Offshore Renewable Energy

Offshore Renewable Energy

Editors

Jijian Lian
Qiang Fu
Lin Cui
Run Liu
Bingyong Guo

MDPI

Basel • Beijing • Wuhan • Barcelona • Belgrade • Novi Sad • Cluj • Manchester

Editors

Jijian Lian
Tianjin University of Technology
Tianjin
China

Qiang Fu
Tianjin University of Technology
Tianjin
China

Lin Cui
National Ocean Technology Center
Tianjin
China

Run Liu
Tianjin University
Tianjin
China

Bingyong Guo
Northwestern Polytechnical University
Taicang
China

Editorial Office
MDPI
St. Alban-Anlage 66
4052 Basel, Switzerland

This is a reprint of articles from the Special Issue published online in the open access journal *Journal of Marine Science and Engineering* (ISSN 2077-1312) (available at: https://www.mdpi.com/journal/jmse/special_issues/Z82P08W628).

For citation purposes, cite each article independently as indicated on the article page online and as indicated below:

Lastname, A.A.; Lastname, B.B. Article Title. *Journal Name* **Year**, *Volume Number*, Page Range.

ISBN 978-3-7258-1141-0 (Hbk)
ISBN 978-3-7258-1142-7 (PDF)
doi.org/10.3390/books978-3-7258-1142-7

Cover image courtesy of Jijian Lian

Contents

About the Editors

Jijian Lian

Jijian Lian, professor and President of Tianjin University of Technology and professor at Tianjin University. He has been engaged in scientific and technological research and engineering practice in the fields of water conservancy, hydropower, and offshore energy engineering. He is an academic leader in this field, a recipient of the National Science Fund for Distinguished Young Scholars, a distinguished professor of Yangtze River Scholars, and a leading scientific and technological innovation talent of the National Ten Thousand Talents Plan. He has led more than 70 national and major engineering research projects and systematically developed engineering safety theories and control technologies under the coupling effects of complex water, gas, and fluid dynamics and structures in the fields of water conservancy, hydropower, and offshore wind and solar energy engineering. His research has been successfully applied or industrialized in the construction and operation of over 60 high dam hubs, long-distance water transfer, and offshore wind and solar energy projects. He has received one National Science and Technology Progress special award, one National Science and Technology Progress first prize (ranking second), three second prizes (ranking first, first, and second), and fifteen provincial- and ministerial-level special and first prizes (ranking first out of seven), obtained 148 invention patents, including achievements in eight national industry standards, published over 400 papers in SCI and EI journals, and published five monographs.

Qiang Fu

Qiang Fu is an associate professor at Tianjin University of Technology. He received his Ph.D. in Intelligent Mechanical Systems Engineering from Kagawa University in Japan in 2017. He is a Tianjin candidate for the third level of 131 talents, middle-aged and young backbone teachers, and a member of the professional committee of Intelligent Robot. His research interests concern robot structure design, control algorithms, human–machine interaction, and other related fields. He has proposed a marine maintenance and operation robot with independent intellectual property rights, which can be used to solve the maintenance and operation of offshore floating photovoltaic power generation devices. He has published more than 30 high-level papers and has been granted three patents.

Lin Cui

Lin Cui is a professor, was a member of the working group for offshore renewable energy at IOC/WESTPAC from 2012 to 2014, and has been an organization member of the Asian Wave and Tidal Energy Conference (AWTEC) since 2020. He served as the alternative representative for China of the IEA OES-TCP from 2011 to 2016. At present, he serves as the Chinese lead of the UK–China Centre of Offshore Renewable Energy, which is jointly funded by Newton Funds and the NSFC, a member of Technical Committee 114 (Marine energy—wave, tidal, and other water current converters) of the International Electro-technical Commission as the liaison of subcommittee TC114 with TC82, and a core member of the IEC/TC114 Strategic Business Plan Group. He has published more than 40 journal papers in high-level international academic journals such as *Renewable and Sustainable Energy Reviews*, *Applied Energy*, *Energy Conversion and Management*, *Desalination*, *Micromachines*, and others.

Run Liu

Run Liu is a distinguished professor and doctoral advisor at Tianjin University. Her scholarly endeavors predominantly focus on the intricate challenges of marine soil strength degradation and recovery under complex stress paths, as well as the interactions between such soils and structural components. Her pioneering research has yielded novel insights into the engineering properties and quantitative assessment methods of marine soils, load-bearing capacities and sinking resistance of cylindrical bases in offshore wind installations, predictive and evaluative methodologies for ultra-large and ultra-long steel tubular piles, and theoretical analyses of subsea pipe–soil interaction and its consequent buckling and axial movement. Her findings have been integral to the development of several major marine engineering projects, including offshore wind foundations, deep-water pile foundations in the South China Sea, and subsea tunnels across the Pearl River Delta, Dalian Bay, and Bohai Sea's underwater oil and gas systems.

Bingyong Guo

Bingyong Guo is a professor at the Ocean Institute, Northwestern Polytechnical University. He graduated from Northwestern Polytechnical University, China, with B.Eng. and M.Sc. degrees in Information Countermeasure Technology and Underwater Acoustics in 2010 and 2013, respectively. He received his Ph.D. in Electronic Engineering from the University of Hull, UK, in 2017. He worked as a post-doctoral researcher in the Nonlinear Dynamics and Control Lab, University of Exeter, UK, from 2019 to 2021 and as a Marie Curie Postdoctoral Research Fellow at the Centre for Ocean Energy Research, Maynooth University, Ireland, from 2019 to 2021. His research interests lie in the numerical modelling, experimental verification, and optimal control system design of wave energy conversion devices. He has published around 50 peer-reviewed papers in the field of ocean energy.

Journal of
Marine Science and Engineering

:ditorial

Offshore Renewable Energy

ijian Lian [1,2], Qiang Fu [3,*], Lin Cui [4], Run Liu [2] and Bingyong Guo [5]

1 Institute of Ocean Energy and Intelligent Construction, Tianjin University of Technology, Tianjin 300384, China
2 School of Civil Engineering, Tianjin University, Tianjin 300072, China; liurun@tju.edu.cn
3 School of Electrical Engineering and Automation, Tianjin University of Technology, Tianjin 300384, China
4 National Ocean Technology Center, Ministry of Natural Resources, Tianjin 300112, China; cuilin_oceanenergy@126.com
5 Ocean Institute, Northwestern Polytechnical University, Taicang 215400, China; bingyongguo@outlook.com
* Correspondence: fuqiang@email.tjut.edu.cn

1. Introduction

Offshore renewable energy is an abundant clean energy source that includes offshore solar energy, offshore wind power, wave energy, ocean thermal energy conversion (OTEC), and tidal energy. These energy forms have the characteristics of abundant resources, a wide distribution, no occupation of land space, sustainable utilization, and relatively small environmental impact, and are important directions for the global response to climate change and promoting energy transformation. Compared with terrestrial renewable energy, offshore renewable energy has irreplaceable advantages [1–3]. On the one hand, offshore renewable energy has abundant reserves, high energy density, and is suitable for large-scale and centralized development, which plays a more significant supporting role in the development of the coastal economy. On the other hand, the multi-level distribution characteristics of different forms of marine energy are more obvious, and diversified energy development and utilization can be carried out [4–6].

At present, the development of offshore wind power technology is relatively mature and has entered the stage of large-scale development. With advancements in technology, the scale and efficiency of offshore wind turbines are constantly increasing and the cost is gradually decreasing, making offshore wind power more competitive. Meanwhile, other forms of marine energy such as offshore photovoltaic power, tidal current energy, and wave energy are also actively being researched and developed [7–9]. Although their commercialization level is not as high as that of offshore wind power, technological progress and cost reductions are driving these energy forms to gradually enter the market. The complex marine environment poses greater challenges to the development and utilization of renewable energy at sea. Due to the harsh marine environmental conditions, offshore renewable energy development platforms need to have better corrosion resistance and wind and wave resistance, which leads to difficulties in platform construction as well as operation and maintenance. In addition, the development of offshore renewable energy has an impact on the ecological environment, such as marine pollution and ecological damage. Effective measures need to be taken to prevent and control such impacts.

In summary, offshore renewable energy is in a rapid development stage, but it is worth noting that, with the development and gradual maturity of various technologies, offshore renewable energy will become the mainstream new energy in the future.

2. An Overview of the SI and Published Articles

Aiming to provide updates regarding the development and utilization of offshore renewable energy, the *Journal of Marine Science and Engineering (JMSE)* proudly presents this SI, entitled "Offshore renewable energy". The SI aims to provide updates regarding novel

:itation: Lian, J.; Fu, Q.; Cui, L.; Liu, ₹.; Guo, B. Offshore Renewable :nergy. *J. Mar. Sci. Eng.* **2024**, *12*, 749. ıttps://doi.org/10.3390/jmse12050749

ₗeceived: 15 April 2024
Accepted: 27 April 2024
ᵖublished: 30 April 2024

concepts, control strategies, modeling methods, intelligent equipment, and monitoring technology applications.

This SI discusses the key technologies of the offshore renewable energy system: a behavior analysis of the Bucket Foundation (BF) with an offshore wind turbine during the in-water sinking process (contribution 1); the mooring design of floating wind turbines (contribution 8); wave energy converters (contributions 2 and 7); ocean thermal energy conversion (contribution 4, contribution 11); the performance analysis of a floating wind-wave power generation platform (contribution 6); and the tidal stream turbine attachment fault detection method (contribution 3). The AD-LADRC circulation suppression strategy is proposed based on the traditional circulation suppression scheme (contribution 5) in order to improve the circulation suppression effect, increase the stability of high-voltage direct current (HVDC) transmission systems, and provide certain theoretical and application support for the subsequent development and large-scale utilization of offshore wind power. Additionally, this SI discusses the suitable installation position and the optimum strut height of T-foil for an offshore wind power operation and maintenance vessel. A longitudinal control system for ship multiple-degrees-of-freedom-coupled motion is established for severe sea states (contribution 10). A novel three-DOF turbine access system (TAS) with active motion compensation for the Fujian coastal area is designed. A proposed new stacking compensation algorithm and a fuzzy controller with feedforward compensation are used to achieve TAS end position compensation (contribution 9). An electromagnetic generator with a circular structure is proposed (contribution 7) that can quickly charge capacitors and light up LED lights under various sea conditions. It has the ability to provide power for small ocean monitoring equipment and has good application prospects. These studies provide valuable insights for the development and application of offshore renewable energy.

3. Conclusions

We sincerely thank the *Journal of Marine Science and Engineering (JMSE)* for inviting us to collaborate in creating the SI "Offshore Renewable Energy", and the authors for providing high-quality papers. Firstly, we would like to express our sincere gratitude to the reviewers. They carefully reviewed the content of the paper with a rigorous scientific attitude and provided insightful criticism and guidance. Their strict requirements for the quality of the paper and suggestions for further improvement have enabled us to examine our research more deeply and enhance the scientific approach of the article. Their in-depth review and constructive modification suggestions have deeply inspired us. Secondly, we would like to express our deepest gratitude to the diligent editors. They not only carefully reviewed and revised our paper with professional insight, but also provided many valuable suggestions with selfless spirit and patience. Their organization and optimization of the structure and language of the paper have improved the fluency of our article. Meanwhile, their guidance and assistance during the paper revision process benefited us greatly. Finally, we would like to thank the editors and reviewers for providing us with professional knowledge and technical support. Their knowledge and experience have enabled our paper to be improved and perfected at a higher level. Thank you again to all of the professionals who participated in the development of this Special Issue. Your hard work and selfless dedication have laid a solid foundation for the successful publication of this SI. We sincerely hope that, in the future, we can continue to work together to promote the progress and development of research related to offshore renewable energy.

Author Contributions: Conceptualization, J.L.; writing—original draft preparation, Q.F.; review and editing, J.L., Q.F., L.C., R.L. and B.G. All authors have read and agreed to the published version of the manuscript.

Conflicts of Interest: The authors declare no conflict of interest.

List of Contributions

1. Ye, F.; Lian, J.; Xiao, T.; Xiong, D.; Wang, H.; Guo, Y.; Shao, N. Behavior Analysis of a Bucket Foundation with an Offshore Wind Turbine during the In-Water Sinking Process. *J. Mar. Sci. Eng.* **2024**, *12*, 494. https://doi.org/10.3390/jmse12030494.
2. Wang, C.; Guo, L.; Chen, P.; Fu, Q.; Cui, L. Annular Electromagnetic Generator for Harvesting Ocean Wave Energy. *J. Mar. Sci. Eng.* **2023**, *11*, 2266. https://doi.org/10.3390/jmse11122266.
3. Song, D.; Liu, R.; Zhang, Z.; Yang, D.; Wang, T. IRNLGD: An Edge Detection Algorithm with Comprehensive Gradient Directions for Tidal Stream Turbine. *J. Mar. Sci. Eng.* **2024**, *12*, 498. https://doi.org/10.3390/jmse12030498.
4. Lu, B.; Liu, Y.; Zhai, X.; Zhang, L.; Chen, Y. Design and Experimental Study of 50 kW Ocean Thermal Energy Conversion Test Platform Based on Organic Rankine Cycle. *J. Mar. Sci. Eng.* **2024**, *12*, 463. https://doi.org/10.3390/jmse12030463.
5. Xu, X.; Wang, D.; Zhou, X.; Tao, L. Suppression of Negative Sequence Current on HVDC Modular Multilevel Converters in Offshore Wind Power. *J. Mar. Sci. Eng.* **2024**, *12*, 383. https://doi.org/10.3390/jmse12030383.
6. Chen, M.; Deng, J.; Yang, Y.; Zhou, H.; Tao, T.; Liu, S.; Sun, L.; Hua, L. Performance Analysis of a Floating Wind–Wave Power Generation Platform Based on the Frequency Domain Model. *J. Mar. Sci. Eng.* **2024**, *12*, 206. https://doi.org/10.3390/jmse12020206.
7. Qin, J.; Zhang, Z.; Song, X.; Huang, S.; Liu, Y.; Xue, G. Design and Performance Evaluation of an Enclosed Inertial Wave Energy Converter with a Nonlinear Stiffness Mechanism. *J. Mar. Sci. Eng.* **2024**, *12*, 191. https://doi.org/10.3390/jmse12010191.
8. Chen, M.; Jiang, J.; Zhang, W.; Li, C.B.; Zhou, H.; Jiang, Y.; Sun, X. Study on Mooring Design of 15 MW Floating Wind Turbines in South China Sea. *J. Mar. Sci. Eng.* **2024**, *12*, 33. https://doi.org/10.3390/jmse12010033.
9. Wang, J.; Zhang, S.; Cheng, J.; Li, Y.; Shen, Y.; Wu, Z. Modeling and Simulation of a Turbine Access System with Three-Axial Active Motion Compensation. *J. Mar. Sci. Eng.* **2023**, *11*, 2237. https://doi.org/10.3390/jmse11122237.
10. Yuan, J.; Liu, Z.; Geng, H.; Zhang, S.; Liang, L.; Zhao, P. Design Longitudinal Control System Using Suitable T-Foil Modeling for the Offshore Wind Power Operation and Maintenance Vessel with Severe Sea States. *J. Mar. Sci. Eng.* **2023**, *11*, 2182. https://doi.org/10.3390/jmse11112182.
11. Zhao, D.; Li, S.; Shi, W.; Zhou, Z.; Guo, F. Design and Optimization of the Teardrop Buoy Driven by Ocean Thermal Energy. *J. Mar. Sci. Eng.* **2024**, *12*, 661. https://doi.org/10.3390/jmse12040661.

References

1. Zhao, Y.; Yuan, H.; Zhang, Z.; Gao, Q. Performance analysis and multi-objective optimization of the offshore renewable energy powered integrated energy supply system. *Energy Convers. Manag.* **2024**, *304*, 118232. [CrossRef]
2. Weiss, C.V.; Guanche, R.; Ondiviela, B.; Castellanos, O.F.; Juanes, J. Marine renewable energy potential: A global perspective for offshore wind and wave exploitation. *Energy Convers. Manag.* **2018**, *177*, 43–54. [CrossRef]
3. Cullinane, M.; Judge, F.; O'Shea, M.; Thandayutham, K.; Murphy, J. Subsea superconductors: The future of offshore renewable energy transmission? *Renew. Sustain. Energy Rev.* **2022**, *156*, 111943. [CrossRef]
4. Konispoliatis, D.N.; Katsaounis, G.M.; Manolas, D.I.; Soukissian, T.H.; Polyzos, S.; Mazarakos, T.P.; Voutsinas, S.G.; Mavrakos, S.A. REFOS: A renewable energy multi-purpose floating offshore system. *Energies* **2021**, *14*, 3126. [CrossRef]
5. Raileanu, A.; Onea, F.; Rusu, E. An overview of the expected shoreline impact of the marine energy farms operating in different coastal environments. *J. Mar. Sci. Eng.* **2020**, *8*, 228. [CrossRef]
6. Pryor, S.C.; Barthelmie, R.J.; Shepherd, T.J. Wind power production from very large offshore wind farms. *Joule* **2021**, *5*, 2663–2686. [CrossRef]
7. Wan, L.; Moan, T.; Gao, Z.; Shi, W. A review on the technical development of combined wind and wave energy conversion systems. *Energy* **2024**, *294*, 130885. [CrossRef]
8. Li, G.; Zhu, W. Tidal current energy harvesting technologies: A review of current status and life cycle assessment. *Renew. Sustain. Energy Rev.* **2023**, *179*, 113269. [CrossRef]
9. Wang, G.; Chao, Y.; Chen, Z. Facilitating developments of solar photovoltaic power and offshore wind power to achieve carbon neutralization: An evolutionary game theoretic study. *Environ. Sci. Pollut. Res.* **2023**, *30*, 45936–45950. [CrossRef] [PubMed]

Journal of
Marine Science
and Engineering

Article

Behavior Analysis of a Bucket Foundation with an Offshore Wind Turbine during the In-Water Sinking Process

Fangdi Ye [1,2], Jijian Lian [1,2,3], Tianrun Xiao [1,2], Dongzhi Xiong [1,2], Haijun Wang [1,2], Yaohua Guo [1,2] and Nan Shao [2,4,*]

1 State Key Laboratory of Hydraulic Engineering Intelligent Construction and Operation, Tianjin University, Tianjin 300072, China; fdye@tju.edu.cn (F.Y.); jjlian@tju.edu.cn (J.L.); xiaotr@tju.edu.cn (T.X.); xdongzhi@tju.edu.cn (D.X.)
2 School of Civil Engineering, Tianjin University, Tianjin 300072, China
3 School of Computer Science and Engineering, Tianjin University of Technology, Tianjin 300384, China
4 School of Water Conservancy and Hydroelectric Power, Hebei University of Engineering, Handan 056038, China
* Correspondence: shaonan@tju.edu.cn

Abstract: To enhance the safety of the in-water sinking operation for an integrated system, including a bucket foundation (BF), tower, and rotor nacelle assembly (RNA), in complex marine environmental conditions, a model test of in-water sinking for an offshore wind turbine and bucket foundation (OWT–BF) is conducted. The motion behavior of the OWT–BF and the sling tensions during the in-water sinking process are investigated, and the numerical method is validated through test results. The results demonstrate a positive correlation between the wave height and motion response of the OWT–BF. The most critical stage of the in-water sinking operation occurs when the top cover of the bucket is fully submerged, resulting from the substantial cross-sectional difference between the bucket base and the transition section. Furthermore, the closer the OWT–BF is to the seabed, the less it is affected by waves in terms of motion response. It is advisable to conduct the in-water sinking operation of the OWT–BF in sea states with wave heights below 1.5 m. Simultaneously, slings can efficiently control the motion response of the OWT–BF, thereby enhancing the safety of the sinking operation.

Keywords: offshore wind power; air-floating structure; bucket foundation; wave action; in-water sinking

Citation: Ye, F.; Lian, J.; Xiao, T.; Xiong, D.; Wang, H.; Guo, Y.; Shao, N. Behavior Analysis of a Bucket Foundation with an Offshore Wind Turbine during the In-Water Sinking Process. *J. Mar. Sci. Eng.* **2024**, *12*, 494. https://doi.org/10.3390/jmse12030494

Academic Editor: Eva Loukogeorgaki

Received: 26 January 2024
Revised: 12 March 2024
Accepted: 13 March 2024
Published: 15 March 2024

1. Introduction

At present, the proportion of offshore wind capacity is still relatively small compared to that of onshore wind capacity [1]. The complete utilization of offshore wind energy resources will effectively improve energy supply problems and generate significant economic benefits [2,3]. However, technical issues such as the construction, transportation and installation of offshore wind power structures have been important factors restricting the development of offshore wind power. The traditional method of installing an OWT and its supporting foundations usually involves sea sweeping, positioning, transportation, piling, lifting, etc., and the entire construction process is operated at sea [4,5]. This method is difficult, costly, and time-consuming, and it is also limited by the marine environment and window period. Therefore, to accelerate the development of offshore wind power, there is an urgent need for a new and efficient program to improve the efficiency of offshore wind power construction [6].

To promote the development of offshore wind power, Tianjin University proposed a novel supporting structure for offshore wind power, i.e., a bucket foundation, which has the advantages of strong capsize resistance, bucket-top bearing, self-floating towing, and negative-pressure sinking [7], and it is mainly applied to soft clay (or silty clay) and fine sand (or sandy silt) foundations [8]. In recent years, the BF has been rapidly popularized and applied in the field of offshore wind power. In addition, the research team of the BF also developed a set of integrated for the construction, transportation, and installation of the OWT–BF, which mainly includes the processes of onshore prefabrication, air-floating towing, in-water sinking, and in-soil sinking. The integrated technology of the OWT–BF avoids the use of large-scale machinery such as lifting equipment at sea, simplifies the construction process, reduces the difficulty of offshore operation and the risk of damage to the wind turbine, and easily realizes the goal of immediate installation and use. The technology is not only safe, efficient, environmentally friendly, and economical, but also accelerates the low-cost, large-scale development and application of offshore wind power. The in-water sinking operation of the OWT–BF is one of the key steps of the integrated technology. The motion and dynamic characteristics of the OWT–BF during the sinking process are directly related to the safety of the installation, so these need to be focused on by research.

The BF can be regarded as an air-floating structure during the in-water sinking process, which is mainly provided with buoyancy by the gas inside the bucket and realizes in-water sinking through dynamic regulation of the gas. The concept of an air-floating structure was put forward as early as the 1980s, and it was subsequently studied by many scholars [9,10]. Compared to ordinary floating bodies, air-floating structures are characterized by the fact that the structure is supported by gas, and the gas inside the structure can absorb wave energy and reduce the motion response of the structure under the action of waves, which improves the safety of the structure [11–13]. Cheung et al. [14] analyzed the dynamic response of an air-floating platform consisting of a series of vertical cylinders with bottom openings by numerical methods, and the numerical results were better verified by model tests. The oscillations of the water column and the motion characteristics of the platform were discussed in detail. Lee and Newman [15] analyzed a very large floating structure (VLFS) supported by an air cushion. They assumed the air–water interface to be a free liquid surface and derived the extended equations of motion for rigid body motion and generalized modes. Pinkster et al. [16] analyzed the hydrodynamic characteristics of a large mobile offshore platform supported by gas. Based on the three-dimensional potential flow theory, the response transfer function of the structure under wave action was derived from numerical calculations and model tests. The results showed that the numerical method was able to better predict the motion and the bending moments of the structure in waves. Maeda et al. [17] proposed a pontoon-type mega floating structure supported by an air cushion. The wave pressure was analyzed by applying the potential flow theory using the pressure distribution method, and the pressure and volume changes of the air cushion were linearized to derive the basic characteristics of the elastic deflection of the structure. Ikoma et al. [18–21] investigated the motion characteristics and hydroelastic properties of air-floating structures under linear wave action. Regarding the in-water sinking of air-floating structures, T. Næss et al. [22] investigated the wave-slamming force on a four-bucket suction anchor during lifting and sinking into the water. The study reveals the importance of the suction-anchor-top cover passing through the splash zone and shows that the slamming force will not damage the structure and crane when the suction-anchor-top cover passes through the splash zone. Zhang et al. [23,24] conducted a model test on the in-water sinking of a four-bucket jacket foundation and revealed the effect of waves on the foundation during the sinking process. In addition, DNV-GL [25] also published a special recommendation manual on modeling and analysis methods for the offshore operation of multiple-bucket foundations.

From the above analysis, it can be seen that the current research mainly focuses on the in-water sinking of the foundation, none of which involves the tower or RNA. Due to the presence of the tower and RNA, the center of gravity of the OWT–BF is significantly higher compared to the foundation alone. The stability of the structure is significantly weaker, and the installation system is more complex, making the integrated installation of OWT–BF more difficult. The current OWT–BF integrated-installation programs were applied at water depths of less than 15 m, as shown in Figure 1. When the water depth is shallow, the in-water sinking process of the OWT–BF is fast, and it is correspondingly less difficult to install. Furthermore, the height of the bucket base is generally within the range of 5–15 m. Throughout the sinking process, when the bottom of the bucket contacts the seabed, the top cover is not submerged yet. Once the bucket base contacts the seabed the OWT–BF is constrained by the soil, and the safety of the in-water sinking operation is significantly improved. Therefore, the integrated installation of an OWT–BF is relatively easy in shallow waters. In cases where the height of the bucket base is less than the actual water depth, the bottom of the bucket remains uncontacted with the seabed even when the bucket base is fully submerged. Due to the substantial difference in shape between the bucket base and the transition section, corresponding measures must be taken to assist in the completion of the in-water sinking operation of an OWT–BF. With the development of offshore wind power in China in deeper water, the actual water depth will be much greater than the height of the bucket base. Although the applicable water depth of BF can be up to 50 m [26,27], the installation difficulty increases significantly with the increase in water depth, so the integrated technology of OWT–BF will face serious challenges in deep water. Therefore, it is crucial to explore the motion law of the OWT–BF during the in-water sinking and to determine suitable conditions for the sinking environment for practical engineering applications. In this paper, considering the weight and shape of the OWT–BF model tests are carried out based on the prototype 30 m-water-depth condition to study the motion law of the OWT–BF during the in-water sinking under wave action. The control strategy based on the use of slings and other auxiliary equipment to assist the operation of the OWT–BF's in-water sinking is proposed and verified, which provides the theoretical support and technical guidance for the integrated technology of the OWT–BF.

Figure 1. Integrated installation of an OWT–BF.

2. Experimental Techniques

2.1. Similarity Theory

The model test is an important means to study the hydrodynamics of ocean engineering, and the similarity principle is an important theory to guide the study of the model test as well as to forecast the hydrodynamic performance of the entity. The hydrodynamic model test in ocean engineering aims to examine the motion and forces acting on the structure within the oceanic environment, with gravity and inertia forces being the predominant factors influencing its behavior. Consequently, the model test must adhere to the Froude similarity criterion, indicating that the Froude number of the model and prototype must be equal, ensuring an accurate similarity in gravity and inertia forces between the model and the prototype. Additionally, the motion and forces exerted on the structure under wave action exhibit periodicity, requiring the model and prototype to maintain equality in the Strouhal number. So,

$$V_m/\sqrt{gL_m} = V_s/\sqrt{gL_s}, \tag{1}$$

$$V_m T_m/L_m = V_s T_s/L_s \tag{2}$$

where V_m, L_m, and T_m are the velocity, wavelength, and period of the model, respectively; V_s, L_s, and T_s are the velocity, wavelength, and period of the prototype, respectively.

In actual engineering, the structure is in a seawater environment. However, the model test is carried out in the laboratory, and the test model is in a freshwater environment. Therefore, the density of seawater and freshwater should be maintained at a fixed ratio, and the ratio is generally γ = 1.025. According to the similarity theory, the proportional relationship between the physical quantities of the model and prototype is shown in Table 1.

Table 1. Proportional relationship between the physical quantities of the model and prototype.

Items	Symbol	Ratio	Items	Symbol	Ratio
Line	L_s/L_m	λ	Period	T_s/T_m	$\lambda^{0.5}$
Area	A_s/A_m	λ^2	Frequency	f_s/f_m	$\lambda^{0.5}$
Volume	V_s/V_m	λ^3	Mass	$\Delta_s/\Delta\Delta_m$	$\gamma\lambda^3$
Linear acceleration	a_s/a_m	1	Force	F_s/F_m	$\gamma\lambda^3$
Angle	α_s/α_m	1	Moment	M_s/M_m	$\gamma\lambda^4$
Water density	ρ_s/ρ_m	γ	Moment of inertia	I_s/I_m	$\gamma\lambda^5$

λ is the geometric similarity ratio.

2.2. Model Description

The test relies on a prototype of a 6.45 MW wind turbine situated within an offshore wind farm in the Jiangsu Sea area, as depicted in Figure 2. The prototype primarily comprises the BF (bucket base, transition section), tower, and RNA, and the test is conducted at a 1:50 scale. Based on the similarity theory, the principal parameters of the prototype and model of the OWT–BF are listed in Table 2.

(a) (b)

Figure 2. Test model: (**a**) OWT–BF; (**b**) BF.

Table 2. Principal parameters of both the prototype and model of the OWT–BF.

Parameter	Units	1:50 Model	Full Scale
Bucket diameter	m	0.72	36.0
Bucket height	m	0.24	12.0
Transition-section height	m	0.42	21.0
Tower height	m	1.83	91.7
Bucket foundation mass	kg	23.11	2,889,000
Tower mass	kg	3.96	494,500
RNA mass	kg	3.39	423,400

Figure 3 illustrates the test layout of the OWT–BF. The BF is segmented into seven compartments, comprising six side compartments and a central hexagonal compartment. Each compartment is individually connected to an air pump using a dedicated gas pipe. A pressure sensor is employed to monitor the gas pressure within each compartment. Three slings, distributed at 120°, are positioned above the top cover of the bucket, with the tension of each sling measured using a tension sensor. In practical engineering, the OWT–BF is situated within the U-channel of the transportation and installation vessel. To reduce the motion of the OWT–BF during transportation and installation, three constraints are established within the U-channel. One end of each constraint is affixed to the inner wall of the U-channel, while the other end is in close contact with the bucket wall. The three constraints are positioned on both sides and in the middle of the U-channel. A spring device is employed to simulate the constraint during the test, and the bolts are adjusted to position the spring precisely in the middle of the bucket wall and in direct contact with the wall. The constraint frame is positioned in the middle of the tower and surrounds the tower. The inclinometer and acceleration sensor are positioned on the top of the nacelle. According to the right-hand spiral rule, the x-axis corresponds to the wave-spread direction representing the surge direction of the OWT–BF. The y-axis aligns with the width of the wave flume, which signifies the sway direction of the OWT–BF. The z-axis is defined as the vertical direction of the water surface, which represents the heave of the OWT–BF.

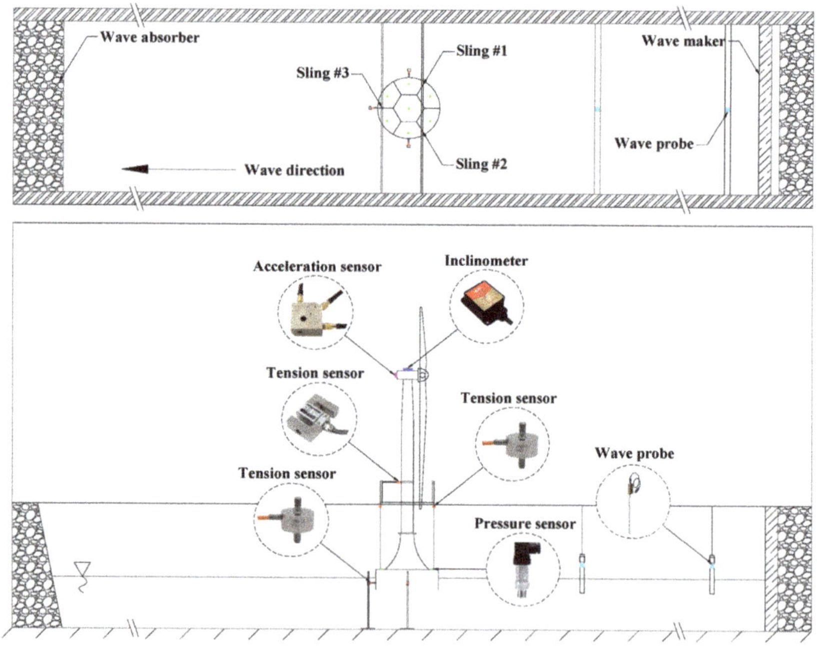

Figure 3. The test layout of the OWT–BF.

2.3. Test Conditions

Based on engineering experience, it is recognized that the in-water sinking process of the OWT–BF is relatively slow, resulting in a slow change in the liquid level inside the bucket. To facilitate the test analysis, the dynamic process of continuous sinking is decomposed into quasi-static processes with varying sinking depths. To prevent air leakage from the BF during the wet towing process and to account for the draft depth of the transportation and installation vessel in practical engineering, the initial sinking depth was established at 14 cm (prototype 7 m), with the sinking depth representing the height of the bottom of the bucket from the water surface. The maximum sinking depth is determined based on the model mass, leading to the establishment of a test water depth of 60 cm (30 m for the prototype). Considering the height of the bucket wall (12 m for the prototype) and the condition where the water surface aligns with the top cover of the bucket, the entire sinking process is divided into five depths following the principle of equal division, with a depth interval of 10 cm (5 m for the prototype). Generally speaking, the installation operation of offshore platforms is usually carried out in seasons with calm sea states, roughly equivalent to level 3–4 sea states, and the corresponding wave height is 0.5–2.5 m. Therefore, the wave heights of 0.5 m, 1.5 m, and 2.5 m are selected for the test. Currently, the installation of a few dozen OWT–BFs is completed by the integration technology of OWT–BF. According to the feedback from the actual project, the sea state of a wave period of 7–8 s has a more obvious influence on the in-water sinking operation of OWT–BF. Therefore, the wave period of 7.8 s is selected for the test. Simultaneously, adhering to the principle of a single variable and the requirement for force balance, the initial tension of the slings at each sinking depth is set to 10 N (prototype 1250 kN) by adjusting the gas pressure. The test conditions are detailed in Table 3.

Table 3. Test conditions.

Conditions	Sinking Depth		Wave Height	
	Model (cm)	Prototype (m)	Model (cm)	Prototype (m)
Condition 1-A~C	14	7	1/3/5	0.5/1.5/2.5
Condition 2-A~C	24	12	1/3/5	0.5/1.5/2.5
Condition 3-A~C	34	17	1/3/5	0.5/1.5/2.5
Condition 4-A~C	44	22	1/3/5	0.5/1.5/2.5
Condition 5-A~C	54	27	1/3/5	0.5/1.5/2.5

3. Experimental Results and Discussion

3.1. Motion Responses

Recognizing the importance of the motion response of the wind turbine during in-water sinking, the top of the nacelle is equipped with an acceleration sensor and inclinometer for measuring the surge acceleration, heave acceleration, and pitch angle of the OWT–BF [28]. Comprising the bucket base and the transition section, the BF exhibits significant differences in body shape. To represent the relationship between the sinking depth and the height of the bucket base more intuitively, the relative sinking depth k is employed to indicate various sinking states. This involves expressing k as the ratio of the sinking depth to the height of the bucket base. If k is 1, the water surface aligns precisely with the top cover of the bucket. If k is less than 1, the water surface contacts the bucket base. If k is greater than 1, the water surface contacts the transition section, as depicted in Figure 4.

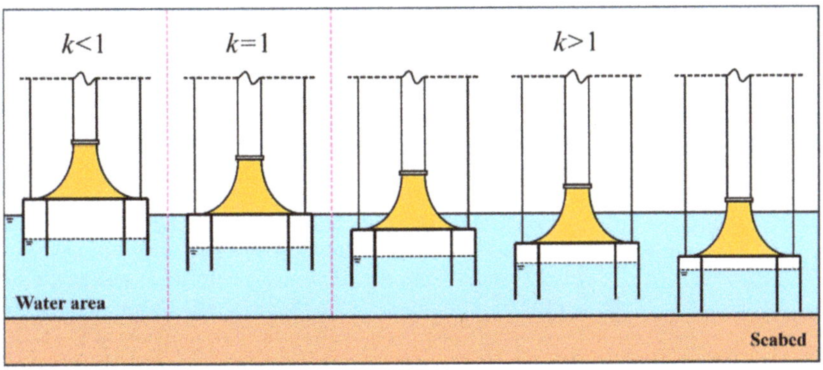

Figure 4. In-water sinking process of the OWT–BF.

Figure 5 illustrates the surge acceleration, heave acceleration, and pitch angle of the OWT–BF under varying wave-height conditions at each sinking depth. It is evident that the surge acceleration, heave acceleration, and pitch angle exhibit an increase with the rising wave height at each sinking depth. In identical conditions, the surge acceleration significantly surpassed the heave acceleration. For wave heights less than 5 cm, pitch angles remained within 2°.

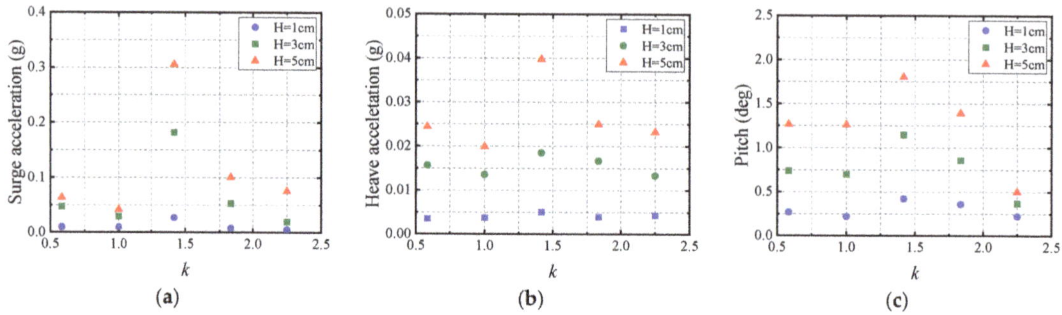

Figure 5. The motion responses of the OWT–BF under varying wave-height conditions at each sinking depth: (**a**) surge acceleration; (**b**) heave acceleration; (**c**) and pitch angle.

For k values less than 1, the surge acceleration decreased with an increase in k. This occurs because the height difference between the liquid surface inside and outside the bucket remains unchanged until the water surface reaches the top cover of the bucket. As k increased, the gas volume in the compartment decreased, and, concurrently, the liquid volume and surge radiation damping increased, resulting in a decrease in surge acceleration. Two factors contribute to the decrease in the pitch motion during this stage. On the one hand, both the additional mass and the radiation damping of the pitch increase with the growing sinking depth. On the other hand, the reduction in the distance between the center of gravity and the center of buoyancy enhances stability.

At k values slightly above one, the surge acceleration increased from 0.01 g, 0.03 g and 0.04 g to 0.03 g, 0.19 g, and 0.31 g for wave heights of 1 cm, 3 cm, and 5 cm, while the pitch angles increased from 0.21°, 0.71°, and 1.43° to 0.43°, 1.15°, and 1.83°, respectively. Evidently, both the surge acceleration and the pitch angle experienced a substantial increase in this stage. At k equal to 1.4, both the surge acceleration and pitch angle attained their maximum values during the entire sinking process. Concurrently, with a wave height of 5 cm, the surge acceleration of 0.31 g surpassed the permissible maximum horizontal acceleration of 0.25 g for the wind turbine. The sudden increase in motion during this stage can be attributed to the water surface surpassing the top cover of the bucket. Consequently, the wave exerted pressure on both the top cover of the bucket and the transition section, leading to the generation of vertical wave forces. Due to the impact of the vertical wave, the motion of the OWT–BF intensified, leading to an augmentation in both the surge acceleration and pitch angle during this stage.

With the ongoing increase in k, both the surge acceleration and pitch angle exhibited a decreasing trend, and the magnitude of the decrease was substantial. The surge acceleration decreased to less than 0.1 g, and the pitch angle reduced to less than 0.6°. This can be attributed to the low rate of change in the cross-sectional area of the transition section with the sinking depth when the water surface reached the transition section. Consequently, there was a slow change in gas volume in the compartments during this sinking phase. Additionally, with the decreasing volume of gas in the bucket, the influence of the gas on the motion of the OWT–BF gradually diminished, resulting in a decrease in both the surge acceleration and pitch angle. Simultaneously, the heave acceleration remained within 0.05 g, which is significantly below the permissible heave acceleration of 0.2 g for the wind turbine. Consequently, the impact of heave acceleration on the in-water sinking of the OWT–BF was relatively negligible.

3.2. Tension on Sling

The in-water sinking operation of the OWT–BF primarily relies on the collaboration between gas pressure and slings. Since slings serve as crucial auxiliary tools, variation in tensions on slings can, to some extent, reflect the interaction between the OWT–BF and the installation vessel. Figure 6 shows the tensions on slings #1, #2, and #3 under different wave heights at each sinking depth. The tensions on slings #1 and #2 closely matched each other as they were situated on the wave-facing side and symmetrically arranged. The tension on sling #3, positioned on the wave-back side, was notably larger than that on sling #1 and #2.

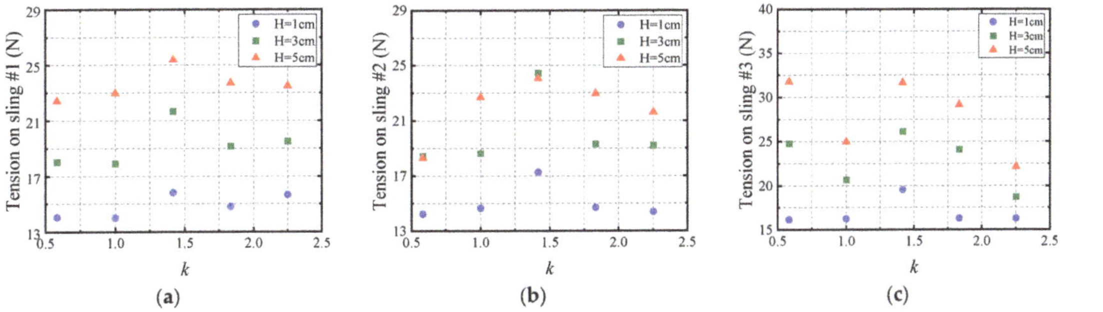

Figure 6. The tensions on slings under different wave heights at each sinking depth: (**a**) sling #1; (**b**) sling #2; and (**c**) sling #3.

At k values less than 1, the tensions on slings #1, #2, and #3 exhibited a tendency to decrease as k increased. This is attributed to the fact that, before the water surface exceeds the top cover of the bucket, the waterline surface area remains unchanged, resulting in the water-spring stiffness remaining constant. With an increase in sinking depth, the gas pressure remains constant, and concurrently, the gas volume in the compartment decreases. This results in an increase in the stiffness of the gas spring. Consequently, the heave stiffness of the gas spring and water spring in series increases. Additionally, the increase in the additional mass and radiation damping of the heave also causes a deceleration of the vertical motion of the OWT–BF. Consequently, these factors contribute to the decrease in the tension on slings #1, #2, and #3 at this stage.

At the point where k exceeded 1, the tensions on slings #1, #2, and #3 experienced an increase. This is because, at the point where the water surface exceeded the top cover of the bucket, the surface wave acted directly on the top cover of the bucket and the transition section, significantly intensifying the vertical effect of the wave. Additionally, it was observed during the test that, due to the close proximity between the water surface and the top cover of the bucket, a water splash occurred above the top cover of the bucket. This splash would intensify the pitch and heave motion of the OWT–BF, leading to increased tension on slings #1, #2, and #3 during this stage.

As k continued to increase, the tensions on slings #1, #2 and #3 decreased. During this stage, no splashing occurred. This is because an increase in the height of the top cover of the bucket from the water surface weakened the vertical action of waves. Additionally, as the submerged volume of the transition section increased, the distance between the center of gravity and the center of buoyancy of the OWT–BF decreased, which led to increased stability. Consequently, the tensions on slings #1, #2, and #3 all decreased during this stage.

4. Numerical Analysis

4.1. Modeling

Figure 7 illustrates the in-water sinking system of the OWT–BF, comprising RNA, tower, bucket, three slings above the top cover, three constraints at the periphery of the bucket wall, and the limit frame of the tower. The focus of this paper is on the motion behavior of the OWT–BF during the in-water sinking process, excluding consideration of the influence of the vessel on the OWT–BF. Consequently, slings, constraints, and limit frames are simulated in a simplified manner. According to the model test device, three slings are positioned above the top cover of the bucket and distributed at 120°. The three constraints are positioned at the left, right, and rear of the bucket, respectively, with a distribution of 90 degrees. Given that the collision of the OWT–BF with the constraints during the in-water sinking process results in an axial deformation of the constraint, the constraints are simulated using the fender model in SIMA [29]. The limit frame surrounds the tower, and the motion of the tower induces axial deformation of the limit frame, so the limit frame is modeled with springs.

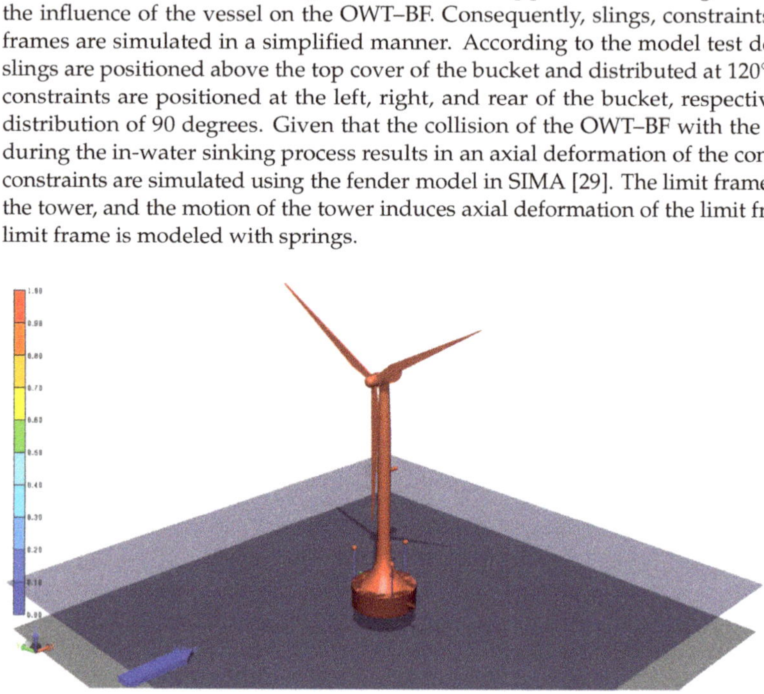

Figure 7. Numerical model of the in-water sinking system of the OWT–BF.

4.2. Validation of Numerical Results

4.2.1. Pitch

Figure 8 displays the time trajectories and statistical values of the pitch angle for the OWT–BF in both the model tests and numerical simulations at various sinking depths with a wave height of 1.5 m. The results indicate that the standard deviation values obtained through the numerical method align with the test results. As the sinking depth increased from 7 m to 27 m, the maximum values of pitch angle varied as follows: 0.890°, 0.709°, 1.148°, 0.859°, and 0.414° for the test results, and 0.713°, 0.605°, 0.980°, 0.801°, and 0.355° for the simulation results, respectively. The increases in the test results over the simulation results were 19.9%, 14.7%, 14.6%, 6.7%, and 14.3%, respectively. This suggests that the test results for the pitch angle were consistently higher than the simulation results. This difference is attributed to the model test being conducted in a 2 m wide and 70 m long flume, where the OWT–BF experienced sidewall effects and wave reflection, introducing a disparity between the model test and the numerical simulation. Nonetheless, the disparity between the two was relatively minor, and the trends in the pitch angle with sinking depth in both the tests and simulations were consistent.

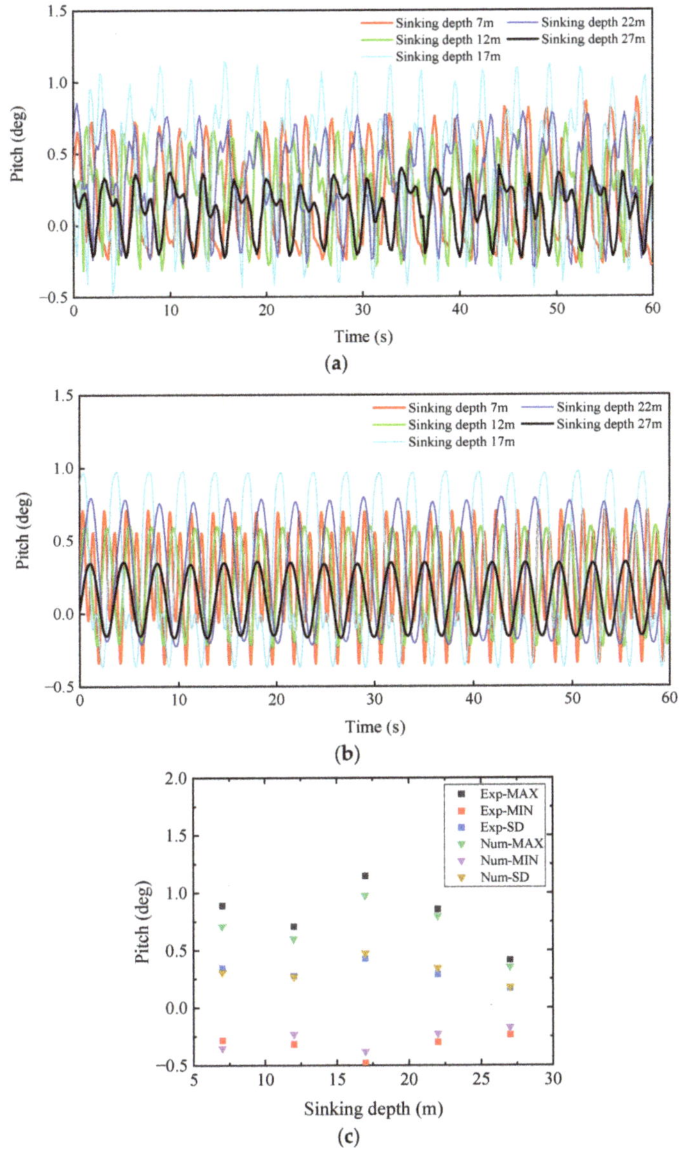

Figure 8. The pitch angle of the OWT–BF at various sinking depths with a wave height of 1.5 m: (**a**) time trajectories of the model tests; (**b**) time trajectories of the numerical simulations; and (**c**) statistical values of the model tests and the numerical simulations.

4.2.2. Surge Acceleration

Figure 9 illustrates the time trajectories and statistical values of surge acceleration for the OWT–BF in both the model tests and numerical simulations at various sinking depths with a wave height of 1.5 m. The variation trend of surge acceleration with changing sinking depth was consistent between the test and simulation results. As the sinking depth increased from 7 m to 27 m, the maximum values of surge acceleration were 0.113 g, 0.060 g, 0.184 g, 0.110 g, and 0.040 g for the test results, and 0.092 g, 0.050 g, 0.017 g, 0.092 g, and 0.034 g for the simulation results, respectively. The test results were slightly larger than the simulation results, which can be attributed to the model test being conducted in a wave flume, while the numerical method accurately simulates the marine environment. Additionally, the numerical method uses a sealing approach to simulate the air–liquid surface in the bucket, in contrast to the compressible gas in the test. This led to a slightly larger test result than the simulation results, but the difference was not significant.

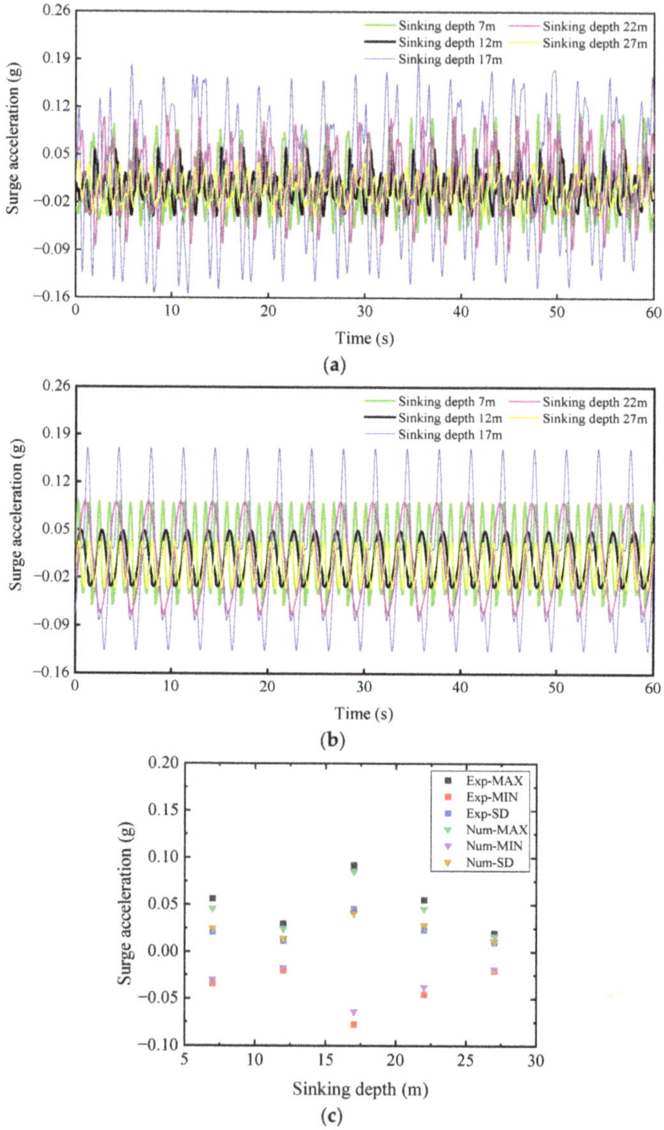

Figure 9. The surge acceleration of the OWT–BF at various sinking depths with a wave height of 1.5 m: (**a**) time trajectories of the model tests; (**b**) time trajectories of the numerical simulations; and (**c**) statistical values of the model tests and the numerical simulations.

4.2.3. Tension on Slings

Compared to the numerical methods, ensuring exact consistency in length and stiffness for each sling during the test is challenging due to the layout of the slings, potentially influencing the test results. Figure 10 illustrates the time trajectories and statistical values of the tension on sling #1 in both the model tests and numerical simulations at various sinking depths with a wave height of 1.5 m. As the sinking depth increased from 7 m to 27 m, the maximum values of the tension on sling #1 were 1357 kN, 1518 kN, 1358 kN, 1375 kN, and 1593 kN for the test results, and 1402 kN, 1541 kN, 1411 kN, 1468 kN, and 1541 kN for the simulation results, respectively. The increases in the test results over the simulation results were 3.2%, 1.5%, 3.7%, 6.3%, and 3.4%, respectively. The average values of the sling tensions in the test and simulation results fell between 1200 kN and 1600 kN, closely aligning with the initial values of the sling tensions and demonstrating good agreement. Furthermore, the variation trends for the sling tensions in the test and simulation results were fundamentally the same with the increasing sinking depth.

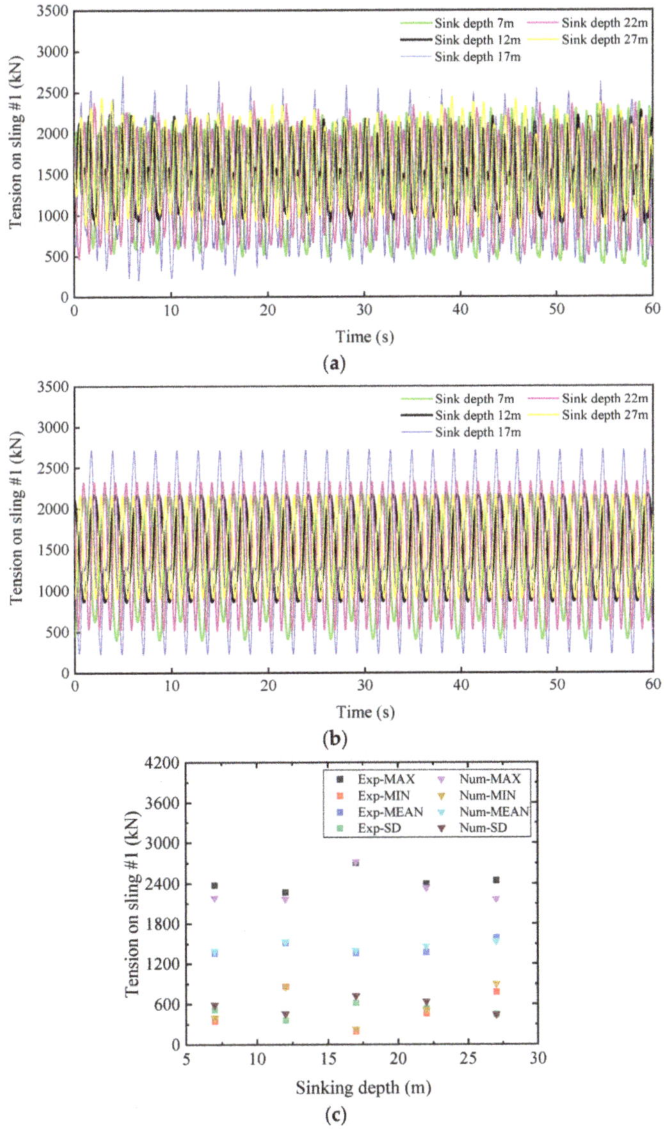

Figure 10. The tension on sling #1 at various sinking depths with a wave height of 1.5 m: (**a**) time trajectories of the model tests; (**b**) time trajectories of the numerical simulations; (**c**) statistical values of the model tests and the numerical simulations.

4.3. Determination of Dangerous Conditions

From both the test and simulation results, it can be observed that as the sinking depth increased from 12 m to 17 m, the sling tension, pitch angle, and surge acceleration all exhibited an increase, reaching their maximum values throughout the sinking process. Due to the small size of the test model, subdividing the sinking depth resulted in insignificant changes in the test results. Therefore, the dynamic characteristics of the OWT–BF at sinking depths from 12 m to 17 m were analyzed by numerical methods at 1 m intervals.

From the above analysis, the variation trend of the tensions on the three slings with changes in sinking depth is consistent. Under the same conditions, the tension on sling #3 was larger than those on slings #1 and #2. Therefore, sling #3 was analyzed in this section. Figure 11 illustrates the tension on sling #3 at sinking depths from 12 m to 17 m. The results indicated that when the sinking depth increased from 12 m to 17 m, the maximum values of the sling tensions were 2204 kN, 2808 kN, 3613 kN, 3500 kN, 3479 kN, and 3357 kN, respectively. When the sinking depth increased from 12 m to 14 m, the tension kept increasing by 64.0%. Subsequently, when the sinking depth increased from 14 m to 17 m, the tension tended to decrease by 7.1%. Consequently, the tension on the slings was maximal at a sinking depth of 14 m.

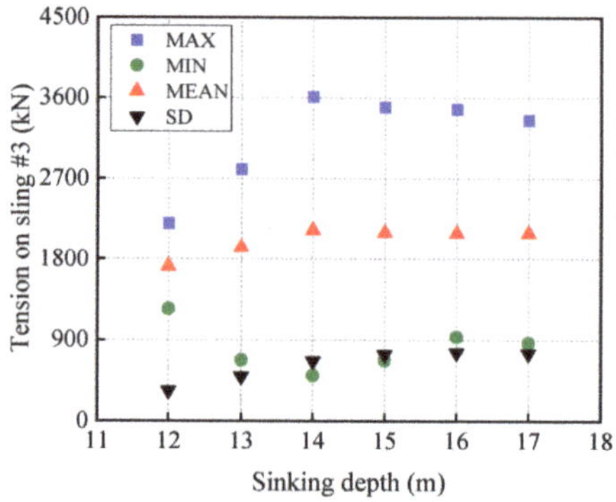

Figure 11. The tension on sling #3 at sinking depths from 12 m to 17 m.

Figure 12 presents the pitch angle at sinking depths from 12 m to 17 m. The results show that when the sinking depth increases from 12 m to 14 m, the pitch angle exhibits a large increasing trend. The maximum, minimum, and average values of the pitch angle increased from 0.60°, 0.23°, and 0.23° to 1.18°, 0.55°, and 0.31°, representing increases of 95.3%, 138.6%, and 32.5%, respectively. When the sinking depth increased from 14 m to 17 m, the pitch angle showed a slower decreasing trend. The maximum, minimum, and average values of the pitch angle decreased to 0.98°, 0.38°, and 0.28°, with decreases of 17.0%, 30.8%, and 9.4%, respectively. Consequently, the pitch angle reached its maximum when the sinking depth was 14 m.

Figure 13 shows the surge acceleration at sinking depths from 12 m to 17 m. The results indicate that when the sinking depth increased from 12 m to 14 m, the surge acceleration kept increasing, with the maximum and minimum values of the surge acceleration increasing from 0.050 g and 0.032 g to 0.186 g and 0.160 g, respectively. When the sinking depth increased from 14 m to 17 m, the surge acceleration showed a decreasing trend and the maximum and minimum values decreased to 0.170 g and 0.126 g, respectively. Consequently, the trends of sling tension, pitch angle, and surge acceleration with sinking depth were consistent at sinking depths from 12 m to 17 m. Moreover, they all reached a maximum at a sinking depth of 14 m, where the relative sinking depth k was 1.2.

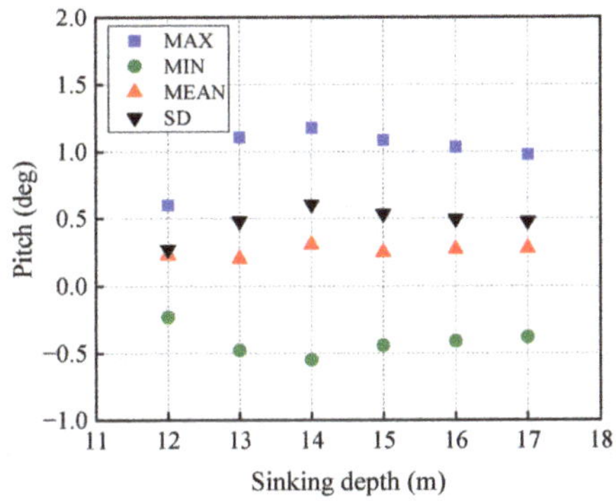

Figure 12. The pitch angle at sinking depths from 12 m to 17 m.

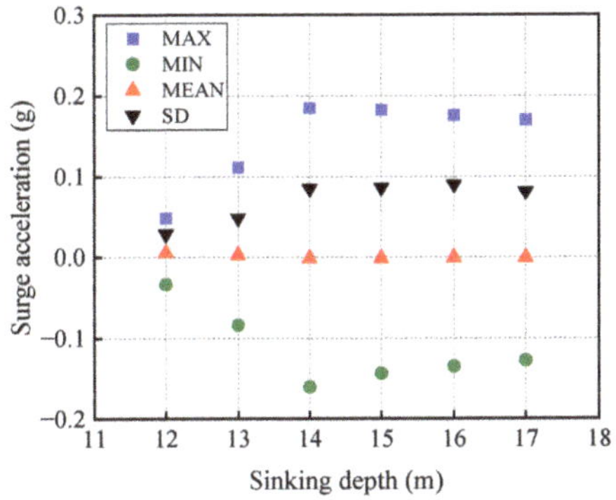

Figure 13. The surge acceleration at sinking depths from 12 m to 17 m.

4.4. Optimization of Sling Layout

The sling is an important auxiliary tool during the in-water sinking process of the OWT–BF, and it is essential to study the layout of the sling. According to the analysis in Section 4.3, the maximum values of the sling tension, pitch angle, and surge acceleration all appeared at sinking depths between 12 m and 17 m, so this section focuses on the motion behaviors of the OWT–BF in this interval of sinking depth. The in-water sinking operation of the OWT–BF was conducted with the assistance of the transportation and installation vessel. The OWT–BF was located in the U-channel of the transportation and installation vessel, and the slings were positioned above the top cover of the bucket. As the layout of the slings is limited by the U-channel, considering that the opening side of the U-channel cannot be arranged with slings, only the inner side of the U-channel is considered to be arranged with slings. In summary, three layouts were selected, and the numbers of slings were 3, 4, and 5, respectively, as shown in Figure 14.

Figure 15 shows the pitch angle and surge acceleration of the OWT–BF for the three layouts of slings. In Figure 15a, when the sinking depth increased from 12 m to 17 m, the pitch angles varied as follows: 0.60°, 1.11°, 1.18°, 1.09°, 1.04°, and 0.98° for layout 1; 0.87°, 1.53°, 1.52°, 1.51°, 1.42°, and 1.23° for layout 2; and 0.46°, 1.06°, 1.26°, 1.15°, 1.00°, and 0.91° for layout 3. The results indicate that the pitch angles of layout 1 and layout 3 were smaller than those of layout 2 under identical conditions, with the difference between the pitch angles of layout 1 and layout 3 being less pronounced. Figure 15b reveals that the surge acceleration of layout 2 surpassed that of layout 1 and layout 3 for sinking depths from 12 m to 17 m. At a sinking depth of 14 m, the surge acceleration values for layout 1, layout 2, and layout 3 were 0.186 g, 0.246 g, and 0.184 g, respectively. The surge acceleration of layout 2 exceeded that of layout 1 and layout 3 by 32.3% and 33.7%, respectively. Consequently, both the pitch angle and surge acceleration of layout 2 with four slings surpassed those of layout 1 with three slings and layout 3 with five slings. Simultaneously, there were minimal disparities in both pitch angle and surge acceleration between layout 1 and layout 3. Considering safety and economic factors, layout 1, comprising three slings, emerges as the most reasonable option.

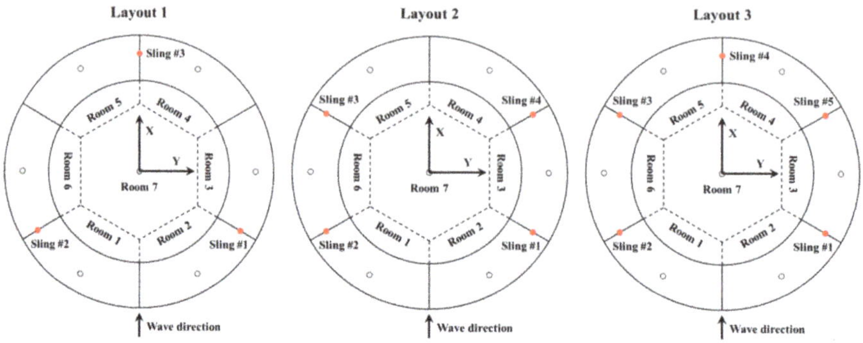

Figure 14. Three layouts for slings.

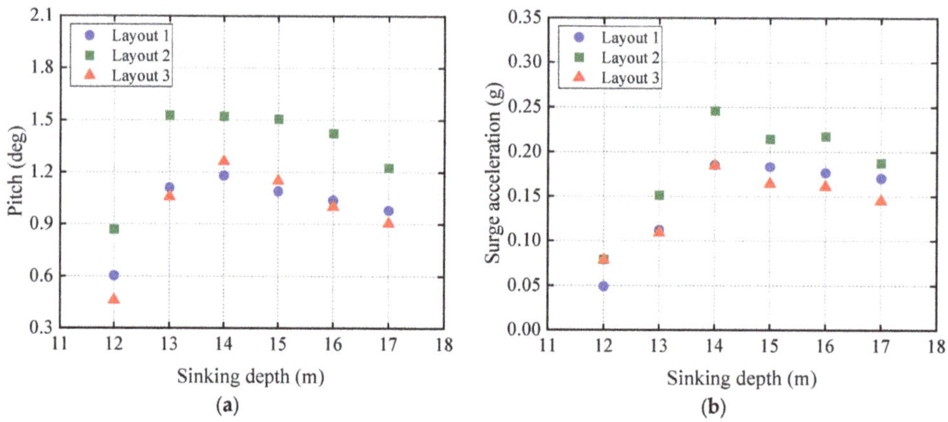

Figure 15. The motion responses of the OWT–BF for the three layouts of slings: (a) pitch angle and (b) surge acceleration.

5. Control Strategies for the In-Water Sinking Process of the OWT–BF

The presence of superstructures, such as towers and RNAs, significantly raises the center of gravity of the entire structure. This leads to the OWT–BF failing to meet stability requirements, even though the BF itself satisfies self-floating stability requirements. Additionally, when the structure tilts under external loads, the gas inside the bucket compresses and fails to provide sufficient restoring force. Consequently, auxiliary equipment and control strategies are essential for the in-water sinking operation of the OWT–BF. The coordinated action of slings will effectively control the motion response of the OWT–BF, facilitating the sinking operation under more complex environmental conditions.

According to the above analysis, there is a similarity in the change rule of the motion response of the OWT–BF and the sling tensions with the sinking depth. Consequently, the entire sinking process is divided into three stages as follows:

Stage 1: Rapid Sinking Stage. Before the water surface reaches the top cover of the bucket, given the columnar structure of the bucket base, the sinking operation at this stage is performed by discharging the gas in the bucket evenly. Simultaneously, the sinking speed of the OWT–BF and the release speed of the sling are determined by the gas discharge speed, ensuring uniformity throughout the entire sinking process.

Stage 2: Hovering Stage. As the water surface approaches the top cover of the bucket, the sinking operation is halted. During this stage, filling the compartments of the top cover with water prevents the rapid inflow of seawater immediately after the water surface surpasses the top cover. This precaution avoids an imbalance of force, which could result in a sudden increase in the sling tension. Subsequently, the tension on the slings is heightened to enhance their constraints on the OWT–BF, limiting the motion response during the most dangerous stage.

Stage 3: Slow Sinking Stage. Once the water surface surpasses the top cover, the OWT–BF accelerates its sinking when the gas is evenly discharged due to the decreasing cross-sectional area of the transition section. However, excessive sinking speed may lead to the loss of sling constraint, resulting in the OWT–BF losing balance. Hence, this stage involves sinking to a specific depth and then pausing to make real-time adjustments in sling tension and air pressure to achieve equilibrium. This process is iterated until the entire sinking operation is completed.

6. Conclusions

The BF has broad application potential in the field of offshore wind power. To enhance the integrity, safety, and reliability of the integrated technology of the OWT–BF, it is of great significance to study the in-water sinking process of the OWT–BF. In this paper, the behavior of the OWT–BF during the in-water sinking process under wave action is investigated through both model tests and numerical simulations. The primary conclusions are as follows:

(1) During the in-water sinking process, the motion of the OWT–BF and the sling tensions increase with an increase in the wave height. Under identical conditions, the surge acceleration is obviously larger than the heave acceleration, establishing the surge acceleration as the predominant controlling factor of the in-water sinking process of the OWT–BF.
(2) A relative sinking depth of 1.2 is the most dangerous condition in the whole sinking process, as the motion of the OWT–BF and the sling tensions reach their maximums at this depth. At a wave height of 2.5 m, the surge acceleration reaches 0.31 g, surpassing the allowable horizontal acceleration of 0.25 g. Consequently, requisite measures should be implemented to mitigate the motion of the OWT–BF at this stage, and it is advisable to conduct the in-water sinking operation of the OWT–BF when the wave height is below 1.5 m.
(3) As the relative sinking depth exceeds 1.2, the motion of the OWT–BF and the sling tensions are significantly reduced by wave influences as the sinking depth increases, thereby enhancing the safety of the sinking operation.
(4) The change rules of the OWT–BF motion and sling tensions with the sinking depth remain relatively consistent during the in-water sinking process. The sinking strategy controlled by the sling and gas pressure proves effective in controlling the motion of the OWT–BF. Therefore, the slings serve as crucial auxiliary equipment for accomplishing the installation of an air-floating structure.
(5) The layout of slings exerts a more significant influence on the motion of the OWT–BF than the quantity of slings.

Author Contributions: Conceptualization, J.L., H.W., Y.G. and N.S.; methodology, J.L., H.W., Y.G. and N.S.; software, F.Y., T.X. and D.X.; validation, F.Y.; formal analysis, F.Y.; resources, J.L.; data curation, F.Y.; writing—original draft, F.Y., T.X. and D.X.; writing—review and editing, F.Y., T.X. and D.X.; supervision, J.L., H.W., Y.G. and N.S.; project administration, J.L. and N.S.; funding acquisition, J.L. and N.S. All authors have read and agreed to the published version of the manuscript.

Funding: This research was funded by the following grants: a Science and Technology Project of the Hebei Education Department, grant number BJK2023099; Hebei Natural Science Foundation grant number E2022402074; and Hebei Natural Science Foundation grant number E2020402074.

Institutional Review Board Statement: Not applicable.

Informed Consent Statement: Not applicable.

Data Availability Statement: Data are contained within the article.

Conflicts of Interest: The authors declare no conflicts of interest.

References

1. Global Wind Energy Council (GWEC). *Global Wind Report 2023*; GWEC: Brussels, Belgium, 2023.
2. Li, Y.; Huang, X.; Tee, K.F.; Li, Q.S.; Wu, X.P. Comparative study of onshore and offshore wind characteristics and wind energy potentials: A case study for southeast coastal region of China. *Sustain. Energy Technol. Assess.* **2020**, *39*, 13. [CrossRef]
3. Li, J.J.; Wang, G.D.; Li, Z.H.; Yang, S.L.; Chong, W.T.; Xiang, X.B. A review on development of offshore wind energy conversion system. *Int. J. Energy Res.* **2020**, *44*, 9283–9297. [CrossRef]
4. Castro-Santos, L.; Filgueira-Vizoso, A.; Lamas-Galdo, I.; Carral-Couce, L. Methodology to calculate the installation costs of offshore wind farms located in deep waters. *J. Clean. Prod.* **2018**, *170*, 1124–1135. [CrossRef]
5. Lacal-Arántegui, R.; Yusta, J.M.; Domínguez-Navarro, J.A. Offshore wind installation: Analysing the evidence behind improvements in installation time. *Renew. Sust. Energ. Rev.* **2018**, *92*, 133–145. [CrossRef]
6. Guo, Y.H.; Wang, H.J.; Lian, J.J. Review of integrated installation technologies for offshore wind turbines: Current progress and future development trends. *Energy Convers. Manag.* **2022**, *255*, 25. [CrossRef]
7. Lian, J.J.; Ye, F.D.; Wang, P.W.; Guo, Y.H.; Wang, H.J.; Xiao, T.R.; Xiong, D.Z. Integrated transportation of offshore wind turbine and bucket foundation based on a U and K shaped assembled platform. *Ocean Eng.* **2023**, *275*, 16. [CrossRef]
8. He, B.; Jiang, J.; Cheng, J.; Zheng, J.B.; Wang, D. The capacities of tripod bucket foundation under uniaxial and combined loading. *Ocean Eng.* **2021**, *220*, 13. [CrossRef]
9. Liu, X.Q.; Le, C.H.; Zhao, M.J.; Ding, H.Y.; Zhang, P.Y.; Lv, N.; Luo, S. Experimental Study on Influencing Factors of Motion Responses for Air-Floating Tetrapod Bucket Foundation. *China Ocean Eng.* **2022**, *36*, 258–267. [CrossRef]
10. Chakrabarti, S.K. Scale effects on a unique launch sequence of a gravity-based structure. *Appl. Ocean Res.* **1995**, *17*, 33–41. [CrossRef]
11. Van Kessel, J.L.F. Aircushion Supported Mega-Floaters. Ph.D. Thesis, Technische Universiteit Delft, Delft, The Nederlands, 2010.
12. Van Kessel, J.L.F.; Pinkster, J.A. The Effect of Aircushion Division on the Structural Loads of Large Floating Offshore Structures. In Proceedings of the ASME 2007 26th International Conference on Offshore Mechanics and Arctic Engineering, San Diego, CA, USA, 10–15 June 2007.
13. Van Kessel, J.L.F.; Pinkster, J.A. Wave-induced Structural Loads on Different Types of Aircushion Supported Structures. In Proceedings of the Seventeenth International Offshore and Polar Engineering Conference, Lisbon, Portugal, 1–6 July 2007.
14. Cheung, K.F.; Phadke, A.C.; Smith, D.A.; Lee, S.K.; Seidl, L.H. Hydrodynamic response of a pneumatic floating platform. *Ocean Eng.* **2000**, *27*, 1407–1440. [CrossRef]
15. Lee, C.H.; Newman, J.N. Wave effects on large floating structures with air cushions. *Mar. Struct.* **2000**, *13*, 315–330. [CrossRef]
16. Pinkster, J.A.; Meevers Scholte, E.J.A. The behaviour of a large air-supported MOB at sea. *Mar. Struct.* **2001**, *14*, 163–179. [CrossRef]
17. Maeda, H.; Ikoma, T.; Masuda, K.; Rheem, C.K. Hydroelastic Behaviors of Large Aircushion Supported Elastic Floating Structures in Regular Waves. In Proceedings of the MTS/IEEE Oceans 2002 Conference, Biloxi, MI, USA, 29–31 October 2002.
18. Ikoma, T.; Masuda, K.; Maeda, H.; Rheem, C.K. Hydroelastic Behavior of Air-Supported Flexible Floating Structures. In Proceedings of the ASME 2002 21st International Conference on Offshore Mechanics and Arctic Engineering, Oslo, Norway, 23–28 June 2002.
19. Ikoma, T.; Masuda, K.; Maeda, H.; Rheem, C.K. Effects of Aircushion Division to Hydroelastic Responses of an Aircushion Type Very Large Floating Structure. In Proceedings of the ASME 2003 22nd International Conference on Offshore Mechanics and Arctic Engineering, Cancun, Mexico, 8–13 June 2003.
20. Ikoma, T.; Masuda, K.; Rheem, C.-K.; Maeda, H. Response Reduction of Motion and Steady Wave Drifting Forces of Floating Bodies Supported by Aircushions in Regular Waves. In Proceedings of the 25th International Conference on Offshore Mechanics and Arctic Engineering, Hamburg, Germany, 4–9 June 2006.
21. Ikoma, T.; Kobayashi, M.; Masuda, K.; Rheem, C.-K.; Maeda, H. A Prediction Method of Hydroelastic Motion of Aircushion Type Floating Structures Considering with Draft Effect into Hydrodynamic Forces. In Proceedings of the ASME 2008 27th International Conference on Offshore Mechanics and Arctic Engineering, Estoril, Portugal, 15–20 June 2008.
22. Næss, T.; Havn, J.; Solaas, F. On the importance of slamming during installation of structures with large suction anchors. *Ocean Eng.* **2014**, *89*, 99–112. [CrossRef]
23. Zhang, P.; Li, Y.e.; Le, C.; Ding, H.; Yang, Z.; Qiang, L. Dynamic characteristics analysis of three-bucket jacket foundation lowering through the splash zone. *Renew. Energy* **2022**, *199*, 1116–1132. [CrossRef]
24. Zhang, P.Y.; Li, Y.E.; Ding, H.Y.; Le, C.H. Response analysis of a lowering operation for a three-bucket jacket foundation for offshore wind turbines. *Renew. Energy* **2022**, *185*, 564–584. [CrossRef]
25. DNV GL. *DNVGL-RP-N103 Modelling and Analysis of Marine Operations*; Det Norske Veritas: Oslo, Norway, 2017.
26. Kim, S.R.; Hung, L.C.; Oh, m. Group effect on bearing capacities of tripod bucket foundations in undrained clay. *Ocean Eng.* **2014**, *79*, 1–9. [CrossRef]

7. Wang, L.Z.; Wang, H.; Zhu, B.; Hong, Y. Comparison of monotonic and cyclic lateral response between monopod and tripod bucket foundations in medium dense sand. *Ocean Eng.* **2018**, *155*, 88–105. [CrossRef]
8. Zhang, P.; Han, Y.; Ding, H.; Zhang, S. Field experiments on wet tows of an integrated transportation and installation vessel with two bucket foundations for offshore wind turbines. *Ocean Eng.* **2015**, *108*, 769–777. [CrossRef]
9. DNV. *SIMO User Manual Version 4.0 Rev 0*; Norwegian Marine Technology Research Institute, Marintek: Trondheim, Norway, 2012.

Journal of
Marine Science and Engineering

MDPI

Article

Design and Optimization of the Teardrop Buoy Driven by Ocean Thermal Energy

Danyao Zhao, Shizhen Li, Wenzhuo Shi *, Zhengtong Zhou and Fen Guo

Institute of Marine Science and Technology, Shandong University, Qingdao 266237, China; zhaody@mail.sdu.edu.cn (D.Z.); lishizhen@sdu.edu.cn (S.L.); ztsdu@mail.sdu.edu.cn (Z.Z.); guofen@sdu.edu.cn (F.G.)

* Correspondence: wzshi@sdu.edu.cn; Tel.: +86-0532-58633267

Abstract: With the inception of the Argo program, the global ocean observation network is undergoing continuous advancement, with profiling buoys emerging as pivotal components of this network, thus garnering increased attention in research. In efforts to enhance the efficiency of profiling buoys and curtail energy consumption, a teardrop-shaped buoy design is proposed in this study. Moreover, an optimization methodology leveraging neural networks and genetic algorithms has been devised to attain an optimal profile curve. This curve seeks to minimize drag and drag coefficient while maximizing drainage, thereby improving hydrodynamic performance. Simulation-based validation and analysis are conducted to assess the efficacy of the optimized buoy design. Results indicate that the drag of the teardrop-shaped buoy with a deflector decreased by 9.2% compared to pre-optimized configurations and by 22% compared to buoys lacking deflectors. The hydrodynamic profile devised in this study effectively enhances buoy performance, laying a solid foundation for ocean thermal energy generation and buoyancy regulation control. Additionally, the optimized scheme serves as a valuable blueprint for the design of ocean exploration devices.

Keywords: buoy; ocean thermal energy; multi-objective optimization; neural network; genetic algorithm; shell optimization

Citation: Zhao, D.; Li, S.; Shi, W.; Zhou, Z.; Guo, F. Design and Optimization of the Teardrop Buoy Driven by Ocean Thermal Energy. *J. Mar. Sci. Eng.* **2024**, *12*, 661. https://doi.org/10.3390/jmse12040661

Academic Editor: Luca Martinelli

Received: 19 March 2024
Revised: 6 April 2024
Accepted: 14 April 2024
Published: 16 April 2024

1. Introduction

In recent years, the scope of human exploration of the oceans has expanded significantly. The introduction of the Argo program, coupled with rapid advancements in underwater exploration technology, has created an enabling environment for the advancement of exploration buoys [1]. Currently, there are over 3000 profiling buoys actively operating in the ocean, making significant contributions to marine environmental observation, weather forecasting, and monitoring atmospheric changes, among other critical areas [2,3]. Conventional buoys rely on batteries for power, which cease functioning once their power is depleted. Consequently, leveraging ocean energy as a sustainable power source for buoys offers a promising solution to address their endurance limitations. Among the various ocean energy sources, ocean thermal energy stands out as one of the most abundant. It encompasses the thermal energy stored in the temperature differential between surface seawater and deep seawater, stemming from solar radiation. As a perpetual and readily accessible renewable energy source, ocean thermal energy holds significant potential for powering buoys and addressing their energy needs [4,5]. Therefore, the exploration of profiling buoys propelled by ocean thermal energy has emerged as a prominent research focus among scientists.

At present, a variety of profiling buoys have been developed. For instance, Sea-Bird Scientific has introduced NAVIS, which is a sleek cylindrical buoy outfitted with a suite of sensors including CTD sensor, optical sensor, and nitrate ultraviolet analyzer. Additionally, it features a built-in power supply enabling it to autonomously conduct 300 profile detections. However, this buoy has been susceptible to malfunctions due to flaws in the design of its hydraulic system [6]. The Scripps Institution of Oceanography has pioneered the development of the SOLO buoy series with various models enhancing efficiency and functionality. Notably, the SOLO-II model stands out as a lighter and more efficient iteration compared to its predecessors; it can carry out detection tasks in water depths of up to 2000 m. Furthermore, the SOLO-DEEP, an upgraded version of the SOLO-II, has achieved an impressive detection depth of 6000 m. Featuring an approximately spherical design, the SOLO-DEEP has successfully completed numerous missions in sea areas as deep as 4000 m off the coast of the United States of America [7,8]. Currently, APEX-type buoys, developed by the Webb company, stand as the most prevalent choice within the Argo program. Characterized by their spherical shape and platforms situated at each end for sensor mounting, these buoys have achieved remarkable milestones. They hold the record for ocean sounding depth, reaching an impressive 6000 m, and have consistently retrieved valuable data. [9]. In addition, operational buoys such as the NOVA buoy, a slender cylindrical design developed by MetOcean company [10], the NINJA buoy, featuring a cylinder with multiple segments rounded on one side and developed by the Agency for Marine-Earth Science and Technology and TSK company [11], and the ARVOR buoy, another slender cylindrical model developed by the Ocean Development Institute, have been successfully deployed [12]. The Ocean Profiling Explorer-COPEX, developed by the National Ocean Technology Center, has been employed in numerous ocean experiments, yielding satisfactory data results [13]. However, most buoy products are commercialized with limited public research, resulting in a scarcity of studies focused on optimizing buoy shapes. Most buoys in operation exhibit spherical or cylindrical shapes, which are prone to high resistance and consequently impact the carrying capacity of the buoy.

Similarly, the shape of underwater gliders driven by ocean thermal energy has also received attention in research. Nonetheless, there are references to various studies on the shape of underwater gliders, which also driven by ocean thermal energy that could serve as valuable resources. For instance, the Spray Glider developed by Sherman at the Scripps Institute of Oceanography features a slender, low-resistance streamlined shell [14] and boasts a maximum submerged depth of 1500 m [15]. The Slocum developed by WEBB measures 1.8 m in length and features a streamlined bow and tail. It is capable of operating at speeds of approximately 0.5 m/s in waters up to 200 m deep [16]. The Sea Glider developed by Erisken et al. at the University of Washington utilizes a spindle-like form factor with a smaller size, and it has also performed well in experiments [17]. Bertram and Alvarez conducted an in-depth discussion on the overall design simulation of the Autonomous Underwater Vehicle's (AUV) shell. Collaborating with naval hydrodynamics experts, they put forth comprehensive guidelines for the shell shape, which are accompanied by empirical coefficients to optimize the maneuverability of the torpedo-like geometry [18]. Ye and Pan employed an enhanced ensemble of surrogates-based global optimization method (IESGO-HSR), which incorporates a hierarchical design space reduction strategy, to optimize the airfoils of the novel flying-wing configuration underwater glider [19]. Sun et al. used the efficient global optimization (EGO) method to achieve a higher maximum lift-to-drag ratio of the blender-wing-body underwater glider [20]. Fu et al. designed a slender ellipsoidal underwater glider and used a genetic algorithm to carry out multi-objective optimization of the hull drag and hull surface pressure for the underwater glider [21].

At present, most buoys in operation exhibit spherical or cylindrical shapes, which are prone to high resistance and consequently impact the carrying capacity of the buoy. Li introduced a streamlined buoy design that effectively enhanced float performance [22]. However, this design failed to account for the reciprocating motion characteristics of profiling buoys. Therefore, this paper proposes a water teardrop float inspired by the motion patterns of profiling buoys. This design incorporates a bow-like structure on both sides of the float to optimize its hydrodynamic performance. The key innovations of this study include: (a) The proposal and design of a teardrop profiling buoy, driven by ocean thermal energy, which considers buoy motion characteristics to enhance hydrodynamic efficiency during both ascent and descent phases. (b) Development of an optimization scheme for the buoy shell, with objectives focused on minimizing drag and drag coefficient while maximizing drainage volume. This approach aims to attain the optimal buoy shape for superior performance and improves buoy energy efficiency which will be crucial to maintaining and improving the Argo array [23]. These findings provide a solid foundation for the self-sustaining generation of ocean thermal energy and buoyancy regulation control for underwater buoys. Furthermore, they offer valuable insights for the development of underwater detection devices.

The remainder of this paper is organized as follows. Section 2 elucidates the theory employed in the design, computational fluid dynamics (CFD) simulation, and optimization process. Section 3 presents the optimization results following discussion, accompanied by an analysis of the hydrodynamic performance of the buoy. Finally, Section 4 offers concluding remarks.

2. Methods

To improve the hydrodynamic performance of the buoy, a teardrop buoy is proposed, designed, optimized and verified in this paper. This section primarily outlines the theoretical framework employed in the design of the buoy, CFD simulation, and optimization process. It is divided into three key parts: buoy structural design, numerical model development, and optimization scheme.

2.1. Buoy Structural Design

The shape of ocean exploration buoys typically resembles a rotational body. This rotational body is often described by a curve that rotates around an axis to form its surface which is known as the generatrix. In this context, we introduce the teardrop curve as the generatrix of the rotational body. The teardrop curve is a streamlined curve resembling the edge of a water droplet, which is described as follows:

$$\begin{cases} y_f = b\left[1-\left(\frac{X_f}{L_f}\right)^{n_f}\right]^{\frac{1}{n_f}} \\ y_a = b\left[1-\left(\frac{X_a}{L_a}\right)^{n_a}\right] \end{cases}, \quad (1)$$

where b represents the maximum radius, L_f denotes the length of the front part, L_a indicates the length of the ahead part, n_f signifies the index of the front part, and n_a represents the index of the ahead part.

The movement direction of a submarine in water aligns with its bow section, necessitating the design of the bow section curve to enhance control performance and reduce drag coefficient. Conversely, the stern section curve is typically designed to mitigate vortex formation at the stern. As ocean exploration buoys move vertically in seawater, both ends of the buoy alternate as the bow. Hence, this study proposes that both ends of the buoy feature teardrop-shaped curves with the design of the generatrix as follows:

$$y = b\left[1-\left(\frac{X}{l}\right)^{n_1}\right]^{\frac{1}{n_2}}. \quad (2)$$

where n_1 and n_2 determine the line shape of the teardrop curve. The curve is a straight line when $n_1 = n_2 = 1$ and is nearly square when $n_1 = n_2 = 3$. To ensure that the research is accurate and effective and the scope of optimization is sufficient, the optimized range of values for n_1 and n_2 is constrained within the range of $[0.8\ ,\ 3]$.

The ocean exploration buoy driven by thermal energy needs a certain amount of internal storage space. Therefore, the design process begins with confirming the internal space of the buoy shell before proceeding with the external structure design. Combined with the heat exchange system of the buoy, it is determined that the shell of the buoy will be a cylinder with a diameter of 246 mm and a height of 1400 mm. The heat exchange system requires 6 heat exchange tubes distributed around the shell. Comprehensive considerations of the buoy shape, water displacement, overall mass, and density of the deflector material lead to the adoption of a transitional structure for the buoy, which is segmented into front, middle, and back sections. The rotating busbar of the front and back sections of the deflector serves as the optimization object. The model of the buoy is depicted in Figure 1. The buoy is designed for a maximum diving depth of 1000 m, a maximum sailing speed of 1 knot (0.514 m/s), and a power generation rating of 300 W. The buoy's internal capacity to carry equipment such as power generation systems, wireless communications, sensors, etc., as well as its external pressure-resistant capacity are considered; detailed dimensions are provided in Table 1.

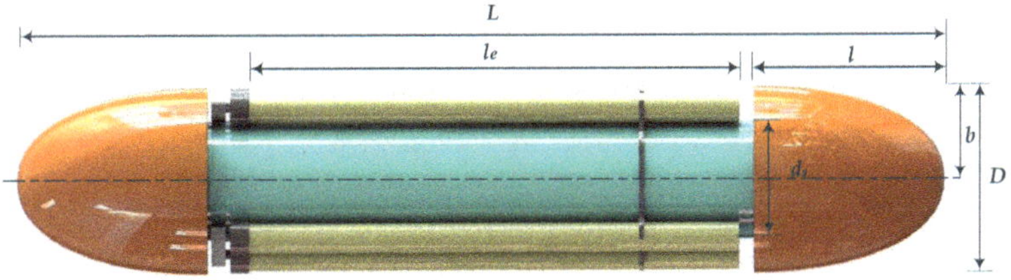

Figure 1. The shape of the teardrop buoy.

Table 1. Model parameters of the designed buoy.

Parameters	Sign	Data
length of the buoy	L	1940 mm
diameter of the buoy	D	380 mm
length of deflector	l	400 mm
radius of deflector	b	190 mm
length of heat exchanger tube	l_e	1050 mm
length of shell	l_s	1400 mm
diameter of shell	d_s	246 mm

2.2. Numerical Model

For the investigation of buoy hydrodynamic performance, accounting for fluid rotation and viscosity, the CFD method is employed. The CFD software Fluent is utilized for simulation purposes. This section outlines the governing equations of the CFD computational method, turbulence model selection, boundary conditions, and verification of mesh convergence.

2.2.1. Turbulence Model

As the buoy moves at a speed of 0.5 m/s, the Reynolds number reaches 1.9×10^5, surpassing a critical value. At this point, laminar flow will be disrupted and turbulence will be generated. In the turbulence model provided by ANSYS 2020 R2, the turbulent SST $k-\omega$ model exhibits better accuracy and stability in the near-wall region, allowing for a more accurate description of the flow field around the buoy. Therefore, the SST $k-\omega$ model has been chosen for this study. The SST $k-\omega$ is an optimized two-equation model that incorporates the equations for turbulent kinetic energy k and turbulence frequency ω [24].

$$\frac{\partial(\rho k)}{\partial t} + \frac{\partial(\rho \vec{u} k)}{\partial x_j} + \rho C_\mu k\omega = P_k + \frac{\partial}{\partial x_j}\left[\left(\mu + \frac{\mu_t}{\sigma_k}\right)\frac{\partial k}{\partial x_j}\right] \tag{3}$$

$$\begin{aligned}\frac{\partial(\rho \omega)}{\partial t} + \frac{\partial(\rho \vec{u} \omega)}{\partial x_j} + C_{\omega 2}\rho\omega^2 = C_{\omega 1}\frac{\omega}{k}P_k + \frac{\partial}{\partial x_j}\left[\left(\mu + \frac{\mu_t}{\sigma_{\omega 1}}\right)\frac{\partial \omega}{\partial x_j}\right] \\ +(1-F_1)\frac{2\rho}{\sigma_{\omega 2}\omega}\frac{\partial k}{\partial x_j}\frac{\partial \omega}{\partial x_j}\end{aligned} \tag{4}$$

$$F_1 = tanh\left\{\left[min\left(max\left(\frac{\sqrt{k}}{C_\mu \omega d}, \frac{500\nu}{d^2\omega}\right), \frac{2k}{d^2 CD_{K\omega}}\right)\right]^4\right\} \quad (5)$$

where P_k is the rate of turbulence generation, $C_\mu = 0.09$, $C_{\omega1} = 5/9$, $C_{\omega2} = 0.75$, $\sigma_k = 2$, $\sigma_{\omega1} = 2$, $\sigma_{\omega2} = 1.17$, ν is kinematic viscosity, d is the distance to the wall, and μ_t is turbulent viscosity, as follows:

$$\mu_t = \rho \min\left(\frac{k}{\omega}, \frac{\alpha_1 k}{SF_2}\right) \quad (6)$$

$$F_2 = tanh\left\{\left[max\left(\frac{2\sqrt{k}}{C_\mu \omega d}\right), \frac{500\nu}{d^2\omega}\right]^2\right\} \quad (7)$$

$$S = \frac{1}{2}(\nabla\vec{u} + (\nabla\vec{u})^{\mathrm{T}}) \quad (8)$$

2.2.2. Boundary Conditions

The boundary conditions and watershed dimensions for the CFD simulation are shown in Figure 2, while the dimensions of the buoy are shown in Table 1. This simulation is designed as a three-dimensional model with the computational domain measuring 6.94 m in length, 2.38 m in width, and 2.38 m in height. The left boundary is designated as the velocity inlet, while the right boundary is assigned as the pressure outlet. The buoy is positioned at the center of the computational domain with its surface defined as the wall. The fluid medium is seawater with a density of 1024.7 kg/m^3 and a dynamic viscosity coefficient of 1.08×10^{-3} Pa·s at 20 °C.

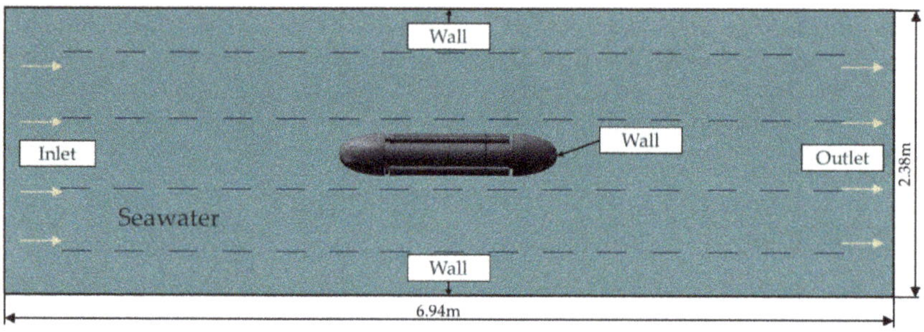

Figure 2. CFD model condition.

2.2.3. Mesh-Independent Verification

To ensure accurate calculation of the flow field around the buoy, automatic inflation is employed in the automatic meshing process of ANSYS 2022 R2. Inflation is applied to the surface of the buoy with the number of layers set to 10. The first layer thickness is set to 2.5 with a growth rate of 1.2. Furthermore, the number and refinement of the mesh significantly impact the accuracy and computational speed of the calculations. Generally, finer meshes yield more accurate results, but an increase in mesh density also escalates computational workload, consequently reducing calculation speed. Hence, it is imperative to determine an appropriate mesh density. Four models with varying mesh sizes are investigated, which are each scaled by a factor of $\sqrt{2}$. The number of meshes and corresponding simulation results are summarized in Table 2:

Table 2. Parameters of meshes and simulation result.

Models	Total Number of Elements	Drag (F_r)	Drag (C_d)
1	162.0 w	8.23 N	13.11
2	210.8 w	7.52 N	11.94
3	321.1 w	7.05 N	11.23
4	605.5 w	6.98 N	11.19

With all other parameters held constant, the computational results gradually converge as the number of meshes increases. When the number of grids reaches 321.1 w, both resistance and resistance coefficient have converged. Further increasing the number of meshes has minimal impact on result convergence. The difference in calculations between model 3 and model 4 is less than 1%. Considering computational efficiency, model 3 is selected as the mesh configuration for computation. The grid configuration is depicted in Figure 3.

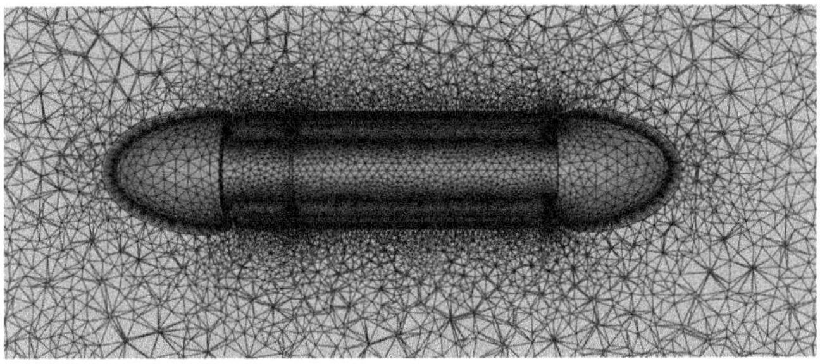

Figure 3. Mesh of simulation with teardrop buoy.

2.2.4. Numerical Simulation Validation

To verify the CFD model presented in this paper, experiments conducted by Gao et al. [25] on an AUV in a circulating water tunnel are simulated. The model parameters are kept consistent with the aforementioned study except for the dimensions and shape of the submarine. The external dimensions of the AUV are depicted in Figure 4. Figure 5 illustrates that the CFD results closely align with the experimental findings with a maximum simulation error of 5.84%. This demonstrates the accuracy of the numerical simulation. Consequently, the simulation results and optimization scheme based on this numerical model can be considered reliable.

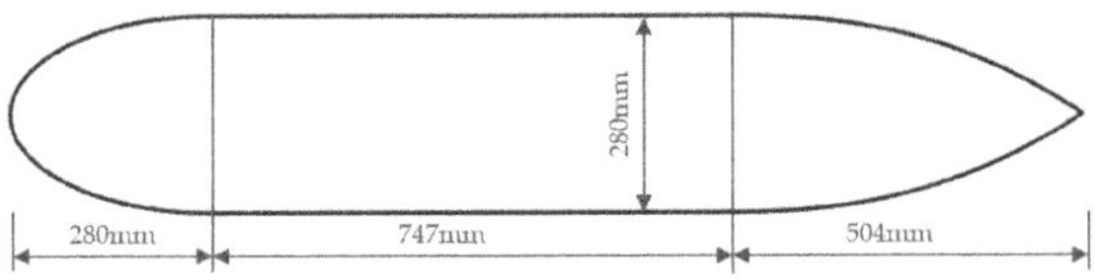

Figure 4. The dimension and shape of the AUV by Gao et al. [25].

2.3. *Optimization Scheme*

When optimizing the shape of the buoy, the primary objective is to reduce water resistance while the buoy moves through seawater. Additionally, the deflector is crafted using lightweight materials to provide extra buoyancy while minimizing buoy resistance. However, an excessively large deflector may impede buoy operations and increase the complexity of buoy deployment. Therefore, the optimization objective is to achieve lower resistance and increased drainage volume under these constraints. Figure 6 illustrates the optimization process: initial parameters obtained through Latin hypercube sampling are simulated to create an initial sample, which is followed by training the samples with a neural network to establish a fitted functional relationship. Subsequently, a genetic algorithm is employed to identify Pareto optimal solutions and plot the Pareto front. Finally, the optimal buoy shape is determined following CFD verification.

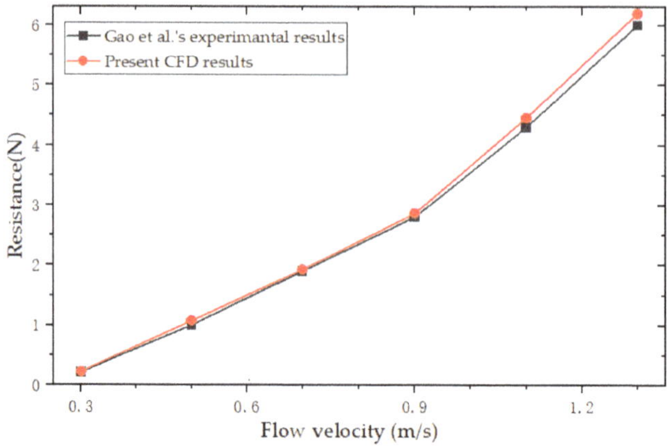

Figure 5. Comparison of resistance between CFD simulation and Gao et al.'s experiment result [25].

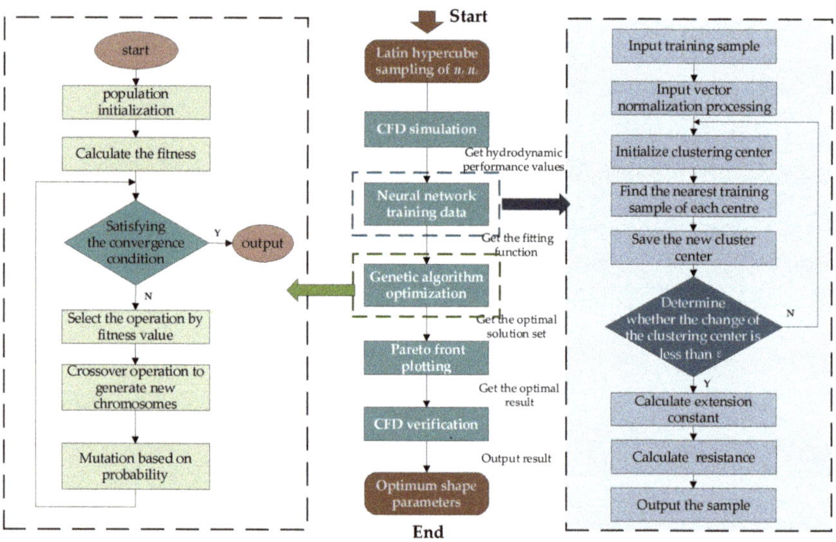

Figure 6. Optimization solution flowchart.

2.3.1. Building Performance Analysis Functions with Neural Networks

Based on the optimization objectives outlined above, this paper will employ a radial basis neural network to examine the relationship between the values of n_1 and n_2 and the water resistance, resistance coefficient, and drainage volume of the buoy during its movement. This analysis aims to explore the connection between the generatrix of the buoy and its hydrodynamic performance.

Neural networks are mainly composed of input, hidden and output layers. The neural network training process is shown on the right side of Figure 4. The training principle is shown in Algorithm 1; Sample A is used as a training set, from which vectors are selected and used as the center of the radial function base. The mean square error is minimized through gradient descent, and iterations continue until the error value meets satisfaction. Subsequently, the functional relationship between n_1, n_2 and resistance are outputted.

Algorithm 1: Neural network fitting the functional relationship between n_1, n_2 and resistance.

INPUT: Training set D = {(x_i, y_i)} i = 1, 2, ...30
Learning rate η = 0.001
Radial basis network based on (x_i, y_i)
REPEAT
FOR ALL (x_k, y_k) IN D
Calculate the error from the network parameters to the output layer;
Calculate the descending gradient of the network parameters based on the error;
Update the parameters according to the descending gradient;
END FOR
UNTIL mean square error < 0.01
OUTPUT the functional relationship between n_1, n_2 and resistance

The input layer x is:

$$x = \begin{bmatrix} n_1 & n_2 \end{bmatrix} \tag{9}$$

The Gaussian function is chosen as the radial basis function for the hidden layer due to its fast approximation and simple structure. The hidden layer function $\phi_i(x)$ can be expressed as

$$\phi_i(x) = \exp\left(-\frac{\| x - c_i \|^2}{\sigma_i^2}\right) \tag{10}$$

where i is the implied layer node number, c_i denotes the radial basis function center vector at the ith node, and σ_i is the radial basis function node width. Combining the above equations and introducing the weight vector w, the output layer y is

$$y = \sum_{i=1}^{N} w_i \exp\left(-\frac{\| x - c_i \|^2}{\sigma_i^2}\right) \tag{11}$$

$$w = [w_1, w_2, \cdots, w_n] \tag{12}$$

2.3.2. Genetic Algorithm

Genetic algorithms are search algorithms that mimic natural evolution and heredity. They search for and improve the optimal solution by performing operations such as selection, crossover, and mutation on individuals within the population. The optimization search principle is outlined in Algorithm 2: after initializing F_r, the fitness is calculated, individuals are selected based on their fitness, and then they are subjected to crossover and mutation. The resulting individuals are added to the new generation of the population, and the process repeats until the number of evolutionary generations reaches 500. The minimum F_r and its corresponding n_1, n_2 values are then outputted. The size of the initial population significantly affects the efficiency of the optimization search. A population that is too small can make it difficult to find the best solution for the optimization problem, whereas a large population may be inefficient and time consuming due to processing numerous unnecessary individuals. Therefore, in this study, the population size is set to 100, the optimal individual coefficient of the genetic algorithm is 0.4, the maximum number of evolutionary generations is 500, and the deviation of the fitness function is set to 10^{-10}.

2.3.3. Pareto Optimal Solution

Under the given constraints, the task of optimizing multiple objectives to attain their maximum (or minimum) values is referred to as a multi-objective optimization problem. The essence of solving such a problem lies in identifying a set of non-inferior solutions, known as Pareto-optimal solutions, which strive to achieve a balanced outcome across individual objectives. The collection of multiple Pareto-optimal solutions plotted in the objective function space forms what is known as the Pareto front.

Algorithm 2: Genetic algorithm for finding the minimum resistance value.

```
BEGIN
    Initialize  Fr(0)  t = 0;  Initialize parameters
    While  (t < 500) do
        Calculate individual fitness;
        Select individuals from the population based on fitness;
        Perform crossover and mutation of selected individuals;
        Add manipulated individuals to the population;
    UNTIL  t = 500  evolutionary generations up to 500
END
OUTPUT  minimum Fr
```

Assuming that the multi-objective optimization problem involves N variables, T optimization objectives, R inequality constraints, and K equality constraints, the expression is as follows:

$$\min\{y_1 = f_1(x),\ y_2 = f_2(x), \ldots, y_T = f_T(x)\} \tag{13}$$

$$\text{s. } t \begin{cases} g_i(x) \leq 0\ i = 0, 1, \ldots, R \\ h_k(x) = 0\ k = 0, 1, \ldots, K \end{cases} \tag{14}$$

where $x = [x_1, x_2, \ldots, x_{N,}]^T \in$ X, X is the space of design variables; $y = [y_1, y_2, \ldots, y_{N,}]^T \in$ Y, Y is the target space; $g(x)$ is the inequality constraint; and $h(x)$ is the equality constraint.

In this paper, the motion data of buoys with varying slewing body buses are utilized as inputs, and a radial basis neural network is trained to establish the relationship between the control parameters of the slewing body buses and the motion performance of the buoys. This relationship serves as the objective function output. Consequently, the shape optimization problem of the buoys is transformed into a multi-objective optimization problem involving the performance parameters of the buoys. The set of Pareto optimal solutions will be determined based on minimizing drag, minimizing drag coefficient, and maximizing drainage volume as the objectives.

3. Results and Discussion

In this section, the simulation results of the buoy's hydrodynamic performance are presented. The parameter optimization process is described in detail, aiming to minimize drag, minimize drag coefficient, and consider drainage volume. The optimized parameters are then simulated in Fluent to verify the hydrodynamic performance of the optimized buoys.

3.1. Sampling and CFD Simulation Results

Firstly, the initial shape of the buoy is simulated and analyzed. The seawater flow velocity is set as 0.1 m/s, 0.2 m/s, 0.3 m/s, 0.4 m/s and 0.5 m/s. The resulting drag force and drag coefficient of the buoy during its profile movement in seawater are shown in the Table 3 below.

Table 3. Kinematic energy parameters of bouy at different speeds.

Velocity	Drag (F_r)	Drag Coefficient (C_d)
0.1 m/s	0.26 N	0.43
0.2 m/s	1.08 N	1.77
0.3 m/s	2.45 N	4.01
0.4 m/s	4.50 N	7.34
0.5 m/s	7.05 N	11.23

When the velocity of the buoy was increased from 0.1 to 0.3 m/s, the drag on the buoy increased by 2.19 N, and from 0.3 to 0.5 m/s, the drag increased by 4.6 N. According to the formula:

$$F_b = \rho g V_b, \tag{15}$$

The skin bag needs to provide a volume change of about 245 mL for the buoy to undergo profiling motion at a velocity of 0.3 m/s, where F_b is the buoyancy on the buoy, and V_b is the buoy drainage volume. When the buoy is moving at a velocity of 0.5 m/s, the skin bag needs to provide a volume change of about 705 mL. However, the size of the skin bag is too large, occupying space inside the shell and requiring additional heat exchangers for operation. Therefore, optimizing the shape of the buoy to minimize drag, skin friction, and energy consumption is essential for enhancing buoy performance.

Secondly, the Latin hypercube sampling method combines random sampling with stratified sampling, providing a strong representation of samples across the overall probability distribution. This method yields statistically accurate results with small sample sizes [26], thereby achieving better sampling accuracy while utilizing smaller computational costs. Moreover, it ensures accurate representation of remote events in the sampling process. Therefore, the Latin hypercube sampling method is employed to generate 30 sets of curves, which are then used to simulate the hydrodynamic performance of the buoy in Fluent at 0.5 m/s. The curve parameters and simulation results are denoted as sample A. The spatial distribution of the 30 sets of samples taken is shown in Figure 7, and several examples are represented in Table 4.

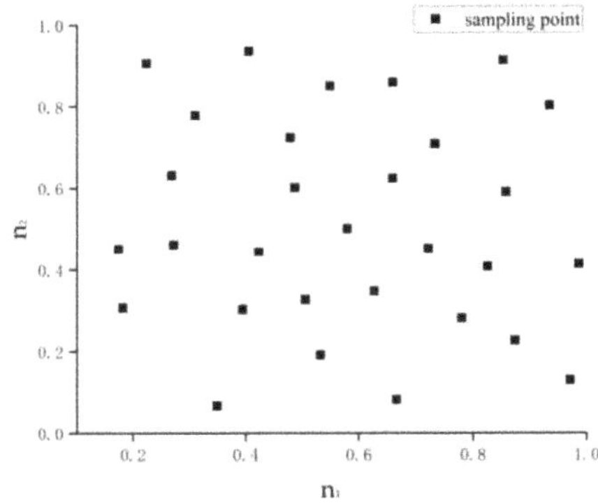

Figure 7. Spatial distribution of the sample.

Table 4. Performance parameters of several samples in sample A.

Sample	n_1	n_2	F_r (N)	C_d
1	2.748	1.498	6.975	11.388
2	2.869	2.766	6.595	10.767
3	2.253	1.765	6.797	11.098
4	2.716	2.300	6.474	10.570
5	1.349	1.992	7.495	12.237
6	2.319	2.891	6.727	10.983
7	2.064	1.421	7.621	12.442
8	2.560	1.619	6.916	11.292
9	1.974	2.323	6.894	11.255
10	2.198	2.871	6.684	10.912

The maximum value in minimum distance $\max\{d_{\min}\}$ and minimum potential energy U are used to discuss the accuracy of the randomization of the Latin hypercube sampling method, as shown below:

$$\max\{d_{\min}\} = \max\left\{ \min_{1 \le i,j \le 30,\ i \ne j} d(x_i, x_j) \right\} \tag{16}$$

$$U = \sum_{i=1}^{29} \sum_{j=i+1}^{30} d(x_i, x_j)^2 \quad (17)$$

where $d(x_i, x_j)$ is the distance between sample i and sample j.

It is calculated that $\max\{d_{\min}\}$ is 0.105, and U is 4041.3. Upon assessing the indicators and the sample distribution shown in Figure 7, it is evident that the Latin hypercube sampling method ensures the randomness of the sample. Given the complex and nonlinear functional relationship between resistance and curve parameters, it is essential to utilize the simulation results from the 30 sets of sample parameters as training data for the neural network to accurately capture this relationship.

3.2. Neural Network Training Results

The relationship between the parameter n_2, n_2 and the drag force on the buoy profile during motion is fitted to the data using the radial basis neural network described in Section 2.3.1. Thirty sets of buoy parameters with different slewing buses from the table are used as training data and inputted into the neural network for iterative computation. The relationship between the parameters n_2, n_2 and the resistance of the buoy is finally obtained as shown in Figure 8. It can be seen that there exists a certain set of defined and unique values within a specific range that minimizes the resistance to movement of the buoy. However numerous saddle points exist in the surface function, making the gradient descent method unsuitable for searching the optimal value as it may lead to falling into local optima.

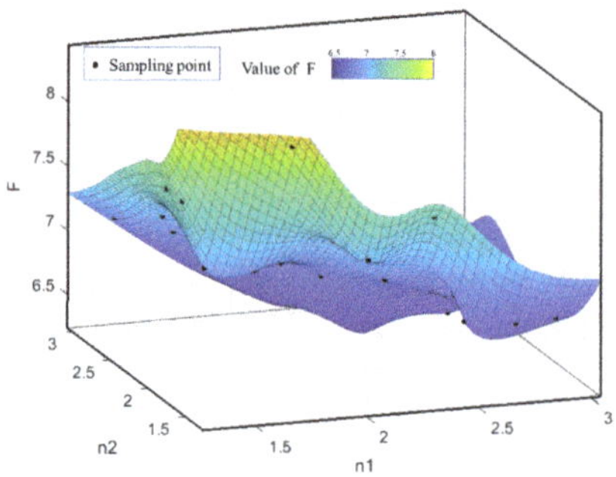

Figure 8. Neural network training result.

3.3. Optimal Solution Selection

Based on the global search results obtained from the genetic algorithm, the Pareto front is plotted, as shown in Figure 9. The Pareto solution set is determined based on criteria such as the minimum resistance, the minimum resistance coefficient, and the maximum drainage volume. Finally, five sets of parameters are selected, as illustrated in Table 5.

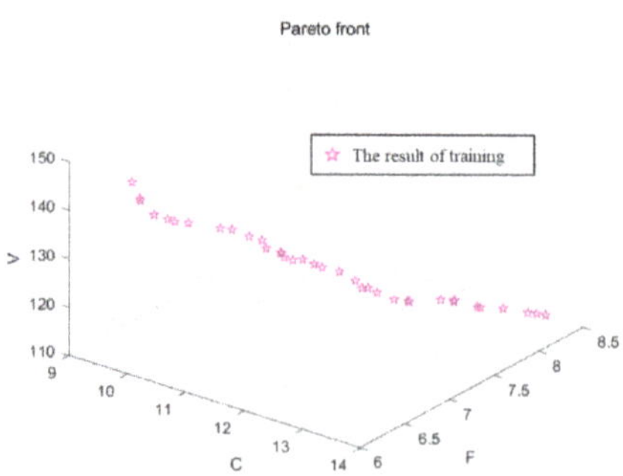

Figure 9. Pareto frontier of the optimization in this study.

Table 5. Optimal solution set of curve parameters.

Number	n_1	n_2	F_r (N)	C_d	V (L)
1	2.963	1.854	6.360	10.308	137.994
2	2.584	2.418	6.401	10.457	141.169
3	2.232	0.853	7.672	12.589	122.543
4	2.707	1.239	7.157	11.494	130.047
5	2.569	1.565	7.013	11.421	132.108

In the optimal solution set, the first group exhibits the smallest drag of 6.36 N with a corresponding resistance coefficient of 10.308. Additionally, its drainage volume measures 137.994 L. On the other hand, the second group records a resistance of 6.401 N and a resistance coefficient of 10.457, albeit with a higher drainage volume of 141.169 L, representing a 3.75 L increase compared to the first group. According to Equation (15), the load of the second group is increased by 3.75 kg compared to the first group. Remarkably, despite this increase in load, the resistance is only raised by 0.041 N. Consequently, the skin bag needs to provide a volume change of about 4 mL for the buoy to execute profiling motion. In summary, while the second group exhibits similar resistance to the first group, with both being significantly lower than the initial value, the second group boasts the highest load capacity. Hence, the parameters of the second group are deemed optimal.

3.4. Simulation Verification

According to the optimization results, the rotary generatrix is obtained as follows:

$$y = 190\left[1 - \left(\frac{X}{400}\right)^{2.824}\right]^{\frac{1}{2.818}}. \tag{18}$$

Fluent simulation is utilized to validate the plausibility of the results obtained from neural network fitting. At a flow velocity of 0.5 m/s, corresponding to the profiling velocity of the buoy, the simulated resistance on the buoy is 6.41 N. Comparatively, the fitted value obtained from the neural network is 6.401 N, resulting in a fitting error of approximately 0.14%. Additionally, Figure 10 illustrates the velocity streamline around the buoy, showcasing uniform flow velocity and stable flow lines. Consequently, the fitting function of the neural network and the optimization scheme are deemed credible. In comparison to the initial sample drag of 7.05 N, the optimized buoy drag experiences a reduction of 9.2%. Moreover, when compared to buoys lacking a deflector, the optimized buoy demonstrates a 22% reduction in resistance. This optimized solution effectively minimizes the resistance of the buoy while ensuring maximum drainage volume and stable flow field around the buoy with uniform flow velocity. Furthermore, in contrast to Li's streamlined buoy [24], the proposed and optimized teardrop buoy better aligns with the reciprocating motion characteristics of the buoy, thereby enhancing its performance during both descent and ascent. This optimized design is more practical and efficient.

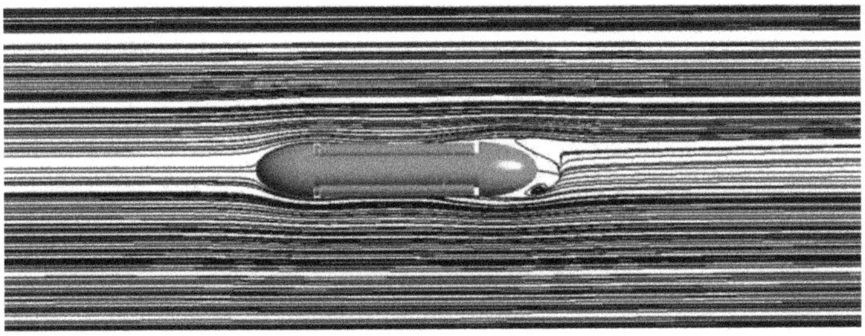

Figure 10. Velocity streamlines at the flow rate of 0.5 m/s.

To visually demonstrate the hydrodynamic performance of the buoy, velocity fields around the buoy are compared at different flow velocities ranging from 0.1 to 0.6 m/s, as depicted in Figure 11. The optimized teardrop buoy maintains a stable flow field even at low velocities, as observed in Figure 11a–d. Moreover, the flow field remains stable as the velocity increases, which is illustrated in Figure 11e,f. Noticeably, a small region of low velocity is observed at the top of the bow with slightly elevated flow velocities at the ends of the deflector. Fluid along the structural path of the device is compelled to form vortices around the edges, resulting in energy dissipation, which is known as form drag. The teardrop-shaped deflector serves to reduce boundary layer separation, mitigate backflow and vortex action, and consequently decrease pressure differential resistance. Additionally, the smooth passage of surface fluid minimizes the creation of vacuum zones on the buoy's surface, effectively reducing pressure differences that may be generated in this area. This reduction in resistance contributes to the superior hydrodynamic performance of teardrop-shaped buoys.

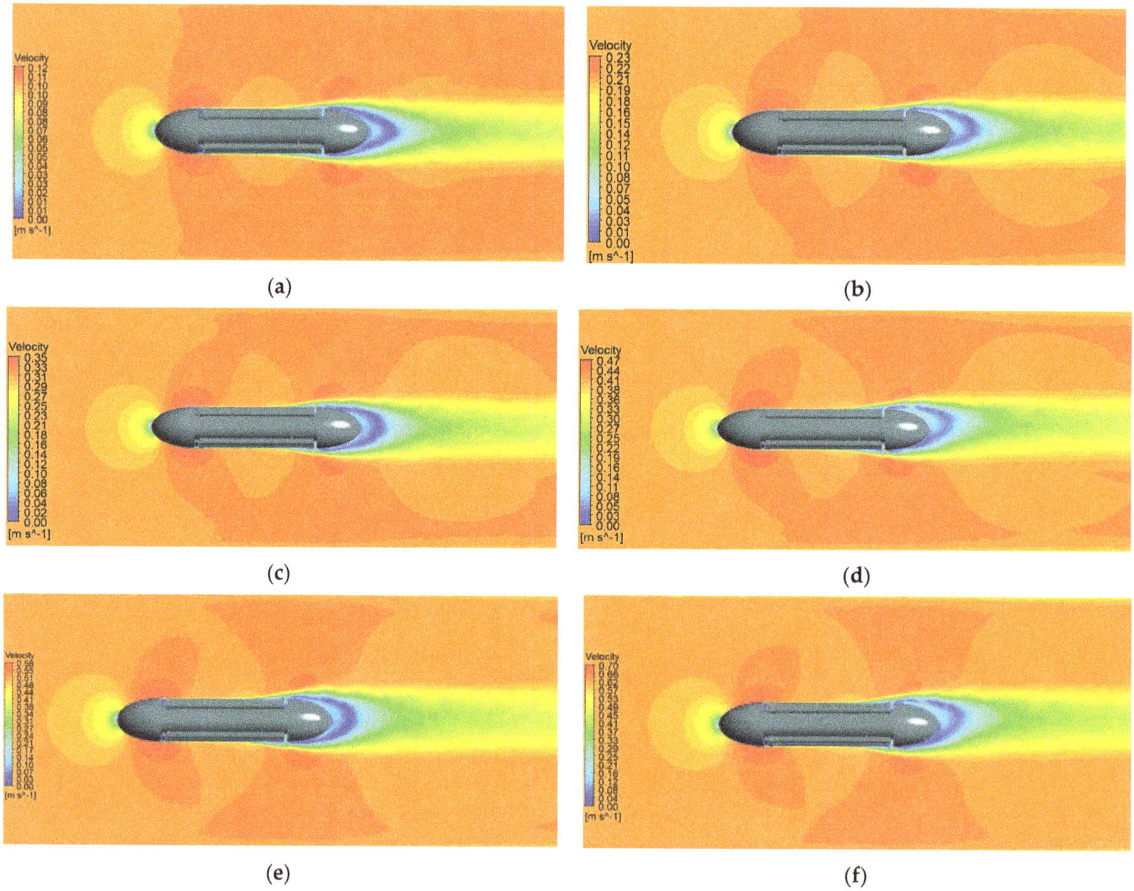

Figure 11. Velocity field around the teardrop buoy: (**a**) 0.1 m·s^{-1}; (**b**) 0.2 m·s^{-1}; (**c**) 0.3 m·s^{-1} (**d**) 0.4 m·s^{-1}; (**e**) 0.5 m·s^{-1}; (**f**) 0.6 m·s^{-1}.

4. Conclusions

This paper investigates the motion characteristics of buoys propelled by ocean thermal energy. Given the significant impact of the deflector shape on buoy performance, which directly influences operational efficiency and payload capacity, a teardrop buoy design is proposed and developed. Subsequently, a multi-objective optimization scheme, integrating genetic algorithms and neural networks, is formulated to minimize drag.

Firstly, a numerical simulation platform was established in Fluent, mirroring wind tunnel tests. Detailed descriptions of boundary conditions and mesh division procedures were provided. Thirty sample sets for n_2, n_2 are generated using the hypercubic sampling method and then simulated in Fluent to analyze buoy resistance and drag coefficients under varied parameters.

Furthermore, data are trained using a radial basis neural network to establish a functional relationship between parameters and drag. Results indicated the presence of a unique n_2, n_2 value that minimizes buoy resistance. Multi-objective optimization of the shell of the buoy, focusing on hydrodynamic performance and drainage volume, is conducted through a combination of genetic algorithms and Pareto front analysis, yielding optimal values for n_2 and n_2 as well as the buoy shell curve.

Finally, optimization results are verified in Fluent. Simulation outcomes demonstrate that the optimized shell reduces buoy resistance by 9.2% during seawater profile movement at 0.5 m/s, and by 22% compared to buoys lacking a deflector. This enhancement in motion performance and reduction in energy consumption in similar operational states provide a strong foundation for self-sustaining ocean thermal energy generation and buoyancy regulation control in underwater buoy applications.

Author Contributions: Conceptualization, D.Z. and W.S.; methodology, D.Z.; software, D.Z.; validation, D.Z. and Z.Z.; formal analysis, D.Z.; investigation, Z.Z.; resources, S.L.; data curation, D.Z. and W.S.; writing—original draft preparation, D.Z.; writing—review and editing, D.Z., W.S., F.G. and S.L.; visualization, Z.Z.; supervision, W.S.; project administration, S.L.; funding acquisition, S.L. All authors have read and agreed to the published version of the manuscript.

Funding: This research was funded by the National Key Research and Development Program of China under Grant 2023YFC2810000, the National Natural Science Foundation of China under Grant 52175054 and the Shandong Province Science and Technology-oriented Small and Medium-sized Enterprises (SMEs) Innovation Capacity Enhancement Project under Grant 2023TSGC0320 and 2022TSGC2172.

Institutional Review Board Statement: Not applicable.

Informed Consent Statement: Not applicable.

Data Availability Statement: Dataset available on request from the authors.

Conflicts of Interest: The authors declare no conflict of interest.

References

1. Xing, X.G.; Claustre, H. Toward deeper development of Biogeochemical-Argo floats. *Atmos. Ocean. Sci. Lett.* **2018**, *11*, 287–290. [CrossRef]
2. Potiris, M.; Mamoutos, I.G.; Tragou, E.; Zervakis, V.; Kassis, D.; Ballas, D. Dense Water Formation in the North—Central Aegean Sea during Winter 2021–2022. *J. Mar. Sci. Eng.* **2024**, *12*, 221. [CrossRef]
3. Hu, Y.; Shao, W.; Li, J.; Zhang, C.; Cheng, L.; Ji, Q. Short-Term Variations in Water Temperature of the Antarctic Surface Layer. *J. Mar. Sci. Eng.* **2022**, *10*, 287. [CrossRef]
4. Abbas, S.M.; Alhassany, H.D.S. Review of enhancement for ocean thermal energy conversion system. *J. Ocean. Eng. Sci.* **2023**, *8*, 533–545. [CrossRef]
5. Herrera, J.; Sierra, S. Ocean Thermal Energy Conversion and Other Uses of Deep Sea Water: A Review. *J. Mar. Sci. Eng.* **2021**, *9*, 356. [CrossRef]
6. Barnard, A.H.; Mitchell, T.O. Biogeochemical monitoring of the oceans using autonomous profiling floats. *Ocean News Technol.* **2013**, *19*, 16–17.
7. Roemmich, D.; Johnson, G.C.; Riser, S. The Argo Program: Observing the global ocean with profiling floats. *Oceanography* **2009**, *22*, 34–43. [CrossRef]
8. Roemmich, D.; Sherman, J.T.; Davis, R.E. Deep SOLO: A full-depth profiling float for the Argo Program. *J. Atmos. Ocean. Technol.* **2019**, *36*, 1967–1981. [CrossRef]
9. Petzrick, E.; Truman, J.; Fargher, H. Profiling from 6000 meter with the APEX-Deep float. In Proceedings of the Oceans 2013 MTS/IEEE San Diego, San Diego, CA, USA, 23–27 September 2013.

10. Kobayashi, T.; Amaike, K.; Watanabe, K. Deep NINJA: A new float for deep ocean observation developed in Japan. In Proceedings of the 2011 IEEE Symposium on Underwater Technology and Workshop on Scientific Use of Submarine Cables and Related Technologies, Tokyo, Japan, 5–8 April 2011.
11. Gould, W.J. From swallow floats to Argo—The development of neutrally buoyant floats. *Deep Sea Res. Part II Top. Stud. Oceanogr.* **2005**, *52*, 529–543. [CrossRef]
12. André, X.; Moreau, B.; Reste, S.L. Argos-3 Satellite Communication System: Implementation on the Arvor Oceanographic Profiling Floats. *J. Atmos. Ocean. Technol.* **2015**, *32*, 1902–1914. [CrossRef]
13. Yu, L.Z.; Zhang, S.Y.; Shang, H.M. Progress of China Argo float. *Ocean. Technol.* **2005**, *2*, 121–129.
14. Sherman, J.; Davis, R.E.; Owens, W.B.; Valdes, J. The autonomous underwater glider "spray". *IEEE J. Ocean. Eng.* **2001**, *126*, 437–446. [CrossRef]
15. Davis, R.E.; Eriksen, C.C.; Jones, C.P. Autonomous buoyancy-driven underwater gliders. In *The Technology and Applications of Autonomous Underwater Vehicles*; Griffiths, G., Ed.; Taylor and Francis: London, UK, 2002; pp. 37–58.
16. Webb, D.C.; Simonetti, P.J.; Jones, C.P. SLOCUM: An underwater glider propelled by environmental energy. *IEEE J. Ocean. Eng.* **2001**, *26*, 447–452. [CrossRef]
17. Eriksen, C.C.; Osse, T.J.; Light, R.D. Seaglider: A long-range autonomous underwater vehicle for oceanographic research. *IEEE J. Ocean. Eng.* **2001**, *126*, 424–436. [CrossRef]
18. Bertram, V.; Alvarez, A. Hydrodynamic Aspects of AUV Design 2006. Available online: https://www.researchgate.net/publication/228395786 (accessed on 5 April 2024).
19. Ye, P.; Pan, G. Shape optimization of a blended-wing-body underwater glider using surrogate-based global optimization method IESGO-HSR. *Sci. Prog.* **2020**, *103*, 0036850420950144. [CrossRef] [PubMed]
20. Sun, C.; Song, B.; Wang, P. Parametric geometric model and shape optimization of an underwater glider with blended-wing-body. *Int. J. Nav. Arch. Ocean Eng.* **2015**, *7*, 995–1006. [CrossRef]
21. Fu, X.; Lei, L.; Yang, G.; Li, B. Multi-objective shape optimization of autonomous underwater glider based on fast elitist non-dominated sorting genetic algorithm. *Ocean Eng.* **2018**, *157*, 339–349. [CrossRef]
22. Li, Z.; Liu, Y.; Guo, F.; Xue, G.; Li, S.; Li, X. Multi-objective optimization of the shell in autonomous intelligent Argo profiling float. *Ocean Eng.* **2019**, *187*, 106176. [CrossRef]
23. Johnson, G.C.; Hosoda, S.; Jayne, S.R.; Oke, P.R.; Riser, S.C. Argo—Two decades: Global oceanography, revolutionized. *Annu. Rev. Mar. Sci.* **2022**, *14*, 379–403. [CrossRef]
24. Liu, S.; Ong, M.C.; Obhrai, C. Influences of free surface jump conditions and different $k-\omega$ SST turbulence models on breaking wave modelling. *Ocean. Eng.* **2020**, *217*, 107746. [CrossRef]
25. Gao, T.; Wang, Y.; Pang, Y. Hull shape optimization for autonomous underwater vehicles using CFD. *Eng. Appl. Comp. Fluid* **2016**, *10*, 601–609. [CrossRef]
26. Yang, W.; Zhang, Y.; Wang, Y.; Liang, K.; Zhao, H.; Yang, A. Multi-Angle Reliability Evaluation of Grid-Connected Wind Farms with Energy Storage Based on Latin Hypercube Important Sampling. *Energies* **2023**, *16*, 6427. [CrossRef]

Journal of
Marine Science and Engineering

Article

IRNLGD: An Edge Detection Algorithm with Comprehensive Gradient Directions for Tidal Stream Turbine

Dingnan Song [1], Ran Liu [2], Zhiwei Zhang [3], Dingding Yang [1] and Tianzhen Wang [1,*]

1 Logistics Engineering College, Shanghai Maritime University, Pudong District, Shanghai 201306, China; songdingnan@126.com (D.S.); 15272965282@163.com (D.Y.)
2 Leshan Shawan Power Supply Branch, State Grid Sichuan Electric Power Company, Leshan 614900, China; rainoc@126.com
3 Shanghai Power Industrial & Commerical Co., Ltd., State Grid Shanghai Municipal Electric Power Company, Huangpu District, Shanghai 200001, China; dianjing@21cn.com
* Correspondence: tzwang@shmtu.edu.cn; Tel.: +86-2138-282-640

Abstract: Tidal stream turbines (TSTs) harness the kinetic energy of tides to generate electricity by rotating the rotor. Biofouling will lead to an imbalance between the blades, resulting in imbalanced torque and voltage across the windings, ultimately polluting the grid. Therefore, rotor condition monitoring is of great significance for the stable operation of the system. Image-based attachment detection algorithms provide the advantage of visually displaying the location and area of faults. However, due to the limited availability of data from multiple machine types and environments, it is difficult to ensure the generalization of the network. Additionally, TST images degrade, resulting in reduced image gradients and making it challenging to extract edge and other features. In order to address the issue of limited data, a novel non-data-driven edge detection algorithm, indexed resemble-normal-line guidance detector (IRNLGD), is proposed for TST rotor attachment fault detection. Aiming to solve the problem of edge features being suppressed, IRNLGD introduces the concept of "indexed resemble-normal-line direction" and integrates multi-directional gradient information for edge determination. Real-image experiments demonstrate IRNLGD's effectiveness in detecting TST rotor edges and faults. Evaluation on public datasets shows the superior performance of our method in detecting fine edges in low-light images.

Keywords: tidal stream turbine; attachment fault detection; edge detection; renewable energy

Citation: Song, D.; Liu, R.; Zhang, Z.; Yang, D.; Wang, T. IRNLGD: An Edge Detection Algorithm with Comprehensive Gradient Directions for Tidal Stream Turbine. *J. Mar. Sci. Eng.* **2024**, *12*, 498. https://doi.org/10.3390/jmse12030498

Academic Editor: Kamal Djidjeli

Received: 21 February 2024
Revised: 9 March 2024
Accepted: 13 March 2024
Published: 17 March 2024

1. Introduction

Climate change and energy shortages are two rising issues on the agenda of the world. Human activities are driving the global warming trend observed since the mid-20th century, and the effects are irreversible and will worsen in the decades to come. Renewable energy sources, which are cleaner and more efficient, could play a pivotal role in sustainability development. The theoretical capacity of tidal energy in China reaches 110 GW [1], indicating vast prospects for development. Its underwater generator, the tidal stream turbine (TST), does not occupy land resources, avoiding noise or visual pollution [2,3]. A TST can provide similar power to a bigger wind turbine [4]. The above characteristics prove that tidal stream energy is attractive for electric power generation. However, TSTs are susceptible to factors such as biofouling [5], sudden changes in instantaneous flow velocity [6], seawater corrosion [7], cavitation, and turbulence. These factors can result in mechanical faults or blade damage, leading to torque imbalances and voltage fluctuations, which significantly affect the quality of power generation, efficiency, and grid stability [8–10]. Biofouling can cause an increase in device maintenance time and structural loading [11]; there is also a biosecurity risk since immersed devices can serve as potential vectors for the introduction of non-native species [12]. Therefore, timely and effectively detection of TST rotor attachment is of significant importance.

Current methods for TST rotor attachment detection primarily rely on electrical and image signals. Electrical signal-based methods, usually utilizing time–frequency or statistical analysis and machine learning, have shown some effect in detecting unbalanced rotor attachments [13–16]. However, these methods are susceptible to sea state and can find it difficult to detect balance attachments. Image-based methods overcome the shortcomings by analyzing the attachment conditions from underwater TST images directly. They focus on classification and semantic segmentation networks. Zheng et al. [17] carried out attachment detection using an improved sparse autoencoder (SAE) and Softmax Regression (SR) method. They collected TST attachment images and divided the data based on the degree of attachment. Their proposed method achieved higher accuracy compared to the traditional principal component analysis (PCA) feature extraction algorithm. However, they only focused on TSTs in a static state, which is an idealized condition not reflective of reality. Therefore, Xin et al. [18] collected TST images under operational conditions to make the dataset more representative of real-world scenarios. Then, the data were classified using a depthwise separable convolutional neural network (CNN), which achieved higher recognition accuracy than SAE+SR and a reasonable computational cost compared to large deep networks such as ResNet. However, the classification-based methods fail to precisely localize and display the biofouling. Peng et al. [19] proposed a semantic segmentation network to identify TST blade attachments. It identified the TST blade edge by constructing two branches of coarse and fine segmentation. However, it requires a large number of labeled data, high computational costs, and a longer training time. Peng et al. [20] proposed an image generation method, which extended and generated labeled data to reduce the workload of manual labeling. The proposed C-SegNet further improved the segmentation accuracy but performed poorly in contour localization. To address the localization misalignment caused by motion blur, Qi et al. [21] combined preceding and succeeding frames for fault detection, significantly improving the localization accuracy on blurred images. However, a remaining challenge is the insufficient diversity in data, leading to inadequate network generalization. The limited data restricts data-driven algorithms to specific environments, posing a challenge for the application of TST attachment fault detection algorithms. Moreover, the severe degradation of underwater images will impact the extraction of features such as edges.

To address the above problem, this paper proposes a non-data-driven edge detection algorithm, Indexed Resemble-Normal-Line Guidance Detector (IRNLGD), for TST attachment fault detection. Focusing on the difficulty of extracting edge features in degraded underwater images, we employ an eight-directional gradient operator to extract image gradients. Next, to better utilize gradient direction information, we introduce the concept of "indexed resemble-normal line direction" and calibrate the gradient directions based on the trend of gradient changes. Finally, edge points are determined through the joint calibration of gradient direction and magnitude. The experimental results on the TST dataset and public datasets demonstrate IRNLGD's effectiveness in dim images. What is more, a TST attachment fault detection method is explored on the basic of IRNLGD. A two-level detection is carried out with a data-driven lightweight classification network and the non-data-driven algorithm IRNLGD. It combines the advantages of both algorithms, providing more reliable and precise fault detection results.

The main contributions of this paper are as follows:

1. An effective edge detection algorithm, IRNLGD, is proposed to extract edge from low-contrast images.
2. A TST attachment fault detection method, MobileNet-IRNLGD, is explored, which combines data-driven and non-data-driven algorithms to strike a balance between limited data and precise detection.
3. The proposed TST attachment fault detection method is applied specifically to the real TST images, demonstrating the feasibility for engineering applications.

The remainder of the paper is organized as follows: In Section 2, we review the related work on edge detection. In Section 3, the proposed method is introduced in detail. Then the experimental results and analysis are presented in Section 4. Finally, we give the conclusions of this study in Section 5.

2. Related Work

The TST images captured by underwater sensors degrade due to the environment. According to the classical Jaffe–McGlamery underwater imaging model [22,23], the received light by the camera can be divided into three components, i.e., a direct, forward-scattered, and back-scattered component. The degradation process can be simplified as follows: an increase in the camera–object distance leads to direct component attenuation, resulting in low contrast; forward scattering causes convolution of the point spread function with scene radiance, causing blurring; and increased background light due to backward scattering amplifies image noise. These factors collectively drive the degradation of underwater images, resulting in a smaller rate of intensity variation, making it challenging to extract features such as edges.

Edge detection can significantly reduce the irrelevant information in an image, allowing essential structures to be preserved. Edge detection has a rich history and can be divided into the following two categories: handcrafted-based methods and machine learning-based methods [24]. The two main streams have achieved satisfactory performance in edge detection tasks. However, they still face challenges when dealing with specific tasks such as blurred or dim images.

2.1. Handcrafted-Based Edge Detection Methods

The existing handcrafted-based edge detection methods extract spatial, geometric, and other features from the image. They perform operations such as gradient calculations or feature statistics to achieve edge detection. Gradient-based detectors, such as Sobel [25], Laplacian [26], and Canny [27], perform first- or second-order gradient calculations on the image and combine the gradient values with directions to perform edge detection. Gonzalez et al. [28] combined gradient and fuzzy logic theory to detect edge of color images, but the gradient information is not fully used. Ranjan et al. [29] carried out guided image filtering before edge detection. Their strategy achieves satisfactory results, but the performance may fluctuate with the manually selected parameters. Ma et al. [30] proposed a two-level edge detection method, combined with a back propagation (BP) network, to measure the concrete surface roughness. However, low-quality images can pose a challenge to the algorithm's accuracy. According to the analysis of the degradation of underwater images, the TST images captured by the camera are severely blurred. The pixel values of a blurred image are represented as a weighted sum of multiple pixels in the neighborhood of the clear image. Therefore, the image gradient decreases, making it difficult to extract edge features. The gradient-based edge detection operators often focus on two or four gradient directions, neglecting the possibility of other gradient directions. When determining edge points, only the information of the maximum gradient direction or the dominant direction is considered, limiting the contribution of gradients in other directions. They do not fully utilize the information about gradient direction, which leads to the failure of edge detection in low-contrast and blurred images, such as underwater images, as shown in Figure 1b–d.

2.2. Machine Learning-Based Edge Detection Methods

The machine learning-based methods excel at learning features from a large number of samples, thereby better capturing edge structures [24,31,32]. Inspired by the idea of edge patches with obvious local structures, P. Dollár et al. [33] introduced structure learning into edge detection, proposing a structured edge (SE) detector. This method demonstrates satisfactory performance in both accuracy and real-time processing. Thanks to the remarkable performance of deep learning on images, it has gained popularity in edge detection tasks. Inspired by fully CNN, Xie et al. [34] proposed an end-to-end edge detection algorithm, Holistically-Nested Edge Detection (HED), which utilized multi-scale and deep supervision to improve the localization accuracy of edge pixels. Building on the HED style, Liu et al. [35] used relaxed deep supervision (RDS) to guide intermediate layer learning, incorporating operators like Canny and SE to facilitate the learning of edge features. Considering the commonality between edge detection and image segmentation, Yang et al. [36] introduced the U-Net [37] structure into edge detection tasks, proposing a fully convolutional encoder-decoder network (CEDN). Zou et al. [38] constructed DeepCrack based on SegNet [39] utilizing multi-scale feature fusion for crack detection. Su et al. [40] combined traditional difference operators with convolution and introduced a Pixel Difference Network (PiDiNet). It focuses on edge-related features, achieving a better trade-off between accuracy and efficiency for edge detection. The machine learning-based methods have demonstrated impressive capabilities, but their performance relies on sufficiently comprehensive datasets. Their accuracy may not be guaranteed on limited samples.

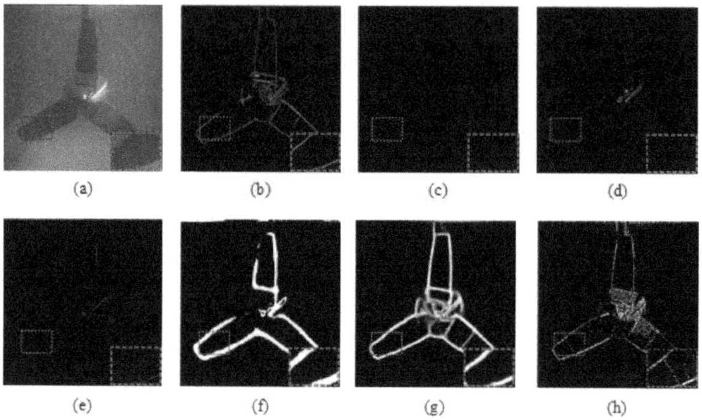

Figure 1. The comparison of different algorithms on underwater TST image: (**a**) original image; (**b**) Canny; (**c**) Roberts; (**d**) Sobel; (**e**) SE; (**f**) HED [34]; (**g**) PiDiNet [40]; (**h**) IRNLGD. The red boxes indicate the dim region.

In summary, current edge detection algorithms have achieved good results. However, they still face challenges when dealing with low-quality images. Aiming at the problems mentioned above, inspired by classical gradient-based operators, we propose an edge detection algorithm that needs no training. It captures additional gradient clues and combines gradient magnitudes with directions to determine edges, allowing for effective preservation of edges in low-contrast images. As shown in Figure 1, most handcrafted-based methods fail to detect the contour of the TST and none of the compared methods are able to detect the edges of the dim regions, as indicated by the red boxes in Figure 1b,f,g. Only the proposed method is capable of detecting the contour of the TST and the edges between the attachment and the blade in the dim region (as shown in Figure 1h). A detailed introduction is given in Section 3.

3. Indexed Resemble-Normal-Line Guidance Detector

The edge detection algorithm proposed in this paper is divided into three steps: (1) The eight-direction gradients of the input image are calculated to form a gradient matrix. (2) The direction is calibrated according to the pairwise gradient magnitude, and the indexed resemble-normal-line direction is obtained according to the calibrated direction matrix. (3) The edge point determination is performed by combining the gradient magnitude and the indexed resemble-normal-line direction. In the following, we will introduce the proposed method in detail and return to its application in Section 4.

3.1. Calculation of Eight-Direction Gradients

Gradients contain rich image information, and edge detection tasks can be performed using gradients. Traditional operators typically focus on two or four directions when computing gradients, thereby losing information from other directions. To obtain comprehensive multi-directional gradient information, an adjustable eight-direction gradient calculation method is proposed. The gradient magnitude for the first direction is calculated by

$$g_1 = \gamma[f(x,y) - f(x-1,y-1)] = \gamma \begin{bmatrix} -1 & 0 & 0 \\ 0 & 1 & 0 \\ 0 & 0 & 0 \end{bmatrix} * \begin{bmatrix} f(x-1,y-1) & f(x,y-1) & f(x+1,y-1) \\ f(x-1,y) & f(x,y) & f(x+1,y) \\ f(x-1,y+1) & f(x,y+1) & f(x+1,y+1) \end{bmatrix}, \quad (1)$$

where $f(x,y)$ indicates the pixel intensity of the pixel (x,y). We design an eight-directional gradient operator, sequentially rotating 45° from the first direction. According to Equation (1), the eight-directional gradient operator is given by:

$$\begin{aligned} \mathbf{C}_1 &= \gamma \begin{bmatrix} -1 & 0 & 0 \\ 0 & 1 & 0 \\ 0 & 0 & 0 \end{bmatrix}, \mathbf{C}_2 = \gamma \begin{bmatrix} 0 & -1 & 0 \\ 0 & 1 & 0 \\ 0 & 0 & 0 \end{bmatrix} \\ \mathbf{C}_3 &= \gamma \begin{bmatrix} 0 & 0 & -1 \\ 0 & 1 & 0 \\ 0 & 0 & 0 \end{bmatrix}, \mathbf{C}_4 = \gamma \begin{bmatrix} 0 & 0 & 0 \\ 0 & 1 & -1 \\ 0 & 0 & 0 \end{bmatrix} \\ \mathbf{C}_5 &= \gamma \begin{bmatrix} 0 & 0 & 0 \\ 0 & 1 & 0 \\ 0 & 0 & -1 \end{bmatrix}, \mathbf{C}_6 = \gamma \begin{bmatrix} 0 & 0 & 0 \\ 0 & 1 & 0 \\ 0 & -1 & 0 \end{bmatrix} \\ \mathbf{C}_7 &= \gamma \begin{bmatrix} 0 & 0 & 0 \\ 0 & 1 & 0 \\ -1 & 0 & 0 \end{bmatrix}, \mathbf{C}_8 = \gamma \begin{bmatrix} 0 & 0 & 0 \\ -1 & 1 & 0 \\ 0 & 0 & 0 \end{bmatrix}, \end{aligned} \quad (2)$$

where γ is the gradient amplification coefficient. It is introduced to amplify the gradient magnitudes without affecting subsequent gradient comparisons and direction calibration. Since the edge pixel intensity is based on the gradient magnitude, it is necessary to appropriately amplify the gradient values to enhance the edge. However, the calculation may overflow if γ is too large. Here, we take $\gamma = 2$.

For the input image $\boldsymbol{I}$, the gradient is obtained by convolution with the gradient operator in the corresponding direction. Therefore, according to Equation (2), the gradient matrix can be written as $\boldsymbol{G} = [\boldsymbol{G}_1, \ldots, \boldsymbol{G}_8]$, where

$$\boldsymbol{G}_k = \boldsymbol{C}_k * \boldsymbol{I} \quad k = 1,2,\ldots,8. \quad (3)$$

Notice that $\boldsymbol{G}_k \in \mathbb{R}^{H \times W}$. $\boldsymbol{G}$ records pixelwise eight-direction gradient magnitudes of the input image, and the sequence indicates the corresponding direction. The gradient matrix $\boldsymbol{G}$ provides the basic information for direction calibration and edge discrimination.

3.2. Indexed Resemble-Normal-Line Direction Calibration

In order to further utilize the cues of the gradient matrix, we introduce gradient directions in the form of recalibration. In classical algorithms, the role of gradient direction in edge detection tasks is to determine the edge direction and filter out pixels outside the specified direction. In this process, only the information of the maximum gradient direction or dominant direction is considered, resulting in the loss of information from other directions. Therefore, a novel gradient direction calibration method is proposed to preserve more useful cues.

According to the eight-directional gradients mentioned above, the calibrated direction is given by

$$\begin{cases} D_{x,y,k} = sgn\left(\left|G_{x,y,k}\right| - \left|G_{x,y,k+4}\right|\right) \\ D_{x,y,k+4} = -sgn\left(\left|G_{x,y,k}\right| - \left|G_{x,y,k+4}\right|\right), \end{cases} \quad (4)$$

where $D \in \mathbb{R}^{H \times W \times 8}$ is the calibrated direction matrix. $D_{x,y,k}$ indicates the $k^{th}(k \leq 4)$ calibrated direction of the pixel (x, y). D consists of $-1, 0$, and 1, indicating the change trend of a pixel spreading to its neighborhood. When the intensity increases, the corresponding direction is calibrated as 1.

More specifically, assume that there exists a neighborhood of size 3×3 for pixel (x, y) and calculate the eight gradient magnitudes $G_1, \ldots G_8$. The gradient direction calibration process is as follows:

1. Compare the magnitude of two gradients at 180° to each other;
2. Calibrate the directions to 1, 0, or -1, respectively, based on Equation (4). The calibration is carried out simultaneously due to the pairwise comparison;
3. Iterate over all gradient magnitudes in the neighborhood.

Following the idea of obtaining multi-directional gradient information, the set of directions calibrated as 1 in $D_{x,y}$ is defined as the indexed resemble-normal-line direction of the pixel (x, y) , marked as $r_{x,y}$, i.e.,

$$r_{x,y} = \left\{k \middle| D_{x,y,k} = 1\right\}. \quad (5)$$

$r_{x,y}$ is an indefinite-length vector, with a length of $l(l \in \{1,2,3,4\})$. It contains the direction information of intensity growth. Different from traditional methods using single direction information, the indexed resemble-normal-line direction records all directions where the intensity has an increasing trend. It preserves more gradient information and is beneficial for extracting edge features from degraded images. To fully utilize differential information across various directions, gradient magnitude and direction are retained to complement edge discrimination where gradient directions are not recorded in the indexed resemble-normal-line direction.

3.3. *Edge Discrimination*

Due to the characteristics of degraded underwater images, edge features such as gradients are suppressed. Traditional edge detection methods fail because they consider information limited to partial gradients in the image. Therefore, a multi-directional-based edge point determination method is proposed. The pixelwise edge discrimination is carried out by combining the gradient magnitude and the indexed resemble-normal-line direction. Inspired by classical Gestalt cues [41]: similarity, continuity, and proximity, the following three criteria are proposed:

1. The gradient magnitude of the edge point is much larger than that of the non-edge.
2. The length of the indexed resemble-normal-line direction of the edge point satisfies $len(r_{x,y}) \geq 3$.
3. Within 3×3 neighborhood of the edge point, at least one pixel has the same indexed resemble-normal-line direction, and the gradient magnitude is about the same.

Criterion 1 follows the principle of abrupt changes in intensity at edges. The gradient reflects the intensity variation, where a larger gradient magnitude indicates a more significant grayscale variation and a higher likelihood of an edge. For criterion 1, the decision threshold is calculated as follows:

For the eight-direction gradient $G_{x,y}$ of pixel (x, y), the quadratic sum matrix $Gs_{x,y}$ can be given as

$$Gs_{x,y} = \sum_{k=1}^{8} G^2_{x,y,k}. \quad (6)$$

To eliminate the effect of too small pixel values on the threshold calculation, a truncation parameter μ is set, and a dynamic truncation threshold $\boldsymbol{M}$ can be obtained as

$$\boldsymbol{M} = \mu \times max(\boldsymbol{Gs}_{x,y}). \tag{7}$$

The pixel is involved if and only if its quadratic sum value is greater than $\boldsymbol{M}$. Through experimental verification, here, we take $\mu = 0.1$.

Then, the threshold λ is calculated as

$$\lambda = \sqrt{\frac{\sum\limits_{x=1}^{H} \sum\limits_{y=1}^{W} \{\boldsymbol{Gs}_{x,y} | \boldsymbol{Gs}_{x,y} \geq M\}}{n}}, \tag{8}$$

where n is the number of pixels involved in the calculation.

Criterion 2 ensures that the detected edge points exhibit intensity variations in as many directions as possible. This helps in locating edge points accurately and also contributes to refining the edge. Criterion 3 aims to maintain edge continuity by controlling the similarity of gradients among neighboring pixels.

In conclusion, the proposed IRNLGD is finally detailed through the following steps:

1. Calculating eight-direction gradients: after graying the input image, the grayscale image is convolved with the eight-direction gradient operator to calculate the gradient matrix $\boldsymbol{G}$.
2. Indexed resemble-normal-line direction calibration: First, the gradients on the opposite direction are compared. The eight directions are calibrated to $-1, 0$, or 1, and saved in the gradient direction calibration matrix $\boldsymbol{D}$. Then, for each pixel, the set of the channel sequence numbers with value 1 in $\boldsymbol{D}$ is denoted as indexed resemble-normal-line direction $\boldsymbol{r}$.
3. Edge discrimination: First, the decision threshold λ is calculated based on the gradient matrix $\boldsymbol{G}$. Then, the edge points are judged by the proposed criteria.

3.4. MobileNet-IRNLGD TST Rotor Attachment Detection Network

In image-based TST attachment detection methods, the classification methods perform category-level detection, with a relatively low dependence on data diversity but insufficient precision in detection. The segmentation algorithm, on the other hand, conducts pixel-level detection, accurately indicating attachment and blade regions, but it requires abundant data under various conditions to ensure accuracy. To balance the limited data and high-precision detection, we explore a two-level TST rotor attachment detection method that combines the strengths of classification and segmentation: a classification network conducts first-level detection for preliminary fault severity assessment, while a non-data-driven IRNLGD edge detection branch performs second-level detection for precise fault localization. The design of the two-level detection combines the advantages of supervised algorithms and non-data-driven methods: the category-level detection in the first stage can fully utilize the existing data priors with the advantages of a deep network to make a preliminary fault assessment, reducing the workload of manual identification; the pixel level in the second stage can further locate the fault on the basics of fault assessment, and conduct secondary detection to prevent possible false alarms in the first step, providing more precise fault detection results for subsequent maintenance operations.

Classification allows for an initial assessment of the severity of faults. The standard convolution-based CNNs can achieve effective feature extraction but often come with a considerable parameter count, leading to constraints on hardware resources. For the convenience of practical applications, a lightweight network is necessary for category-level detection. Considering the significant reduction in computational complexity offered by depthwise separable convolution, we employed MobileNetV1 [42] as the backbone network.

Depthwise separable convolution reduces the network parameter count by decomposing the standard convolution into depthwise convolution and pointwise convolution. The network parameter count mainly depends on the 1×1 pointwise convolution and the fully connected layer. The first layer of standard convolution and depthwise convolution parameters account for less than 2%. The overall parameter count is significantly reduced compared with that of the deep CNN, which can be verified by the experimental results in Section 4.

The proposed IRNLGD is extended to achieve pixel-level attachment detection. The edge detection, offering the semantic output, is carried out to visualize the attachment localization. The edge detection branch is paralleled with MobileNet, and the detection results are given in both branches. The data are input into the trained network branch for classification, and the IRNLGD branch achieves fault localization. The design of two level detection combines the supervised algorithm with the non-data-driven method achieving complementarity.

4. Results and Discussion

In this section, we test the performance of our algorithm. First, the proposed method is applied on the TST dataset. Then, we conducted experiments on the BSDS500 dataset [43] and the UIEB [44] dataset to evaluate the universality of the proposed method. The performance of the proposed method is compared with classical and state-of-the-art methods.

4.1. Experiment on TST Dataset

To demonstrate the effectiveness of the proposed method for TST rotor attachment detection, experiments were conducted on the TST dataset. The dataset is obtained by simulating TST rotor attachment faults on a marine current power generation system experiment platform and collected with an underwater camera.

4.1.1. Data Collection Experiment

To ensure the sample diversity, we add more fault types and different data collection environments on the basis of Ref. [17]. The data collection experiment is performed on the marine current power generation system experimental platform, and the structure and parameters are the same as Ref. [18]. The rope is used to wrap around the TST rotor to simulate the biofouling, as shown in Figure 2. Different degrees of attachment are simulated by employing dry ropes of varying weights and different wrapping methods. Specifically, the degree of attachment is distinguished by the dry rope weight (20 g, 40 g, and 60 g), area, and position. The total number of winding turns is 13, of which there are 3 different configurations for 40 g, namely 13, 3 (near the tip)-10 (near the hub), 6 (near the tip)-7 (near the hub). The detailed attachment degrees are shown in Table 1. The data collection experiment is carried out in two working environments of clean and turbid water. Image data collection is carried out at different shooting angles to simulate the real underwater monitoring situation. The collected data are divided into 11 categories according to the attachment degrees. The training set and test set contain 100/50 images for each category, totaling 1100/550 images.

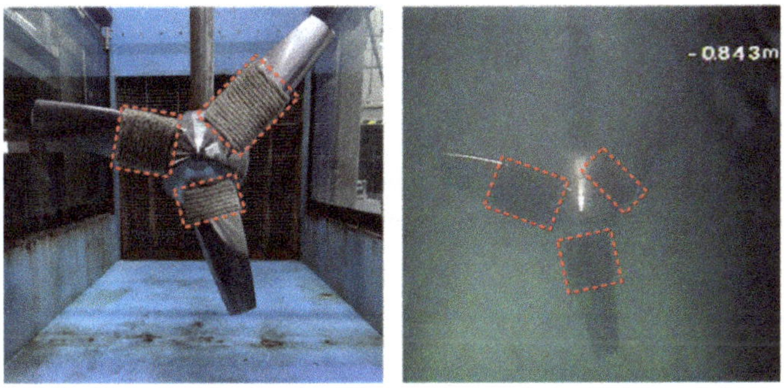

(a) (b)

Figure 2. Examples of TST rotor attachments: (**a**) in air; (**b**) in water. The red boxes indicate the attachment area.

Table 1. Details of attachment degrees of TST.

Category	Classifier Labels	Attachment Details
Attachment degree 0	0	0 g-0 g-0 g
Attachment degree 1	1	0 g-20 g-0 g
Attachment degree 2	2	0 g-20 g-40 g
Attachment degree 3	3	0 g-20 g-60 g
Attachment degree 4	4	20 g-40 g-60 g
Attachment degree 5	5	20 g-40 g (6-7)-60 g
Attachment degree 6	6	40 g (6-7)-40 g (6-7)-60 g
Attachment degree 7	7	40 g-40 g (6-7)-60 g
Attachment degree 8	8	40 g (3-10)-40 g (6-7)-60 g
Attachment degree 9	9	40 g (3-10)-40 g (3-10)-60 g
Attachment degree 10	10	40 g-40 g (3-10)-60 g

4.1.2. Network Implementation Details

The method implementation process and network have been stated in Section 3. All the algorithms are designed by Python 3.9. The classification network is built with open source deep learning framework TensorFlow [45]. The details of the network parameters involved are listed in Table 2.

Table 2. Parameter settings of the lightweight network.

Parameter Name	Value
Learning rate	1.0×10^{-3}
Training epochs	1000
Training batch size	16
Image size	495×495
Width coefficient α	1.0
Resolution coefficient ρ	1.0

4.1.3. Experiment Results

We apply the proposed MobileNet-IRNLGD rotor attachment detection network to the TST dataset to validate the effectiveness. First, we evaluate the performance of MobileNet in first-level detection. Table 3 shows the average results of classification networks under ten repeated experiments. The dataset comes from the data collection experiment mentioned above. It can be seen that MobileNet achieves the best performance in terms of reducing the number of trainable parameters. In the case of costing 13.68% parameters of ResNet50, the accuracy of MobileNet is 2.21% higher. The two-layer CNN and VGG-16 fail to extract attachment features well due to insufficient network layers, so the recognition accuracy is unsatisfying. Moreover, because of the standard convolution, the number of parameters is dozens or even hundreds of times that of MobileNet, even if the depth is not as deep as the same. The results of two different ResNets show that the accuracy of the network does not increase significantly with the depth of the network; it can be analyzed that the TST images are captured underwater, and the contrast decreases compared with the onshore image due to the medium absorption and attenuation of light propagation. As the image intensity decreases, the features will be lost in the deep network, so the accuracy does not improve much with the increase in the network depth.

Table 3. Quantitative experimental results of the five networks.

Name of Network	Trainable Parameters	Accuracy
MobileNet	3,218,251	96.69%
ResNet-101	42,522,699	94.48%
ResNet-50	23,530,571	94.46%
VGG-16	134,305,611	25.00%
Two-layer CNN	63,046,091	12.50%

Next, we assess the performance of IRNLGD. To verify the effectiveness of the proposed algorithm in a complex underwater environment, the operating environment of the TST is divided into clean/turbid water. Each environment is further characterized by strong/weak illumination. The data under strong illumination come from Ref. [17], and the data under weak illumination come from the data collection experiment mentioned above. Four representative edge detection algorithms are selected, including Canny, SE, HED and PiDiNet. The mean squared error (MSE) and average gradient (AG) are employed to evaluate the performance of each method. MSE is used to evaluate the similarity of the edge detection results to the original images, while AG is used to measure the rate of gray change in the results, which reflects the clarity and the expression of detail. The quantitative evaluation results are presented in Table 4. IRNLGD achieves the best results in both MSE and AG, indicating that the edges detected by the proposed method are closest to the original image and possess the richest details. Canny and PiDiNet are ranked second and third, respectively. The outcome is consistent with the characteristics observed in qualitative comparisons. Figure 3 shows some results on the TST dataset, with two samples for each environment. The experimental results indicate that SE hardly detects meaningful edges. The results of HED and parts of Canny are discontinuous, failing to detect closed edges, as shown in the blue boxes in Figure 3. While PiDiNet achieves the most outstanding performance among the comparative methods, its output cannot distinguish between attachments and blades. This leads to a crucial problem: when the blades are completely covered by attachments, the results from PiDiNet fail to provide meaningful information, as indicated in the green boxes in Figure 3. Only the proposed IRNLGD can detect complete blade contours and the texture of attachments, which can achieve attachment localization (the red boxes in Figure 3). What is more, a detailed comparison is conducted to demonstrate the advantages of the proposed method in TST attachment fault detection. The performance evaluation criteria are: (1) the ability to separate the foreground and background; (2) the ability to detect the edge of the TST rotor; and (3) the ability to distinguish between the blades and attachment.

Table 4. The quantitative comparison results of five methods on TST dataset.

Method	MSE	AG
Canny	17,374.37	9.186
SE	17,573.62	0.647
HED	17,813.11	3.299
PiDiNet	17,471.07	3.953
IRNLGD	16,742.77	14.627

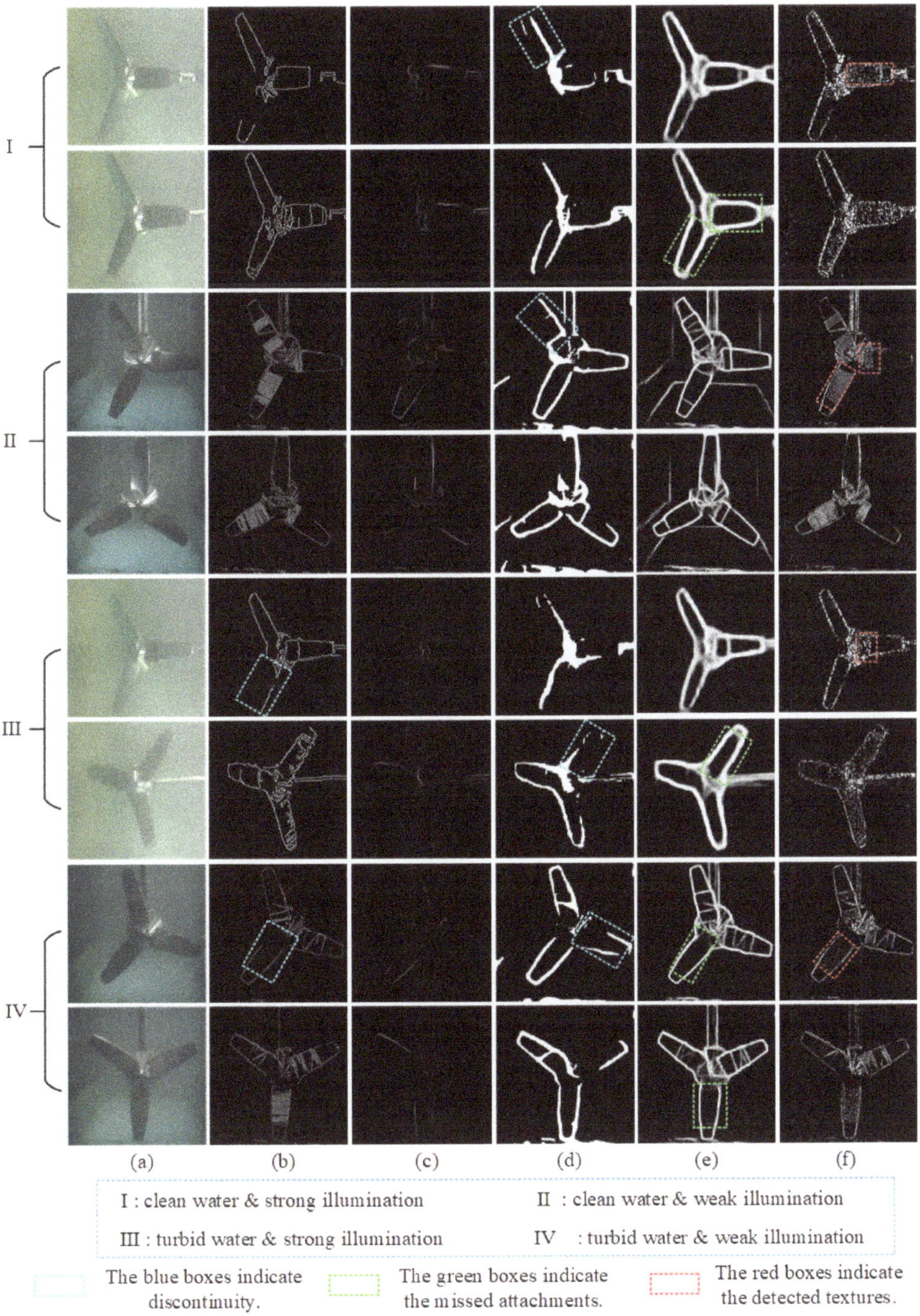

Figure 3. The comparison of five edge detection methods on TST dataset: (**a**) original image; (**b**) Canny; (**c**) SE; (**d**) HED [34]; (**e**) PiDiNet [40]; (**f**) IRNLGD.

Figure 4 shows the edge detection results of the TST image under clean water and strong illumination. Canny is capable of capturing the complete edge of the TST rotor and can partially differentiate the attachment from the blades. However, the distinction may not be very clear. SE and HED fail to obtain the complete edge of the rotor. PiDiNet captures complete edges but fails to differentiate between the blades and attachment, which may affect the subsequent maintenance process. The proposed algorithm not only captures clear and complete edges of the TST rotor but also highlights the distinction between healthy blades and the attachment.

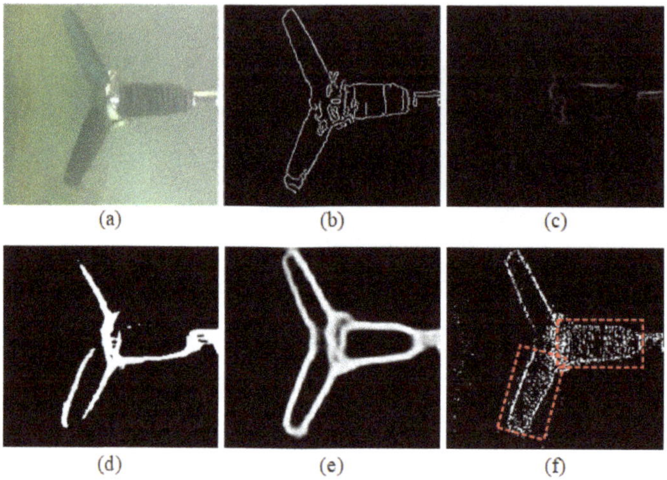

Figure 4. The comparison of five edge detection methods for clean water and strong illumination (**a**) original image; (**b**) Canny; (**c**) SE; (**d**) HED [34]; (**e**) PiDiNet [40]; (**f**) IRNLGD. The red boxes indicate the attachment area.

Figure 5 shows the edge detection results of the TST image under turbid water and strong illumination. Similar to Figure 4, Canny can extract rotor edges but cannot indicate the location of the attachment. SE is almost completely ineffective, and HED detects incomplete edges. PiDiNet can detect complete edges, but they tend to be thick with insufficient refinement. It also ignores the localization of the attachment. IRNLGD is capable of both extracting complete edges and indicating attachment areas.

Figure 6 shows the edge detection results of the TST image under clean water and weak illumination. We use several boxes to highlight the details. Except for SE, most of the algorithms can detect clear edges but some differences exist: Canny and IRNLGD can extract fine edges and differentiate between clean blades and biofouling regions, while the other two can only identify the outline. HED can only detect partial edges and tends to lose edges in details, such as the attachments at the blade tip (left red box in Figure 6d) the detected edges are incomplete, as shown in the dark side of the blade (right green box in Figure 6d). PiDiNet can extract precise rotor edges but fails to distinguish between the blades and attachment. What is more, it also displays some edges of background, which are useless for fault diagnosis. In the comparison of Canny and IRNLGD, the attachment area detected by Canny shows partial loss (left red box in Figure 6b), but IRNLGD can fully display the texture of the attachment (left red box in Figure 6f). Both of them have achieved better results in edge refinement than other methods. Canny performs well in refinement due to non-maximum suppression. However, it is important to note that Canny's high performance relies on manually selected thresholds, while the proposed IRNLGD is adaptive.

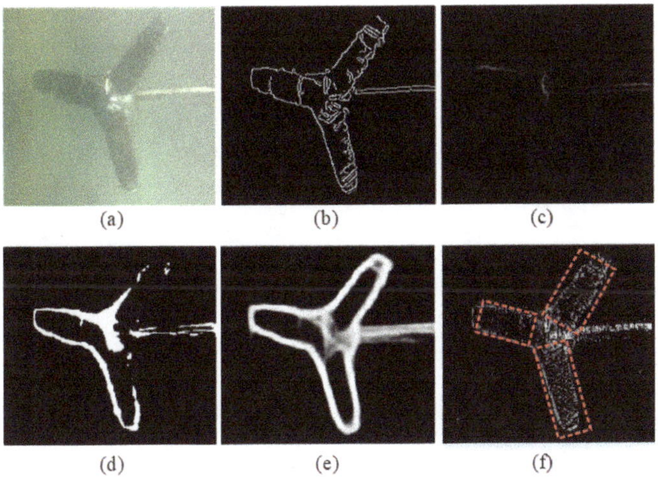

Figure 5. The comparison of five edge detection methods for turbid water and strong illumination: (**a**) original image; (**b**) Canny; (**c**) SE; (**d**) HED [34]; (**e**) PiDiNet [40]; (**f**) IRNLGD. The red boxes indicate the attachment area.

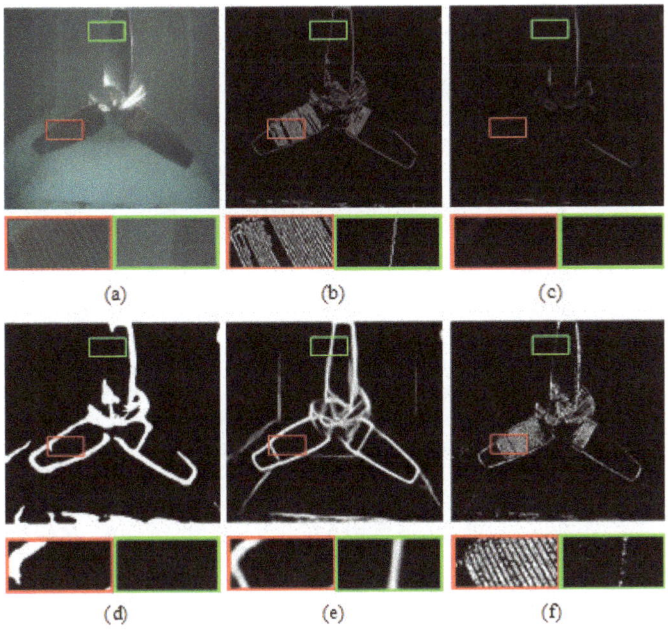

Figure 6. The comparison of five edge detection methods for clean water and weak illumination: (**a**) original image; (**b**) Canny; (**c**) SE; (**d**) HED [34]; (**e**) PiDiNet [40]; (**f**) IRNLGD.

Figure 7 shows the edge detection results of the TST image under turbid water and weak illumination. SE fails to detect a meaningful edge. HED produces incomplete contours (right green box in Figure 7d) and fails to detect the boundary between the attachment and blade near the tip (left red box in Figure 7d). PiDiNet offers advantages in edge extraction, but it cannot indicate the attachment areas. Canny and IRNLGD perform well in distinguishing between the blade and the attachment, but Canny tends to lose details in densely textured areas (right green box in Figure 7b). In contrast, IRNLGD not only detects complete edges but also clearly distinguishes the attachment area from the blades.

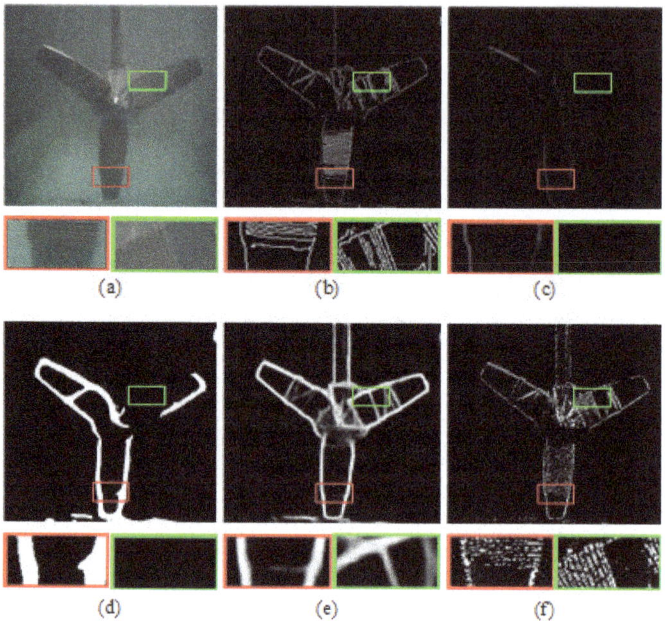

Figure 7. The comparison of five edge detection methods for turbid water and weak illumination. (**a**) original image; (**b**) Canny; (**c**) SE; (**d**) HED [34]; (**e**) PiDiNet [40]; (**f**) IRNLGD.

4.2. Experiment on BSDS500 Dataset

Due to the lack of a publicly available edge detection dataset designed for underwater images, the BSDS500 dataset is used to evaluate the quality of the proposed edge detection algorithm. The dataset is universally adopted for natural edge detection evaluation, consisting of 500 natural images and real human annotations. Each image is segmented by five different experts independently. The five segmentation results are combined with equal weights for objective ground truth. Color channel attenuation and blurring operations are performed on the dataset to simulate the characteristics of an underwater environment.

We employ three quantitative metrics commonly used in edge detection evaluation: the optimal dataset scale (ODS), optimal image scale (OIS), and average precision (AP). Table 5 compares the different edge detection algorithms on BSDS500. The proposed algorithm does not obtain satisfactory results in quantitative evaluation, primarily because IRNLGD can detect fine edges, while human ground truth focuses more on contours and ignores details. This difference causes the detected detailed textures to be penalized in the evaluation. The distinctive feature can be highlighted in visual comparisons. Figure 8 compares the results of IRNLGD with those of other methods. Due to the simulated underwater operation, the images are degraded and feature extraction is compromised, leading to various degrees of failure for the algorithms: SE fails to detect meaningful edges; HED produces a few non-closed contours, losing the most meaningful information, as shown in Figure 8d,e. Canny and PiDiNet achieve better performance among the compared methods but still exhibit incomplete edges. There are discontinuities in low-contrast regions, which is evident in the results of Canny (Figure 8c). The dependence of deep learning on data can result in decreased performance when dealing with new types of samples. For example, in Figure 8f, PiDiNet fails to detect edges or produces weak responses in some dim areas. In contrast, IRNLGD can detect complete edges and capture details not marked in human ground truth, such as the bird wing in sample III in Figure 8g.

The proposed method may have weaker noise suppression capabilities but successfully detects complete edges and fine textures.

Table 5. Quantitative results on the BSDS500 dataset.

Detectors	ODS	OIS	AP
Human	0.80	0.80	-
Sobel	0.563	0.594	0.537
Canny	0.546	0.548	0.004
Roberts	0.553	0.581	0.524
SE	0.668	0.683	0.659
HED [34]	0.557	0.560	0.041
PiDiNet [40]	0.762	0.777	0.753
IRNLGD	0.487	0.489	0.009

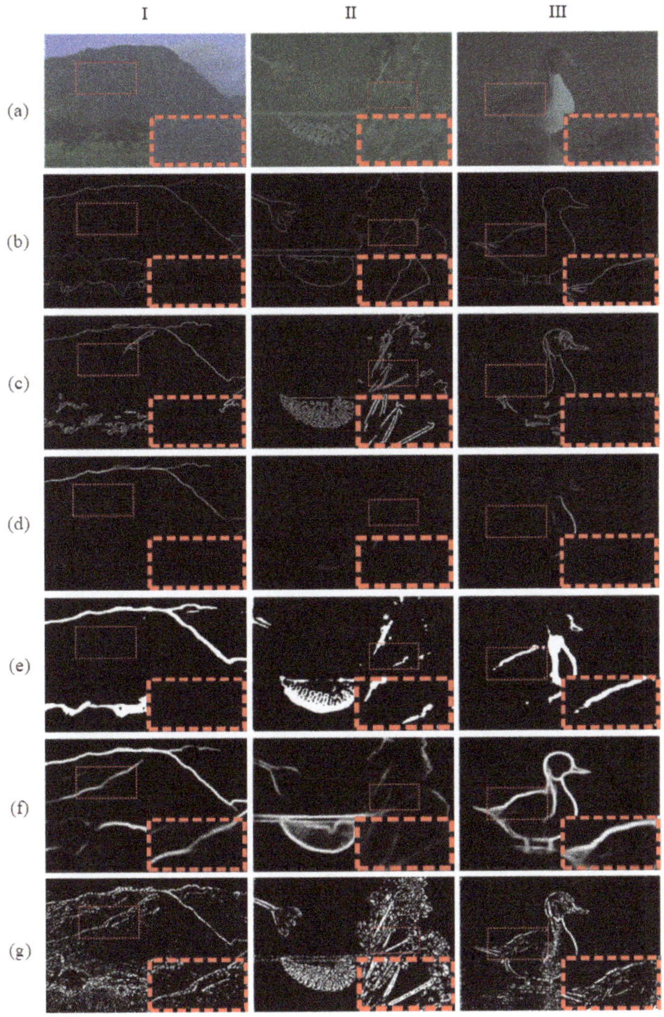

Figure 8. The comparison of five edge detection methods on the BSDS500 dataset: (**a**) original image; (**b**) ground truth; (**c**) Canny; (**d**) SE; (**e**) HED [34]; (**f**) PiDiNet [40]; (**g**) IRNLGD. I, II, and III refer to different samples. The red boxes indicate details.

4.3. Experiment on UIEB Dataset

Experiments were conducted on the UIEB dataset [44] to demonstrate the proposed method's effectiveness on underwater images. The dataset contains 950 original underwater images captured under various illuminations, including natural light, artificial light, or a combination of both. Due to the lack of an objective and publicly available human ground truth, only a qualitative assessment is performed here.

Figure 9 shows some detection results. Due to the degradation of underwater images Canny and SE fail to detect meaningful edges and only prove effective in some high-contrast regions. HED produces a clear contour but can only detect partial edges in a high-turbidity environment. PiDiNet takes into account the idea of pixel differentiation and is sensitive to gradients, achieving more complete edge detection compared to HED. However, it also misses some edges, especially in low-contrast regions, as shown in red boxes in Figure 9e. IRNLGD not only achieves complete edges but also detects fine details, even in dim parts (as shown in the green boxes in Figure 9f). Moreover, its performance does not rely on an extensive amount of data for training.

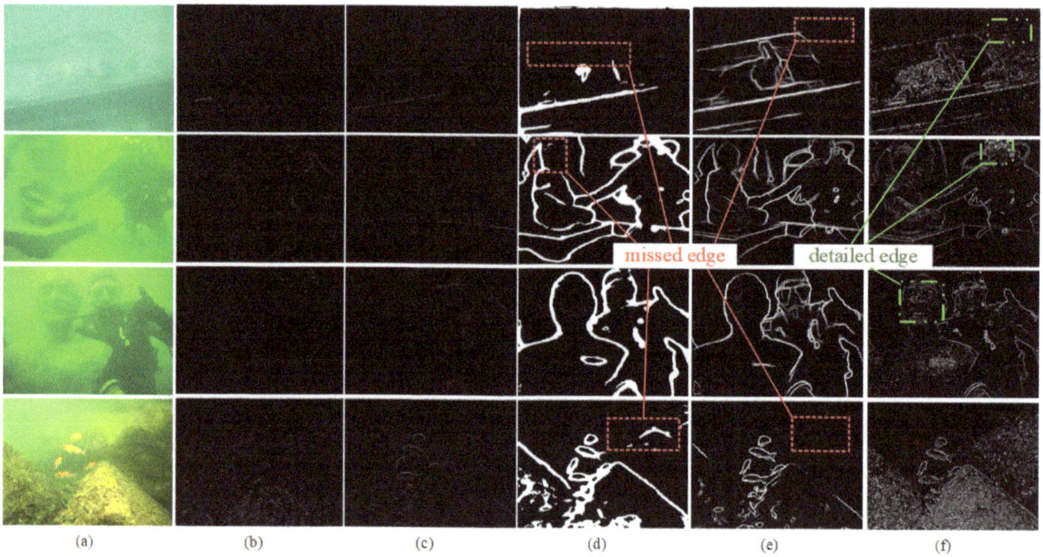

Figure 9. The comparison of five edge detection methods on the UIEB dataset: (**a**) original image; (**b**) Canny; (**c**) SE; (**d**) HED [34]; (**e**) PiDiNet [40]; (**f**) IRNLGD.

4.4. Analysis

Based on the above experimental results, it is evident that the proposed method outperforms other techniques in the detection of fine edges in underwater images. IRNLGD obtains better results on the TST and UIEB datasets than other methods. The failure of the compared methods is due to the scattering during the propagation of light underwater which causes image degradation. Forward scattering weakens the energy of the light blurring the image and defocussing the contour lines; backward scattering allows light reflected by floating particles to enter the camera, resulting in an image with much noise These factors make underwater images characterized by low contrast and attenuated intensity, leading to the failure of various edge detection algorithms that perform well in air images. The pixel values of underwater images do not vary significantly, making some gradient-based operators unable to obtain the desired results. The accuracy of supervised algorithms relies on rich training data. They fail on underwater images with low contrast due to a lack of labeled samples. However, our method considers gradients in multiple directions, allowing for the capture of richer information. Additionally, IRNLGD creatively introduces the concept of "gradient direction calibration", enabling gradient directions to directly participate in edge determination; it contributes to IRNLGD's ability to better capture changes in pixel intensity with limited information, thus facilitating edge detection in degraded images such as those with low light or turbidity.

In the quantitative assessment, the results of IRNLGD are undesirable. The reason is that BSDS500 is annotated for image segmentation and contour detection, and some detailed edges detected by the proposed method are penalized. However, the new edges and textures can also provide valuable information in some cases. The quantitative evaluation results of supervised algorithms are generally better due to the advantage of deep learning-based methods with good feature extraction. They can extract deep feature representations from a large number of training images. A similar feature extraction process does not exist in unsupervised methods. However, our method can still obtain satisfactory results and has no requirement for paired data for training. Therefore, the proposed IRNLGD is competent for edge detection tasks.

While the proposed method has some limitations compared with other methods, these can also be turned into advantages in certain situations. It is undeniable that, in terms of semantic understanding, IRNLGD falls short compared to deep learning-based approaches. However, the performance of deep learning relies on a large amount of labeled data, which is not aligned with the research purposes. The advantage of IRNLGD lies precisely in its ability to perform comprehensive and detailed edge detection without the need of training. Additionally, the proposed IRNLGD exhibits some noise in certain samples; this is acceptable. IRNLGD is designed for fault diagnosis in underwater equipment such as a TST, and this characteristic helps to distinguish between attachments and equipment bodies in the outcome: submerged devices often use smooth materials and anti-fouling coatings to delay biofouling, while the biofouling has a rough surface [46,47]. Experimental results on the TST dataset demonstrate that IRNLGD can fully display the texture of attachments, thereby indicating attachment areas; something that other methods cannot achieve. Therefore, a small amount of noise does not diminish the contribution of the proposed method to the research purposes.

In summary, the proposed method can detect fine edges better, without training or manually selected thresholds, and is more robust than other methods.

5. Conclusions

In this paper, a novel edge detection method IRNLGD based on multi-direction gradient is proposed. IRNLGD enhances the edge determination process by calibrating the gradient directions, allowing a more comprehensive inclusion of gradient from multiple directions. Experiments on a TST dataset, BSDS500, and UIEB demonstrate as follows: (1) IRNLGD can effectively detect edges in low-quality images; and (2) IRNLGD can detect a greater amount of fine texture information. In addition, we explore a two-level TST rotor attachment detection method, MobileNet-IRNLGD. IRNLGD is added as a branch to the classification network MobileNet, which provides fault location while the MobileNet provides attachment degree. The advantages of this method are: (1) It requires less computational resources, reducing hardware demands. (2) It combines supervised algorithms with non-data-driven methods to achieve complementarity, reducing the need for extensive training data. The proposed MobileNet-IRNLGD can provide preliminary fault diagnosis in the first stage, enabling technical staff to take appropriate measures based on the fault severity. During underwater cleaning operations, the precise fault localization in the second stage can provide necessary reference and guidance for operators. Since the proposed edge detection method, IRNLGD, is a non-data-driven algorithm, it can meet the monitoring needs of various types of submerged turbines and other underwater devices. This can be extrapolated from experiments conducted on the UIEB dataset, as our method can also achieve detailed edges and texture detection in degraded underwater images. Overall, the proposed IRNLGD has demonstrated its effectiveness in edge detection, and the MobileNet-IRNLGD has also shown promising results and potential.

IRNLGD can effectively extract edge information from TST images without requiring paired data for training. However, it faces challenges when dealing with images affected by motion blur. In the future, the research will focus on exploring solutions for the motion blurred images.

Author Contributions: Conceptualization, D.S., R.L. and T.W.; methodology, D.S. and R.L.; software D.S. and R.L.; validation, D.S., R.L. and D.Y.; formal analysis, D.S.; investigation, D.S. and R.L. resources, Z.Z. and T.W.; data curation, D.S., R.L., Z.Z., D.Y. and T.W.; writing—original draft preparation, D.S. and R.L.; writing—review and editing, D.S., D.Y., Z.Z. and T.W.; visualization D.S.; supervision, Z.Z. and T.W.; project administration, Z.Z. and T.W.; funding acquisition, T.W. All authors have read and agreed to the published version of the manuscript.

Funding: This research received no external funding.

Institutional Review Board Statement: Not applicable.

Informed Consent Statement: Not applicable.

Data Availability Statement: The data presented in this study are available on request from the corresponding author. The data are not publicly available due to restriction.

Conflicts of Interest: Author Zhiwei Zhang was employed by the company Shanghai Power Industrial & Commerical Co., Ltd., State Grid Shanghai Municipal Electric Power Company. The remaining authors declare that the research was conducted in the absence of any commercial or financial relationships that could be construed as a potential conflict of interest.

References

1. Yang, X.; Liu, N.; Zhang, P.; Guo, Z.; Ma, C.; Hu, P.; Zhang, X. The current state of marine renewable energy policy in China. *Mar. Policy* **2019**, *100*, 334–341. [CrossRef]
2. Elghali, S.B.; Benbouzid, M.; Charpentier, J.F. Marine tidal current electric power generation technology: State of the art and current status. In Proceedings of the 2007 IEEE International Electric Machines & Drives Conference, Antalya, Turkey, 3–5 May 2007 ; IEEE: Piscataway, NJ, USA, 2007; Volume 2, pp. 1407–1412.
3. Lust, E.E.; Luznik, L.; Flack, K.A.; Walker, J.M.; Van Benthem, M.C. The influence of surface gravity waves on marine current turbine performance. *Int. J. Mar. Energy* **2013**, *3*, 27–40. [CrossRef]
4. Goundar, J.N.; Ahmed, M.R. Marine current energy resource assessment and design of a marine current turbine for Fiji. *Renew. Energy* **2014**, *65*, 14–22. [CrossRef]
5. Langhamer, O. Effects of wave energy converters on the surrounding soft-bottom macrofauna (west coast of Sweden). *Mar. Environ. Res.* **2010**, *69*, 374–381. [CrossRef] [PubMed]
6. Chen, H.; Ait-Ahmed, N.; Zaim, E.; Machmoum, M. Marine tidal current systems: State of the art. In Proceedings of the 2012 IEEE International Symposium on Industrial Electronics, Hangzhou, China, 28–31 May 2012; IEEE: Piscataway, NJ, USA, 2012; pp. 1431–1437.
7. Turbines, M.C. *SeaGen Environmental Monitoring Programme*; Final Report; Haskoning Uk Ltd.: Edinburgh, UK, 2011.
8. Titah-Benbouzid, H.; Benbouzid, M. Biofouling issue on marine renewable energy converters: A state of the art review on impacts and prevention. *Int. J. Energy Convers.* **2017**, *5*, 67. [CrossRef]
9. Frost, C.; Morris, C.E.; Mason-Jones, A.; O'Doherty, D.M.; O'Doherty, T. The effect of tidal flow directionality on tidal turbine performance characteristics. *Renew. Energy* **2015**, *78*, 609–620. [CrossRef]
10. Kearney, J. Grid Voltage Unbalance and the Integration of DFIG's. Ph.D. Thesis, Technological University Dublin, Dublin, Ireland, 2013.
11. Nall, C.R.; Schläppy, M.L.; Guerin, A.J. Characterisation of the biofouling community on a floating wave energy device. *Biofouling* **2017**, *33*, 379–396. [CrossRef] [PubMed]
12. Loxton, J.; Macleod, A.; Nall, C.R.; McCollin, T.; Machado, I.; Simas, T.; Vance, T.; Kenny, C.; Want, A.; Miller, R. Setting an agenda for biofouling research for the marine renewable energy industry. *Int. J. Mar. Energy* **2017**, *19*, 292–303. [CrossRef]
13. Freeman, B.; Tang, Y.; VanZwieten, J. Marine Hydrokinetic Turbine Blade Fault Signature Analysis using Continuous Wavelet Transform. In Proceedings of the 2019 IEEE Power & Energy Society General Meeting (PESGM), Atlanta, GA, USA, 4–8 August 2019; pp. 1–5. [CrossRef]
14. Saidi, L.; Benbouzid, M.; Diallo, D.; Amirat, Y.; Elbouchikhi, E.; Wang, T. Higher-Order Spectra Analysis-Based Diagnosis Method of Blades Biofouling in a PMSG Driven Tidal Stream Turbine. *Energies* **2020**, *13*, 2888. [CrossRef]
15. Freeman, B.; Tang, Y.; Huang, Y.; VanZwieten, J. Rotor blade imbalance fault detection for variable-speed marine current turbines via generator power signal analysis. *Ocean Eng.* **2021**, *223*, 108666. [CrossRef]
16. Xie, T.; Li, Z.; Wang, T.; Shi, M.; Wang, Y. An integration fault detection method using stator voltage for marine current turbines. *Ocean Eng.* **2021**, *226*, 108808. [CrossRef]

7. Zheng, Y.; Wang, T.; Xin, B.; Xie, T.; Wang, Y. A sparse autoencoder and softmax regression based diagnosis method for the attachment on the blades of marine current turbine. *Sensors* **2019**, *19*, 826. [CrossRef]
8. Xin, B.; Zheng, Y.; Wang, T.; Chen, L.; Wang, Y. A diagnosis method based on depthwise separable convolutional neural network for the attachment on the blade of marine current turbine. *Proc. Inst. Mech. Eng. Part I J. Syst. Control Eng.* **2021**, *235*, 1916–1926. [CrossRef]
9. Peng, H.; Yang, D.; Wang, T.; Pandey, S.; Chen, L.; Shi, M.; Diallo, D. An adaptive coarse-fine semantic segmentation method for the attachment recognition on marine current turbines. *Comput. Electr. Eng.* **2021**, *93*, 107182. [CrossRef]
0. Peng, H.; Wang, T.; Pandey, S.; Chen, L.; Zhou, F. An Attachment Recognition Method Based on Image Generation and Semantic Segmentation for Marine Current Turbines. In Proceedings of the IECON 2020—The 46th Annual Conference of the IEEE Industrial Electronics Society, Singapore, 18–21 October 2020; IEEE: Piscataway, NJ, USA, 2020; pp. 2819–2824.
1. Qi, F.; Wang, T.; Wang, X.; Chen, L. LAW-IFF Net: A semantic segmentation method for recognition of marine current turbine blade attachments under blurry edges. *Proc. Inst. Mech. Eng. Part M J. Eng. Marit. Environ.* **2023**. [CrossRef]
2. McGlamery, B. A computer model for underwater camera systems. In Proceedings of the Ocean Optics VI, Monterey, CA, USA, 23–25 October 1979; SPIE: Bellingham, WA, USA, 1980; Volume 208, pp. 221–231.
3. Jaffe, J.S. Computer modeling and the design of optimal underwater imaging systems. *IEEE J. Ocean. Eng.* **1990**, *15*, 101–111. [CrossRef]
4. Jing, J.; Liu, S.; Wang, G.; Zhang, W.; Sun, C. Recent advances on image edge detection: A comprehensive review. *Neurocomputing* **2022**, *503*, 259–271. [CrossRef]
5. Kittler, J. On the accuracy of the Sobel edge detector. *Image Vis. Comput.* **1983**, *1*, 37–42. [CrossRef]
6. Szeliski, R. *Computer Vision: Algorithms and Applications*; Springer Nature: Berlin/Heidelberg, Germany, 2022.
7. Canny, J. A Computational Approach to Edge Detection. *IEEE Trans. Pattern Anal. Mach. Intell.* **1986**, *PAMI-8*, 679–698. [CrossRef]
8. Gonzalez, C.I.; Melin, P.; Castillo, O. Edge Detection Method Based on General Type-2 Fuzzy Logic Applied to Color Images. *Information* **2017**, *8*, 104. [CrossRef]
9. Ranjan, R.; Avasthi, V. Edge Detection Using Guided Sobel Image Filtering. *Wirel. Pers. Commun.* **2023**, *132*, 651–677. [CrossRef]
0. Ma, J.; Wang, T.; Li, G.; Zhan, Q.; Wu, D.; Chang, Y.; Xue, Y.; Zhang, Y.; Zuo, J. Concrete surface roughness measurement method based on edge detection. *Vis. Comput.* **2023**, *40*, 1553–1564. [CrossRef]
1. Muntarina, K.; Shorif, S.B.; Uddin, M.S. Notes on edge detection approaches. *Evol. Syst.* **2022**, *13*, 169–182. [CrossRef]
2. Yang, D.; Peng, B.; Al-Huda, Z.; Malik, A.; Zhai, D. An overview of edge and object contour detection. *Neurocomputing* **2022**, *488*, 470–493. [CrossRef]
3. Dollár, P.; Zitnick, C.L. Fast edge detection using structured forests. *IEEE Trans. Pattern Anal. Mach. Intell.* **2014**, *37*, 1558–1570. [CrossRef] [PubMed]
4. Xie, S.; Tu, Z. Holistically-nested edge detection. In Proceedings of the IEEE International Conference on Computer Vision, Santiago, Chile, 7–13 December 2015; pp. 1395–1403.
5. Liu, Y.; Lew, M.S. Learning Relaxed Deep Supervision for Better Edge Detection. In Proceedings of the 2016 IEEE Conference on Computer Vision and Pattern Recognition (CVPR), Las Vegas, NV, USA, 27–30 June 2016; pp. 231–240. [CrossRef]
6. Yang, J.; Price, B.; Cohen, S.; Lee, H.; Yang, M.H. Object Contour Detection with a Fully Convolutional Encoder-Decoder Network. In Proceedings of the IEEE Conference on Computer Vision and Pattern Recognition (CVPR), Las Vegas, NV, USA, 27–30 June 2016.
7. Ronneberger, O.; Fischer, P.; Brox, T. U-Net: Convolutional Networks for Biomedical Image Segmentation. In Proceedings of the Medical Image Computing and Computer-Assisted Intervention—MICCAI 2015, Munich, Germany, 5–9 October 2015; Navab, N., Hornegger, J., Wells, W.M., Frangi, A.F., Eds.; Springer: Cham, Switzerland, 2015; pp. 234–241.
8. Zou, Q.; Zhang, Z.; Li, Q.; Qi, X.; Wang, Q.; Wang, S. DeepCrack: Learning Hierarchical Convolutional Features for Crack Detection. *IEEE Trans. Image Process.* **2019**, *28*, 1498–1512. [CrossRef] [PubMed]
9. Badrinarayanan, V.; Kendall, A.; Cipolla, R. SegNet: A Deep Convolutional Encoder-Decoder Architecture for Image Segmentation. *IEEE Trans. Pattern Anal. Mach. Intell.* **2017**, *39*, 2481–2495. [CrossRef] [PubMed]
0. Su, Z.; Liu, W.; Yu, Z.; Hu, D.; Liao, Q.; Tian, Q.; Pietikäinen, M.; Liu, L. Pixel difference networks for efficient edge detection. In Proceedings of the IEEE/CVF International Conference on Computer Vision, Virtual, 11–17 October 2021; pp. 5117–5127.
1. Elder, J.H.; Goldberg, R.M. Ecological statistics of Gestalt laws for the perceptual organization of contours. *J. Vis.* **2002**, *2*, 5. [CrossRef]
2. Howard, A.G.; Zhu, M.; Chen, B.; Kalenichenko, D.; Wang, W.; Weyand, T.; Andreetto, M.; Adam, H. Mobilenets: Efficient convolutional neural networks for mobile vision applications. *arXiv* **2017**, arXiv:1704.04861.
3. Arbelaez, P.; Maire, M.; Fowlkes, C.; Malik, J. Contour detection and hierarchical image segmentation. *IEEE Trans. Pattern Anal. Mach. Intell.* **2010**, *33*, 898–916. [CrossRef]
4. Li, C.; Guo, C.; Ren, W.; Cong, R.; Hou, J.; Kwong, S.; Tao, D. An underwater image enhancement benchmark dataset and beyond. *IEEE Trans. Image Process.* **2019**, *29*, 4376–4389. [CrossRef]
5. Abadi, M.; Agarwal, A.; Barham, P.; Brevdo, E.; Chen, Z.; Citro, C.; Corrado, G.S.; Davis, A.; Dean, J.; Devin, M.; et al. Tensorflow: Large-scale machine learning on heterogeneous distributed systems. *arXiv* **2016**, arXiv:1603.04467.

46. Gittens, J.E.; Smith, T.J.; Suleiman, R.; Akid, R. Current and emerging environmentally-friendly systems for fouling control in the marine environment. *Biotechnol. Adv.* **2013**, *31*, 1738–1753. [CrossRef] [PubMed]
47. Want, A.; Crawford, R.; Kakkonen, J.; Kiddie, G.; Miller, S.; Harris, R.E.; Porter, J.S. Biodiversity characterisation and hydrodynamic consequences of marine fouling communities on marine renewable energy infrastructure in the Orkney Islands Archipelago, Scotland, UK. *Biofouling* **2017**, *33*, 567–579. [CrossRef] [PubMed]

Journal of
Marine Science and Engineering

MDPI

Article

Design and Experimental Study of 50 kW Ocean Thermal Energy Conversion Test Platform Based on Organic Rankine Cycle

Beichen Lu [1], Yanjun Liu [1,2,*], Xiaoyu Zhai [1], Li Zhang [3] and Yun Chen [1,*]

1 Institute of Marine Science and Technology, Shandong University, Qingdao 266237, China; lubeichen@mail.sdu.edu.cn (B.L.); zhaixiaoyu@mail.sdu.edu.cn (X.Z.)
2 Key Laboratory of High Efficiency and Clean Mechanical Manufacture, Ministry of Education, School of Mechanical Engineering, Shandong University, Jinan 250061, China
3 Southern Marine Science and Engineering Guangdong Laboratory (Zhanjiang), Zhanjiang 524006, China; zhangl@zjblab.com
* Correspondence: lyj111@sdu.edu.cn (Y.L.); chenyuneast@mail.sdu.edu.cn (Y.C.)

Abstract: In recent years, clean and renewable energy sources have received much attention to balance the contradiction between resource needs and environmental sustainability. Among them, ocean thermal energy conversion (OTEC), which consists of surface warm seawater and deep cold seawater, can rely on thermal cycling to generate electricity and has great potential in alleviating the energy crisis. In this paper, the design and experiment study of a 50 kW OTEC platform is proposed. Thermodynamic modeling, calculation, optimization, and engineering calibration of the system were carried out, and the thermal efficiency reached 2.63% to meet the power generation demand. Experiments were also carried out by using a heat pump unit to simulate hot and cold seawater environments, and data on the stable operation of the system were obtained, with the grid-connected power reaching 47.5 kW and a thermal efficiency of 2.46%. The accuracy of the design scheme is verified, and the theoretical basis and data support are provided for the practical development and application of ocean thermal energy conversion.

Keywords: ocean thermal energy conversion; organic Rankine cycle; platform design; experimental research

Citation: Lu, B.; Liu, Y.; Zhai, X.; Zhang, L.; Chen, Y. Design and Experimental Study of 50 kW Ocean Thermal Energy Conversion Test Platform Based on Organic Rankine Cycle. *J. Mar. Sci. Eng.* **2024**, *12*, 463. https://doi.org/10.3390/jmse12030463

Academic Editor: Leszek Chybowski

Received: 8 February 2024
Revised: 3 March 2024
Accepted: 5 March 2024
Published: 7 March 2024

1. Introduction

Most of the energy used in the world is fossil fuels. Coal accounts for 70 percent of major energy consumption. Excessive use of fossil fuels produces large quantities of harmful gases, and wastewater emissions are increasing day by day. In recent years, with the gradual deterioration of the global environment, countries around the world have been changing their energy structures, which rely mainly on fossil fuels such as oil and natural gas, and have been actively advocating and developing the use of new and renewable energy sources with more and more attention being paid to ocean energy [1–4]. In the United Nations sustainable development agenda for 2030, ocean renewable energies have been explicitly identified as a key objective of sustainable development [5].

There are various forms of renewable energy in the ocean, including ocean wind energy, wave energy, tidal current energy, ocean thermal energy, and salt difference energy [6–8]. The ocean thermal energy refers to the heat energy stored between warm seawater (25~30 °C) and deep cold seawater (4~7 °C) on the surface of the ocean that is irradiated by the sun all year round [9]. Compared with other ocean energy sources, the advantages of ocean thermal energy are that it is stable and continuous, has large reserves, and is non-polluting, and the thermal cycle system can be used to complete power generation [10,11]. The South China Sea has huge reserves of ocean thermal energy. Theoretical reserves in China's offshore and adjacent waters are 14.4×10^{21}–15.9×10^{21} J, and the total installed capacity can be developed from 17.47×10^{8}–18.33×10^{8} kW, 90% of which is distributed in the

South China Sea. Actively developing ocean thermal energy can provide an important guarantee for resource development in the South China Sea [12–14]. The development of ocean thermal energy conversion can reduce the dependence on fossil energy and reduce carbon dioxide emissions [15]. While improving the environment, it can also obtain a large amount of high-quality deep clean seawater, which is considered to be one of the energy sources with development potential [16–19].

In 1881, French scientist Arsonval first proposed the feasibility of ocean thermal energy conversion (OTEC) [20]. His student Georges then built the first OTEC factory in 1930 [21]. However, at that time, the technology was not widely used due to its high cost and low energy conversion efficiency. In the 1970s, the oil crisis broke out, and human environmental protection awareness increased. Several countries, mainly France, Japan and the United States, began to attach importance to the development of ocean energy [22]. In 1970, a small offshore power station with a net output of 15 kW was built in Hawaii [23]. In 1981, the Tokyo Electric Power Company built a 100 kW OTEC onshore test plant with R22 as the working fluid in the Republic of Nauru, with a net output power of 31.5 kW [24]. In 1982, Japan's Tokushima company established a 50 kW onshore OTEC power plant with R717 as the working fluid [25]. At the same time, the National Institute of Ocean Technology of India and Saga University of Japan reached a cooperation to build an offshore floating OTEC power plant [26], which provided 493 kW of electricity to the grid. In 2013 the U.S. Makai Offshore Engineering Company designed a 100 kW power plant. In the same year, the French DCNS Group successfully constructed and installed on the French island of Reunion an experimental prototype for OTEC, which is a 1/250th scale prototype compared to the proposed 10 MW plant, with the goal of generating 15 kW of equivalent power. In order to make the prototypes as close as possible to the 10 MW OTEC power plant, they use the same operating pressure [27–29]. In 2019, the 1 MW OTEC power plant was built in South Korea and underwent offshore trials in Busan, where the actual power generation was only 338 kW due to the design temperature difference between the cooling and heating source systems [30]. In China, the First Institute of Oceanography of the Ministry of Natural Resources successively developed 10 kW and 15 kW OTEC system prototypes in 2006 [31]. In general, OTEC technology has been well applied in tropical ocean areas, such as the Pacific Ocean, Indian Ocean, and Caribbean Sea. More and more countries are actively researching and developing related technologies. However, most of the research related to ocean thermal energy conversion stays at the stage of theoretical analysis, and industrialization development is lagging behind, so there is an urgent need to promote the development of technology and equipment related to the development of ocean thermal energy and provide the necessary technical support for the development of the industry.

Ocean thermal energy conversion systems are divided into the open type [32], closed type [33,34], and hybrid type [35,36] due to their different cycle modes. Among them, the closed-type cycle is widely used because of its easy miniaturization. The organic Rankine cycle [37], as the most basic OTEC closed cycle, is one of the typical sustainable energy technologies that use low-grade energy for power generation.

The 50 kW OTEC power generation platform proposed in this paper is equipped with a cold and heat source control module. The working fluid of the system is R134a. Scholars have conducted a large number of studies and analyses on the advantages and disadvantages of the thermodynamic performance of various refrigerants under ocean thermal energy conditions [38–41]. In the demonstration equipment and experimental equipment of various countries, the commonly used refrigerants are mainly ammonia and R134a [42]. Although the performance of ammonia is optimal, it is corrosive, toxic, and easy to volatilize, and the incidence of chemical accidents is very high.

However, R134a has a good comprehensive performance. It is non-flammable, non-explosive, non-toxic, non-irritant, non-corrosive, chemically stable, does not have chemical reactions with metals such as iron, copper, and aluminum, and does not destroy the ozone layer. The safety grade is the safest A1, which is a mainstream refrigerant with very high safety and environmental protection and is used in many demonstration power stations developed in Japan and the United States. The experimental test platform for the cycle system circulating agent dosage is very large, and R134a was finally selected as the circulating agent for safety and environmental considerations. [14,43,44]. In the system, Aspen Plus V10 is used to match a suitable water immersion screw chiller to meet the 4 °C/28 °C cooling and heating source conditions. After equipment selection, the 3D packaging experimental platform is designed by Pro-E 5.0 to improve the space utilization. Then, PLC (Programmable Logic Controller) control cabinets for master and slave stations are added to collect analog signals such as the pressure, flow rate, temperature, and valve opening. Finally, the grid-connected inverter feeds the system power into the grid, which improves the resource utilization efficiency and builds a recycling system.

The development and use of ocean thermal energy is based on the system design. This paper about the South China Sea ocean thermal energy application environment focuses on thermodynamic calculation, working fluid selection, system components selection, and engineering verification of the 50 kW ocean thermal energy conversion test platform to carry out the design work, and the heat pump unit simulation of hot and cold seawater environments for experiments has verified the accuracy of the design scheme.

2. Materials and Methods

2.1. Platform Principle and Thermodynamic Calculation

In the ideal state without considering any loss, the organic Rankine cycle can be decomposed into four thermodynamic processes: the isentropic expansion process (a~b'), isothermal condensation process (b'~c), isentropic compression process (c~d), and constant pressure heat absorption process (d~e~a) [45,46]. A system temperature entropy diagram is shown in Figure 1. R134a belongs to the wet working fluid [47], and the slope of the saturated steam curve (dT/ds) is negative. The principle of ocean thermal energy conversion power generation is shown in Figure 2, numbers 1~8 represent the working fluid state points at the inlet and outlet of the equipment in the diagram. R134a in the evaporator heated by warm seawater with a surface temperature of 28 °C makes it become saturated steam in the turbine and promotes the turbine to work and generate electricity. The exhaust steam from the turbine outlet enters the condenser, exchanges heat with the cold seawater at the deep temperature of 4 °C, and condenses into liquid. After being supercharged by a working fluid pump, it returns to the evaporator to complete the cycle. It is assumed that the working fluid has no leakage to characterize the steady-state performance of the system; the fluid velocity is constant and incompressible; and its kinetic energy, potential energy, and heat loss are ignored. The transparent entropy efficiency is set to 85%, the efficiencies of the motor and pump are both 90%, and the temperature difference between the inlet and outlet of the seawater is set to 2 °C.

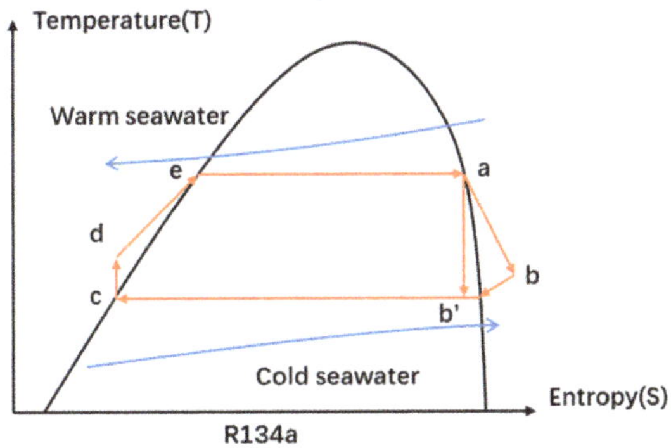

Figure 1. Temperature–entropy diagram of R134a.

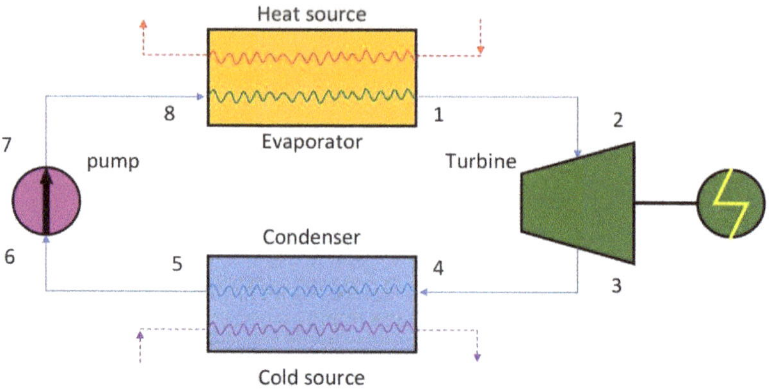

Figure 2. A schematic diagram of organic Rankine cycle.

In thermodynamics, specific enthalpy represents the thermal energy per unit mass of a substance under certain conditions. When a substance passes through a system in the form of flow, the heat transferred is the specific enthalpy of previous and subsequent states multiplied by the mass flow.

The energy conservation equation of the heat exchanger is as follows:

$$Q = \dot{m}\Delta h = m_w c_w \Delta T \tag{1}$$

where Q is the amount of heat exchange (kW), Δh is the specific enthalpy difference between the inlet and outlet of the working fluid side of the heat exchanger (kJ/kg), $\dot{m}$ is the mass flow of the working fluid (kg/s), m_w is the mass flow of seawater (kg/s), c_w is the specific heat capacity of seawater at constant pressure (J/g°C), and ΔT is the seawater side of the heat exchanger temperature rise or drop (°C).

The heat exchange area of the heat exchanger (m^2) is as follows:

$$A = \frac{Q}{k\Delta t_m} \tag{2}$$

where k is the heat transfer coefficient (W/m^2 °C) and Δt_m is the logarithmic heat transfer temperature difference (°C).

The output power of the turbine (kW) is as follows:

$$W = \dot{m}(h_2 - h_3)\eta_g \eta_{is} \tag{3}$$

where h_2 is the specific enthalpy of the inlet of the turbine (kJ/kg), h_3 is the specific enthalpy of the outlet of the turbine (kJ/kg), η_g is the motor efficiency, and η_{is} is the entropy efficiency of the turbine.

The isentropic efficiency of the turbine is as follows:

$$\eta_{is} = \frac{h_2 - h_3}{h_2 - h_{3'}} \tag{4}$$

where $h_{3'}$ is the ideal enthalpy value of the turbine outlet ignoring the isentropic efficiency (kJ/kg).

Since the pressure of the working fluid entering the turbine decreases and becomes exhausted steam, it may cause the liquefaction of the gases in the turbine, which can cause impact damage to the turbine blades. Therefore, it is necessary to prevent the working fluid dryness from being too low. Generally, the dryness x of the working fluid at the outlet of the turbine should be greater than 0.92. The calculation formula is as follows:

$$x = \frac{h_2 - h_5}{h_{5'} - h_5} \tag{5}$$

where $h_{5'}$ and h_5 are the enthalpy values of the saturated gas and the saturated liquid at the outlet of the condenser (kJ/kg), respectively.

The working fluid pump works (kW) as follows:

$$W_P = \frac{\dot{m}(h_7 - h_6)}{\eta_P} \tag{6}$$

where η_P is the pump efficiency.

The net output work (kW) is as follows:

$$W_{net} = W_T - W_P \tag{7}$$

The theoretical cycle thermal efficiency is as follows:

$$\eta_0 = \frac{(h_2 - h_3) - (h_7 - h_6)}{(h_1 - h_8)} \tag{8}$$

The thermal efficiency of the system is as follows:

$$\eta_{net} = \frac{W_{net}}{Q_e} \tag{9}$$

where Q_e is the heat exchange amount of the evaporator (kW).

The optimization process of system thermodynamics calculation is shown in Figure 3. We know the thermodynamic parameters, heat source temperature, cold source temperature, heat transfer end difference, and other working condition settings at each state point. Under the premise of ensuring the dryness of the turbine outlet, firstly, calculate the theoretical cycle of the system efficiency and heat absorption of the evaporator to obtain the working fluid flow; secondly, according to the working fluid flow, enthalpy difference, logarithmic average temperature difference, and heat transfer coefficient, heat exchange capacity, heat exchange area, pump power, working fluid flow, and pipe diameter of cold and hot seawater; finally, select the type according to the pump power of the working fluid and calculate whether the net output power meets 50 kW thermal efficiency. The best working condition setting is obtained through the system's thermal efficiency results.

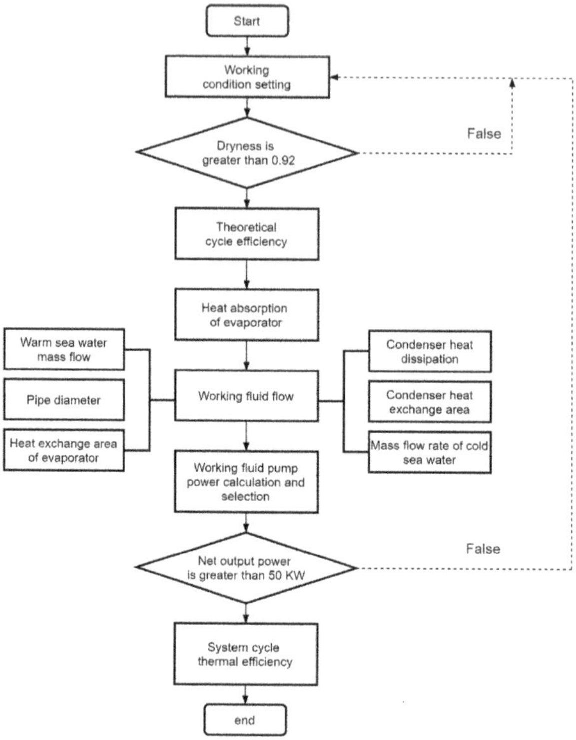

Figure 3. A flow chart of thermodynamic calculation optimization for OTEC.

2.2. Equipment Selection and Engineering Check

After the thermodynamic optimization calculation of the OTEC system, the selection of key components, valve installation, and piping design can be carried out initially. A specific process flow chart is shown in Figure 4. Each section of the pipeline is equipped with a pressure sensor and temperature sensor for easy measurement. The heat exchanger adopts a full-liquid structure [48], the working fluid goes through the shell side, and the water goes through the tube side. Low-fin threaded tubes are immersed in working fluid resulting in a large heat exchange area and high heat transfer efficiency. In order to prevent the turbine's service life from being shortened by the impact of droplets, a gas–liquid separator is added and integrated with the evaporator to increase the space occupancy rate. The turbine adopts a symmetrical and efficient centripetal radial flow type and is directly connected to the generator as an all-in-one machine, with gas bearings support low vibration, low power consumption, a long service life, and a low maintenance cost The working fluid pump adopts a hermetic centrifugal canned motor pump and power frequency control, and the flow is bypassed through the bypass regulating valve. The inlet of the evaporator is equipped with a shut-off valve, which is interlocked with the liquid level control; the turbine inlet is equipped with a pneumatic regulating valve, and the outlet has a shut-off check valve, which is connected in parallel with the bypass regulating valve. The condenser outlet is equipped with a shut-off valve; the inlet of the working fluid pump has two shut-off valves and one filter, and the outlet is equipped with a flowmeter.

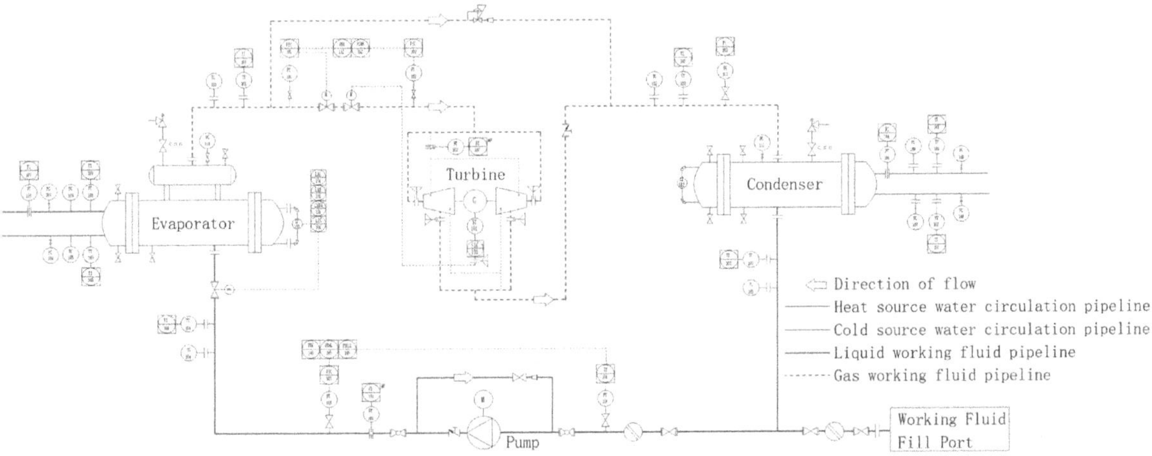

Figure 4. A design diagram of OTEC system.

In actual engineering, the fluid flow resistance caused by the viscous characteristics in the pipeline causes a part of the mechanical energy to be converted into heat energy to be lost. In addition, various valves, flowmeters, filters, and elbows will also cause local energy loss. Therefore, to ensure the smooth implementation of the project, the system is checked, calculated, and corrected.

The head loss along the pipe (m) [49] is as follows:

$$h_f = \frac{\lambda l v^2}{2dg} \tag{10}$$

where l is the pipe length (m), d is the pipe diameter (m), v is the average flow velocity of the cross-section fluid (m/s), g is the acceleration of gravity (m/s^2), and λ is the drag coefficient along the way.

When calculating the local head loss, a certain section of local obstacle resistance is converted into a section of straight pipe along the way resistance with the same diameter as the pipe diameter. The length of this section of pipe is called the equivalent length of the local resistance [50], which is represented by l_q (m), as follows:

$$h_j = \frac{\xi v^2}{2g} = \frac{\lambda l_q v^2}{2dg} \tag{11}$$

$$\sum l_q = \frac{\sum \xi d}{\lambda} \tag{12}$$

Then, the total head loss of the entire pipeline (m) is as follows:

$$\mathrm{h} = h_f + h_j = \lambda \sum \frac{l v^2}{2dg} + \lambda \sum \frac{l_q v^2}{2dg} \tag{13}$$

The total pressure loss (MPa) is as follows:

$$\Delta \mathrm{P} = \Delta P_f + \Delta P_j \tag{14}$$

3. Results and Discussion

3.1. System Design Parameters

The system correction parameters are shown in Tables 1 and 2.

Table 1. Correction of thermodynamic parameters for R134a—organic Rankine cycle.

Status Point	Temperature/T	Pressure/P	Density/ρ	Specific Enthalpy/h	Specific Entropy/s
	°C	kPa	Kg/m^3	kJ/kg	kJ/(kg·K)
1	24	645.78	31.39	411.82	1.7166
2	24	632.23	30.59	412.18	1.7193
3′	8.48	393.96	19.34	402.48	1.7193
3	9.01	393.96	19.18	403.96	1.7246
4	9.01	390.61	19.00	404.05	1.7255
5	8	387.61	1267.90	210.84	1.0388
6	7.79	384.86	1268.60	210.55	1.0378
7	7.90	650.99	1269.40	210.75	1.0378
8	7.90	648.78	1269.40	210.75	1.0378

Table 2. Correction of key parameters for R134a—organic Rankine cycle.

Name	R134a System	
	Initial Value	Check Value
1–2 Working fluid flow rate/(m/s)	15	14.43
3–4 Working fluid flow rate/(m/s)	15	13.25
5–6 Working fluid flow rate/(m/s)	0.5	0.51
7–8 Working fluid flow rate/(m/s)	1.5	1.25
1–2 pipe diameter/mm	144.56	150
3–4 pipe diameter/mm	185.12	200
5–6 pipe diameter/mm	123.90	125
7–8 pipe diameter/mm	87.59	100
1–2 pressure drop loss/MPa	0	0.01355
3–4 pressure drop loss/MPa	0	0.00335
5–6 pressure drop loss/MPa	0	0.00275
7–8 pressure drop loss/MPa	0	0.00221
Diameter of warm seawater pipe/mm	341.29	350
Diameter of cold seawater pipe/mm	333.95	350
Warm seawater mass flow/(kg/s)	182.87	191.49
Cold seawater mass flow/(kg/s)	175.09	183.93
Working fluid mass flow/(kg/s)	7.64	8
Evaporator heat absorption/kW	1536.1	1608.56
Condenser heat dissipation/kW	1470.78	1545.68
Turbine output power/kW	51.08	50.3
System thermal efficiency/%	3.22	2.63

3.2. Overall Structure and Control System

Compared with ordinary dry-type chillers, flooded units have a smaller heat transfer temperature difference and lower outlet superheat, which increases the evaporation temperature and makes full use of the heat exchange area. Therefore, they are more efficient and energy-saving. Screw full-liquid chillers have increasingly become a research and application hotspot of large-scale chillers [51,52].

When the prototype of the power generation system in this paper is running on land, a screw-filled chiller is selected to prepare the cold and heat sources to simulate the 28 °C/4 °C seawater temperature difference to store thermal energy. As shown in Figure 5, the experimental platform consists of an OTEC system and a cold and heat source control system. The cold and heat source control system includes heat pump units (condensers, evaporators, screw compressors), air-cooled auxiliary units (air-cooled condensers, liquid receivers, evaporators, screw compressors), cold and warm water heat exchangers, warm water tanks, a cold water tank, a warm water pump, a cold water pump, an auxiliary water pump, various valves, etc. The physical diagram of the 50 kW OTEC experimental test platform is shown in Figures 6 and 7. Considering factors such as floor space, the organic Rankine cycle module, heat pump unit module, and auxiliary unit module are all skid-mounted, which is convenient for transportation and assembly.

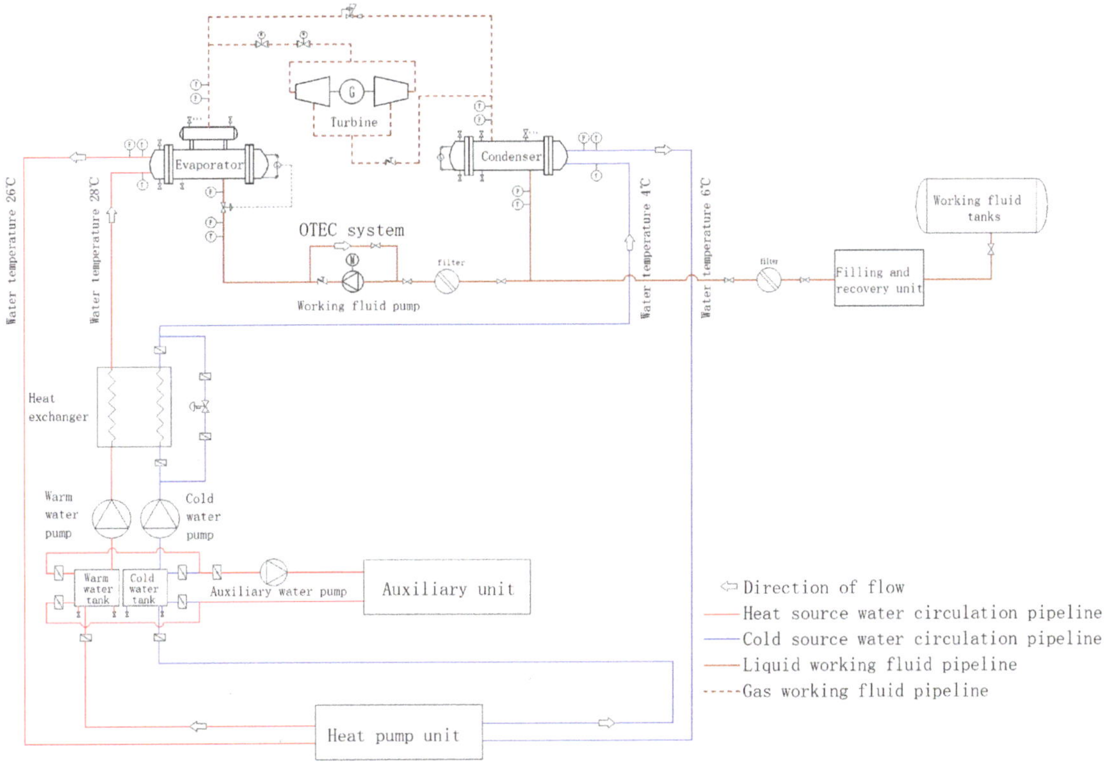

Figure 5. A design diagram of overall experiment setup.

Figure 6. A physical diagram of the 50 kW OTEC analog water recycling system.

Figure 7. A physical diagram of the 50 kW OTEC generation system.

The heat pump unit selects the model LS25LMSA (Moon Environment Technology Co., Ltd., Yantai, China, the same below) screw heat pump refrigerant machine, and the external dimension of the equipment is 6300 × 4000 × 3500 (mm). Among them, the main engine is a twin-screw compressor unit LG25LYA, and the power supply is 3P/50HZ/380V. According to the calculation of Aspen Plus operation, the design working condition of the water-cooled condenser heat transfer is 2187 kW, the waterside inlet temperature is 26 °C and the outlet temperature is 28 °C; the evaporator design working condition heat transfer is 1863 kW, and the waterside inlet temperature is 6 °C. The outlet temperature is 4 °C. Considering that the summer temperature is higher than 28 °C, the air-cooled auxiliary unit LS16SMFA is added. The heat transfer capacity of the air-cooled condenser is 452 kW in the design working condition, and the heat transfer capacity in the design working condition of the evaporator is 387.8 kW.

The temperature of the cold and heat sources of the system should reach 4 °C/28 °C. First, the cold and warm water heat exchanger should make the water temperature reach about 16 °C after heat exchange. Electricity will increase the temperature difference to meet the requirements. The specific control strategy is when the temperature of the water system is lower than 16 °C, the heat pump unit is turned on, and the temperature of the whole system is raised to 16 °C through the cold and warm water heat exchanger, and then the cold and warm water heat exchanger is bypassed. For the cold water system and warm water system, the temperature difference is increased by external power, reaching 4 °C/28 °C; when the temperature of the water system is higher than 16 °C and lower than 28 °C, the auxiliary unit is turned on, and the cold and warm water heat exchanger is used to reduce the temperature of the entire system to 16 °C. The cold and warm water heat exchanger is then bypassed, the heat pump unit is turned on, and work is performed on the cold and warm water system separately to reach a 4 °C/28 °C working condition; when the temperature of the water system is higher than 28 °C, the auxiliary unit is turned on, and the cold and warm water heat exchanger reduces the temperature of the warm water system to 28 °C. Then, the cold and warm water heat exchanger is bypassed, and the temperature of the cold water system is reduced to 4 °C separately. The system operating conditions and ambient conditions are shown in Tables 3 and 4, respectively.

Table 3. OTEC ambient conditions.

Name	Parameter	Value
Electricity supply	Input voltage (V) Frequency (Hz)	380 50
Water supply	Pressure (MPa) Water quality	0.4~0.6 Meet GB/T50050 [53]
Using of equipment	Working fluid Anti-corrosion grade Environmental temperature (°C) Altitude (m)	R134a THWF2 <40 <1000

Table 4. Operating conditions for component.

Name	Type	Parameter	Value
Pump	Shielded	Mass flow (m^3/h) Inlet pressure (MPa) Outlet pressure (MPa)	24.2 0.3~0.5 0.6~0.8
Evaporator	Material side	Inlet/Evaporation temperature (°C) Pressure (kPa)	8/24 646
	Water side	Inlet/outlet temperature (°C) Heat exchange (kW)	28/6 1610
Turbine	Centripetal	Rated speed (rpm) Rating power (kW)	12,500 50
Condenser	Material side	Inlet/condensing temperature (°C) Pressure (kPa)	9/8 388
	Water side	Inlet/outlet temperature (°C) Heat exchange (kW)	4/6 1546

As shown in Figure 8, the prototype in this paper is controlled by a two-level electrical system, with data acquisition and monitoring functions for analog signals, such as the pressure, flow, temperature, and valve switches, and can automatically generate graphs and save data in real time. The main station PLC control cabinet is used as the first-level main console, and the method of PLC with an industrial computer is selected as the core processing component. The PLC uses the Siemens SIMATIC S7-1500 model (Moon Environment Technology Co., Ltd., Yantai, China, the same below), and the upper computer uses the Advantech i7 processor industrial computer (Advantech, Beijing, China). As shown in Figure 9, the master station control system can monitor the control systems of each slave station, control the warm water pump, cold water pump, auxiliary water pump, cold and warm water heat exchanger, and the logical relationship between the actions of the units in the entire system to meet different process adjustments. The OTEC system, heat pump unit, and auxiliary unit have their own Siemens SIMATIC S7-1200, which constitutes the second level as a slave station. All the data from the slave station can communicate with the master station through the integrated PROFINET interface of SIMATIC S7-1200, and the data can be uploaded to the integrated PLC control cabinet in real time, and the master station cabinet SIMATIC S7-1500 will carry out integrated control.

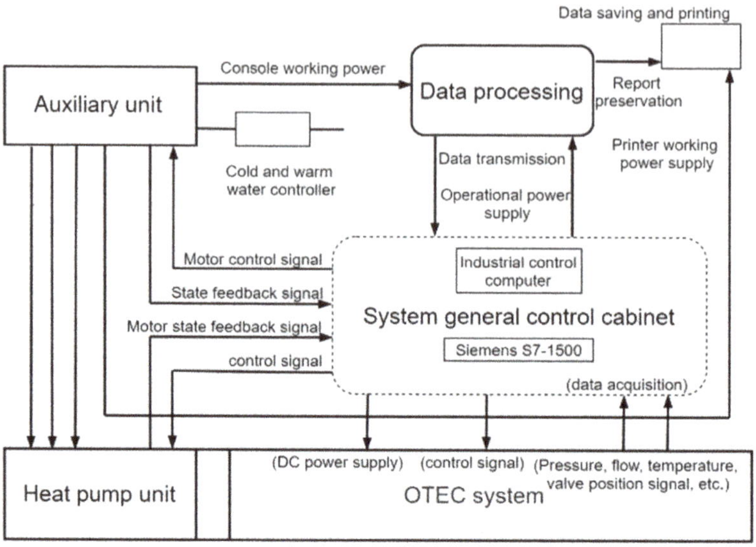

Figure 8. A schematic diagram of control system.

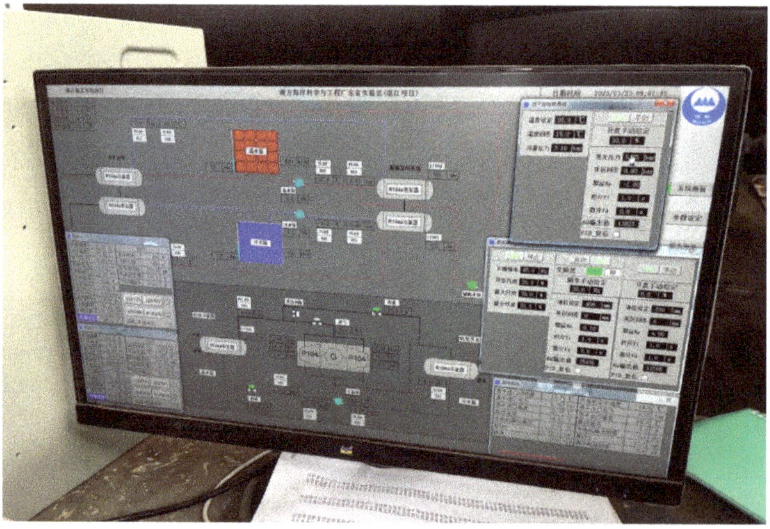

Figure 9. A physical diagram of console with data monitoring and control function.

3.3. On-Grid System

Aiming for power generation with small temperature differences in the ocean, the traditional grid-connected method is improved, and the grid-connected inverter cabinet is adopted. As shown in Figure 10, the system power is converted into DC power through the rectifier equipment, and then the DC power is converted into the same frequency and phase as the grid. Part of the sine wave current can supply power to the local load, and the surplus power can be fed into the grid, eliminating the need for gearboxes, accumulators, and other components. The grid-connected inverter adopts space vector pulse width modulation technology, pure sine wave output, and automatic and synchronous grid tracking; the power factor is close to 1, the current harmonic content is low, there is no pollution to the public grid, there is no impact, and the power generation efficiency is high.

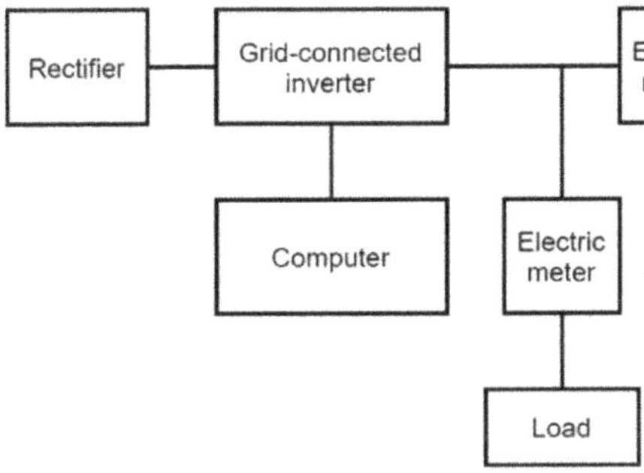

Figure 10. A schematic diagram of grid connection.

3.4. Data Acquisition System

Various states in the OTEC system, such as the temperature, pressure, and mass flow rate, are used as the main monitoring data for equipment operation.

In order to understand the internal operation status of the turbine expander, temperature and pressure sensors are installed at the turbine inlet and outlet, and speed-measuring instruments and vibration sensors are installed in the turbine generator set (Suzhou Xida Low Temperature Equipment Co., Ltd., Suzhou, China). In order to understand the status of system operation, temperature sensors and pressure sensors are installed at the inlet and outlet of the evaporator and condenser, respectively, to monitor the temperature and pressure of the simulated seawater; pressure sensors are installed at the inlet and outlet of the working fluid pump, and a flowmeter is installed at the outlet for the measurement of the working fluid mass flow (Moon Environment Technology Co., Ltd., Yantai, China). Measuring equipment models and parameters are detailed in Table 5.

Table 5. Measuring equipment model and parameters.

Monitoring Instrument	Model Number	Range of Scales	Inaccuracies
Mass flowmeter	DMF-1-U50	0~33 t/h	±0.2%
Temperature sensor	WZP2-83	−50~200 °C	±(0.15 + 0.002 ∣t∣)
Pressure sensor	SIEMENS 7MF0300	−0.1~1.6 MPa	0.075%
RPM sensor	CD-1	0~40,000 r	±0.2%
Vibration sensors	SZ-6	0~2000 μm	±5%

3.5. Experimental Results

Figure 11 shows the variation in the input conditions during the whole system operation, including the temperature and flow rate of warm and cold seawater. Figure 12 reflects the whole process of the system from (1) start-up to (2) stable operation and finally (3) shutdown.

Process (1) is the start-up phase, in which each parameter changes gradually as the opening of the turbine supply valves increases and the opening of the bypass valves decreases, and the output power increases gradually. Process (2) is the stable operation stage: the output parameters of the system can be changed within this range according to the adjustment of the input parameters, maintained in a relatively good control range, without large fluctuations. Process (3) is the shutdown phase, which is the opposite of the valve opening control and start-up process; it can be completed in a relatively short period of time to shut down the experimental equipment.

As can be seen from the experimental process demonstrated in Figures 11 and 12, in the initial stage, the turbine inlet valve is not yet fully open and the bypass pipeline diverts some of the working fluid, resulting in a slow start-up of the turbine. The experiment shows grid-connected power at around 1400 s, and it keeps increasing. When the bypass valve is fully closed, the opening of the turbine inlet valve is about 40%, which leads to a large pressure loss of the working fluid flowing through the valve, so the DC power is kept at a low level. As the opening of the turbine inlet valve gradually reaches 100%, the pressure loss of the mass flowing through the inlet valve decreases and continues in this state for 40 min. Due to the adjustment of the seawater temperature as well as the flow rate, the output power is increasing and the desired design value is reached. The thermal efficiency of the system does not change much as the heat load is also increasing.

Overall, the system is in good operating condition, with gradual changes in the parameters during the start-up phase, and the system can relatively maintain all the parameters within a reasonable range without too much fluctuation during the stable operation phase, and finally a stable shutdown is realized.

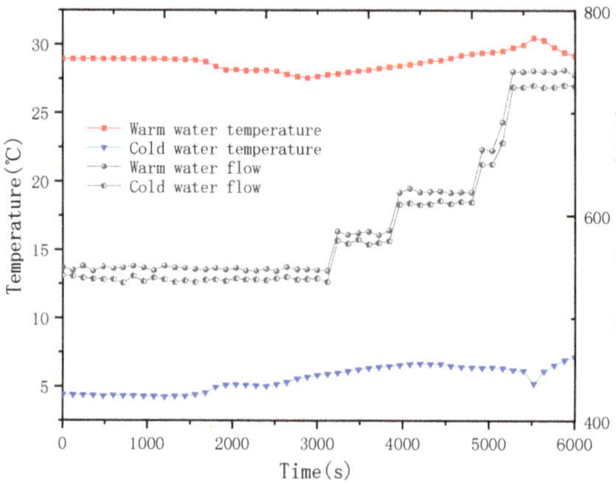

Figure 11. The input working conditions of the power generation system.

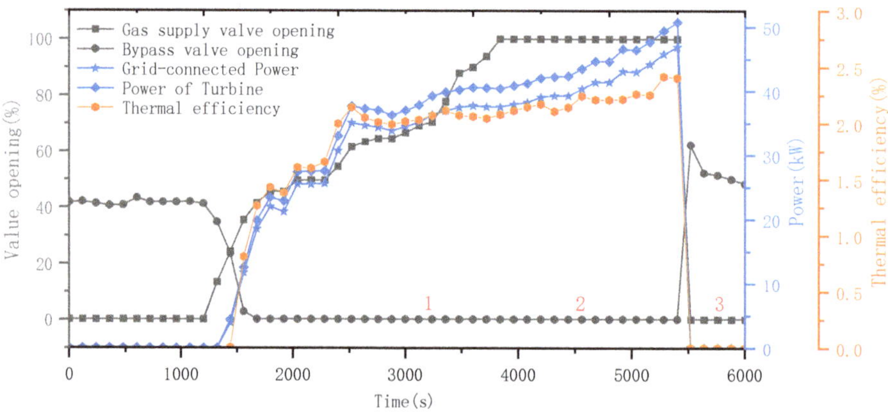

Figure 12. Variation in output parameters during the experiment.

4. Conclusions

In this study, a construction scheme of a land-based prototype simulation system for ocean thermal energy conversion with a cold and heat source control module was proposed and the design and experimental study of a 50 kW ocean thermal energy conversion test platform was carried out in terms of thermodynamic calculations, the selection of the working fluid, the selection of the system components, engineering calibration, and a real-gas test. The following conclusions were obtained:

(1) This study proposes a construction scheme for a land-based prototype simulation system for ocean thermal energy conversion with a heat and cold source control module, which includes an OTEC power generation system, a heat pump unit, and an auxiliary unit, and it is skid-mounted using a modular design to increase the compactness of the system. Equipped with a PLC slave control cabinet, it can monitor the slave pressure, flow, temperature, valve switch, and other analog signal data from the master control cabinet.
(2) The working medium of the system adopts R134a, which has the characteristics of high safety and good thermal conductivity. After engineering calibration calculation, under the design conditions of a 28 °C heat source temperature and 4 °C cold source temperature, the thermal efficiency of the system is 2.63%, and the output power can be up to 50 kW, which meets the design requirements. The grid-connected inverter sends its power into the grid to achieve reasonable conversion and comprehensive utilization of energy.
(3) The experimental analysis shows that the cold and heat source simulation system can stably simulate the seawater environment changes (heat source: 24~28 °C and cold source: 4~8 °C), the power generation system can achieve a smooth start-up and shutdown, there is a continuous power output in the stable operation phase, and the maximum grid-connected power reaches 47.4 kW. The actual thermal efficiency is 2.46%.
(4) As a large ocean thermal energy testing platform in China, the design and experimental research of this experimental platform have achieved good results, which can provide the basis and data support for further research on ocean thermal energy, as well as the experimental simulation basis for further optimization, which is of great significance to the development of ocean thermal energy.

Author Contributions: Conceptualization, B.L. and Y.L.; methodology, X.Z.; software, B.L.; validation, X.Z., B.L. and Y.C.; formal analysis, X.Z. and Y.C.; investigation, B.L.; resources, L.Z.; data curation, Y.C.; writing—original draft preparation, X.Z.; writing—review and editing, Y.L. and Y.C.; visualization, L.Z.; supervision, Y.L.; project administration, Y.L.; funding acquisition, L.Z. All authors have read and agreed to the published version of the manuscript.

Funding: This research was funded by Southern Marine Science and Engineering Guangdong Laboratory (Zhanjiang), grant number: ZJW-2019-05, and the Qingdao Human Resources and Social Security Bureau, the Qingdao Postdoctoral Program: QDBSH20230102007.

Institutional Review Board Statement: Not applicable.

Informed Consent Statement: Not applicable.

Data Availability Statement: Data are contained within the article.

Acknowledgments: Thanks to Yanjun Liu and Yun Chen for their guidance and everyone who contributed to this article.

Conflicts of Interest: The authors declare no conflicts of interest.

References

1. Martínez, M.L.; Vázquez, G.; Pérez-Maqueo, O.; Silva, R.; Moreno-Casasola, P.; Mendoza-González, G.; López-Portillo, J.; MacGregor-Fors, I.; Heckel, G.; Hernández-Santana, J. A systemic view of potential environmental impacts of ocean energy production. *Renew. Sustain. Energy Rev.* **2021**, *149*, 111332. [CrossRef]
2. Feng, C.; Ye, G.; Jiang, Q.; Zheng, Y.; Chen, G.; Wu, J.; Feng, X.; Si, Y.; Zeng, J.; Li, P. The contribution of ocean-based solutions to carbon reduction in China. *Sci. Total Environ.* **2021**, *797*, 149168. [CrossRef]
3. Zereshkian, S.; Mansoury, D. A study on the feasibility of using solar radiation energy and ocean thermal energy conversion to supply electricity for offshore oil and gas fields in the Caspian Sea. *Renew. Energy* **2021**, *163*, 66–77. [CrossRef]
4. Hernández-Fontes, J.V.; Martínez, M.L.; Wojtarowski, A.; González-Mendoza, J.L.; Landgrave, R.; Silva, R. Is ocean energy an alternative in developing regions? A case study in Michoacan, Mexico. *J. Clean. Prod.* **2020**, *266*, 121984. [CrossRef]

5. Tsalis, T.A.; Malamateniou, K.E.; Koulouriotis, D.; Nikolaou, I.E. New challenges for corporate sustainability reporting: United Nations' 2030 Agenda for sustainable development and the sustainable development goals. *Corp. Soc. Responsib. Environ. Manag.* **2020**, *27*, 1617–1629. [CrossRef]
6. Hussain, A.; Arif, S.M.; Aslam, M. Emerging renewable and sustainable energy technologies: State of the art. *Renew. Sustain. Energy Rev.* **2017**, *71*, 12–28. [CrossRef]
7. Ellabban, O.; Abu-Rub, H.; Blaabjerg, F. Renewable energy resources: Current status, future prospects and their enabling technology. *Renew. Sustain. Energy Rev.* **2014**, *39*, 748–764. [CrossRef]
8. Khan, N.d.; Kalair, A.; Abas, N.; Haider, A. Review of ocean tidal, wave and thermal energy technologies. *Renew. Sustain. Energy Rev.* **2017**, *72*, 590–604. [CrossRef]
9. Nihous, G.C. Mapping available Ocean Thermal Energy Conversion resources around the main Hawaiian Islands with state-of-the-art tools. *J. Renew. Sustain. Energy* **2010**, *2*, 043104. [CrossRef]
10. Uehara, H.; Dilao, C.O.; Nakaoka, T. Conceptual design of ocean thermal energy conversion (OTEC) power plants in the Philippines. *Sol Energy* **1988**, *41*, 431–441. [CrossRef]
11. Hashim, H.; Zubir, M.A.; Kamyab, H.; Zahran, M.F.I. Decarbonisation of the industrial sector through greenhouse gas mitigation, offset, and emission trading schemes. *Chem. Eng. Trans.* **2022**, *97*, 511–516.
12. Faizal, M.; Ahmed, M.R. Experimental studies on a closed cycle demonstration OTEC plant working on small temperature difference. *Renew. Energy* **2013**, *51*, 234–240. [CrossRef]
13. Yang, M.-H.; Yeh, R.-H. Analysis of optimization in an OTEC plant using organic Rankine cycle. *Renew. Energy* **2014**, *68*, 25–34. [CrossRef]
14. Sun, F.; Ikegami, Y.; Jia, B.; Arima, H. Optimization design and exergy analysis of organic rankine cycle in ocean thermal energy conversion. *Appl. Ocean Res.* **2012**, *35*, 38–46. [CrossRef]
15. Rau, G.H.; Baird, J.R. Negative-CO2-emissions ocean thermal energy conversion. *Renew. Sustain. Energy Rev.* **2018**, *95*, 265–272. [CrossRef]
16. Esteban, M.; Leary, D. Current developments and future prospects of offshore wind and ocean energy. *Appl. Energy* **2012**, *90*, 128–136. [CrossRef]
17. Faizal, M.; Rafiuddin Ahmed, M. On the ocean heat budget and ocean thermal energy conversion. *Int. J. Energy Res.* **2011**, *35*, 1119–1144. [CrossRef]
18. Wu, Z.; Feng, H.; Chen, L.; Xie, Z.; Cai, C. Pumping power minimization of an evaporator in ocean thermal energy conversion system based on constructal theory. *Energy* **2019**, *181*, 974–984. [CrossRef]
19. Langer, J.; Quist, J.; Blok, K. Recent progress in the economics of ocean thermal energy conversion: Critical review and research agenda. *Renew. Sustain. Energy Rev.* **2020**, *130*, 109960. [CrossRef]
20. D'ARSONVAL, J.A. Utilization des Forces Naturelles, Avenir de l'electricite. *Le Revenue Sci.* **1881**, *17*, 370–372.
21. Nihous, G.C. Ocean Thermal Energy Conversion (OTEC). In *Wind, Water and Fire: The Other Renewable Energy Resources*; World Scientific: Singapore, 2021; pp. 173–196.
22. Aresti, L.; Christodoulides, P.; Michailides, C.; Onoufriou, T. Reviewing the energy, environment, and economy prospects of Ocean Thermal Energy Conversion (OTEC) systems. *Sustain. Energy Technol. Assess.* **2023**, *60*, 103459. [CrossRef]
23. White, H.J. Mini-OTEC. *Int. J. Ambient Energy* **1980**, *1*, 75–88. [CrossRef]
24. Mitsui, T.; Ito, F.; Seya, Y.; Nakamoto, Y. Outline of the 100 kw Otec Pilot Plant in the Republic of Naure. *IEEE Trans. Power Appar. Syst.* **1983**, *PAS-102*, 3167–3171. [CrossRef]
25. Lennard, D. Ocean thermal energy conversion—Past progress and future prospects. *IEE Proc. A (Phys. Sci. Meas. Instrum. Manag. Educ. Rev.)* **1987**, *134*, 381–391. [CrossRef]
26. Kobayashi, H.; Jitsuhara, S.; Uehara, H. *The Present Status and Features of OTEC and Recent Aspects of Thermal Energy Conversion Technologies*; National Maritime Research Institute: Tokyo, Japan, 2020. Available online: https://newsroom.prkarma.com/assets/newsroom/documents/555.svrybz5k.pdf (accessed on 11 December 2023).
27. Sinama, F. *Étude de la Production D'électricité à Partir de L'énergie Thermique des mers à L'île de la Réunion: Modélisation et Optimisation du Procédé*; Université de la Réunion: Saint Denis, France, 2011.
28. Journoud, A.; Sinama, F.; Lucas, F. Experimental Ocean Thermal Energy Conversion (OTEC) project on the Reunion Island. In Proceedings of the 4th International Conference on Ocean Energy, Dublin, Ireland, 17–20 October 2012.
29. Martins, M.; Sinama, F.; Lucas, F. Equivalent Gibbs systems for modelling an onshore OTEC experimental plant on Reunion Island. *Int. J. Energy Res.* **2013**, *37*, 1112–1121. [CrossRef]
30. Kim, H.-J.; Kim, A.S. *Ocean Thermal Energy Conversion (OTEC): Past, Present, and Progress*; IntechOpen: London, UK, 2020.
31. Chen, F.; Liu, L.; Peng, J.; Ge, Y.; Wu, H.; Liu, W. Theoretical and experimental research on the thermal performance of ocean thermal energy conversion system using the rankine cycle mode. *Energy* **2019**, *183*, 497–503. [CrossRef]
32. Heydt, G.T. An assessment of ocean thermal energy conversion as an advanced electric generation methodology. *Proc. IEEE* **1993**, *81*, 409–418. [CrossRef]
33. Kalina, A.I. Combined cycle and waste heat recovery power systems based on a novel thermodynamic energy cycle utilizing low-temperature heat for power generation. In Proceedings of the Turbo Expo: Power for Land, Sea, and Air, Phoenix, AZ, USA, 27–31 March 1983; p. V001T002A003.
34. Zhang, X.; He, M.; Zhang, Y. A review of research on the Kalina cycle. *Renew. Sustain. Energy Rev.* **2012**, *16*, 5309–5318. [CrossRef]

5. Panchal, C.; Bell, K. Simultaneous production of desalinated water and power using a hybrid-cycle OTEC plant. *J. Sol. Energy Eng.* **1987**, *109*, 156–160. [CrossRef]
6. Uehara, H.; Miyara, A.; Ikegami, Y.; Nakaoka, T. Performance analysis of an OTEC plant and a desalination plant using an integrated hybrid cycle. *J. Sol. Energy Eng.* **1996**, *118*, 115–122. [CrossRef]
7. Quoilin, S.; Van Den Broek, M.; Declaye, S.; Dewallef, P.; Lemort, V. Techno-economic survey of Organic Rankine Cycle (ORC) systems. *Renew. Sustain. Energy Rev.* **2013**, *22*, 168–186.
8. Yang, M.-H.; Yeh, R.-H. Investigation of the potential of R717 blends as working fluids in the organic Rankine cycle (ORC) for ocean thermal energy conversion (OTEC). *Energy* **2022**, *245*, 123317. [CrossRef]
9. Ma, Q.; Gao, Z.; Huang, J.; Mahian, O.; Feng, X.; Lu, H.; Wang, S.; Wang, C.; Tang, R.; Li, J. Thermodynamic analysis and turbine design of a 100 kW OTEC-ORC with binary non-azeotropic working fluid. *Energy* **2023**, *263*, 126097.
0. Wahinuddin, M.A.; Mohd, N.A.R.N.; Nasir, M.N.M.; Othman, N.; Mat, S.; Thirugana, S.T. Otec performance evaluation using different working fluids and variations in operating orc conditions. *J. Mek.* **2023**, *46*, 14–26. [CrossRef]
1. Anderson, J.H., Jr. Ocean thermal energy conversion (OTEC): Choosing a working fluid. In Proceedings of the ASME Power Conference, Albuquerque, NM, USA, 21–23 July 2009; pp. 645–653.
2. Yoon, J.-I.; Son, C.-H.; Baek, S.-M.; Kim, H.-J.; Lee, H.-S. Efficiency comparison of subcritical OTEC power cycle using various working fluids. *Heat Mass Transf.* **2014**, *50*, 985–996.
3. Hung, T.; Wang, S.; Kuo, C.; Pei, B.; Tsai, K. A study of organic working fluids on system efficiency of an ORC using low-grade energy sources. *Energy* **2010**, *35*, 1403–1411. [CrossRef]
4. Vera, D.; Baccioli, A.; Jurado, F.; Desideri, U. Modeling and optimization of an ocean thermal energy conversion system for remote islands electrification. *Renew. Energy* **2020**, *162*, 1399–1414. [CrossRef]
5. Xu, W.; Zhao, L.; Mao, S.S.; Deng, S. Towards novel low temperature thermodynamic cycle: A critical review originated from organic Rankine cycle. *Appl. Energy* **2020**, *270*, 115186.
6. Yamada, N.; Hoshi, A.; Ikegami, Y. Performance simulation of solar-boosted ocean thermal energy conversion plant. *Renew. Energy* **2009**, *34*, 1752–1758. [CrossRef]
7. Qyyum, M.A.; Khan, A.; Ali, S.; Khurram, M.S.; Mao, N.; Naquash, A.; Noon, A.A.; He, T.; Lee, M. Assessment of working fluids, thermal resources and cooling utilities for Organic Rankine Cycles: State-of-the-art comparison, challenges, commercial status, and future prospects. *Energy Convers. Manag.* **2022**, *252*, 115055.
8. Hu, B.; Yan, H.; Wang, R. Modeling and simulation of a falling film evaporator for a water vapor heat pump system. *Appl. Energy* **2019**, *255*, 113851. [CrossRef]
9. Kundu, P.K.; Cohen, I.M.; Dowling, D.R. *Fluid Mechanics*; Academic Press: Cambridge, MA, USA, 2015.
0. Elger, D.F.; Williams, B.C.; Crowe, C.T. *Engineering Fluid Mechanics*; John Wiley & Sons: Hoboken, NJ, USA, 2022.
1. Fiaschi, D.; Secchi, R.; Galoppi, G.; Tempesti, D.; Ferrara, G.; Ferrari, L.; Karellas, S. Piston expanders technology as a way to recover energy from the expansion of highly wet organic refrigerants. In Proceedings of the Energy Sustainability, San Diego, CA, USA, 28 June–2 July 2015; p. V002T018A007.
2. Spadacini, C.; Xodo, L.; Quaia, M. Geothermal energy exploitation with Organic Rankine Cycle technologies. In *Organic Rankine Cycle (ORC) Power Systems*; Elsevier: Amsterdam, The Netherlands, 2017; pp. 473–525.
3. *GB/T 50050-2017*; Design Specification for Industrial Circulating Cooling Water Treatment. Ministry of Housing and Urban-Rural Development: Beijing, China, 2017.

Journal of
Marine Science and Engineering

MDPI

Article

Suppression of Negative Sequence Current on HVDC Modular Multilevel Converters in Offshore Wind Power

Xiaoning Xu [1,2], Di Wang [1,2,*], Xuesong Zhou [1,2] and Long Tao [3]

[1] School of Electrical Engineering and Automation, Tianjin University of Technology, Tianjin 300384, China; xxn2523@126.com (X.X.)
[2] Tianjin Key Laboratory of New Energy Power Conversion, Transmission and Intelligent Control, Tianjin 300384, China
[3] School of Electrical Automation and Information Engineering, Tianjin University, Tianjin 300384, China
* Correspondence: wd161718@126.com

Abstract: The High Voltage Direct Current (HVDC) transmission technology employing modula multilevel converters (MMCs) can effectively enhance the transmission efficiency and stability o offshore wind farms, thereby aiding the promotion of large−scale utilization of new energy. Thi holds significant importance for achieving the dual carbon goals. Aiming at the problem of negativ sequence current circulation in MMC−HVDC transmission systems, a circulation suppression strat egy based on augmented order decoupling linear active disturbance rejection control (LADRC) i proposed in this paper. By introducing new state variables into the traditional ADRC structure, th actual output deviation signal and observation gain signal from the disturbance observation value o the system are used. It can not only realize the decoupling control of disturbance and tracking term but also enhance the disturbance immunity, robustness and rapidity of the controller. Finally, an 18−level MMC system model is built based on Matlab (9.12.0.1884302 (R2022a)) & Simulink (R2022a) and the circulation suppression effects of stable operation and voltage sudden change are simulated and compared, which verifies the suppression effect of the improved control strategy on negativ sequence current circulation, which lays a theoretical and application foundation for the sustainabl development of the offshore wind power industry.

Keywords: HVDC; modular multilevel converter; negative sequence current circulation; LADRC circulation suppression

Citation: Xu, X.; Wang, D.; Zhou, X.; Tao, L. Suppression of Negative Sequence Current on HVDC Modular Multilevel Converters in Offshore Wind Power. *J. Mar. Sci. Eng.* **2024**, *12*, 383. https://doi.org/10.3390/jmse12030383

Academic Editor: Yassine Amirat

Received: 4 February 2024
Revised: 18 February 2024
Accepted: 20 February 2024
Published: 23 February 2024

1. Introduction

Global warming and climate change pose an existential threat to humanity. Pursuing industrial development while vigorously expanding renewables is the core strategy of many countries around the world [1]. Renewable energy has been widely used in various industries because of its low cost, environmental friendliness and other advantages. Among them, offshore wind power, ocean energy and solar power are the main new energy forms in the world [2–4]. Offshore wind power has become an important part of the large-scale utilization of new energy because of its rich resource reserves, high power generation efficiency, low water consumption, small land occupation area and close to power load center [5,6]. The development of offshore wind power is helping to build a clean, low carbon, safe and efficient energy future, in order to meet the challenges of climate change and energy security [7]. With the gradual expansion of the scale of offshore wind farms and the continuous increase of offshore distance, the traditional interconnection of offshore wind farms with the onshore substations by high voltage alternating current cables does not find techno-economic feasibility. Advanced schemes and technologies of submarine power transmission are necessary to facilitate the capacity expansion of offshore wind farms [8–10].

In order to improve the large-scale utilization efficiency of offshore new energy and ensure the high quality of electric energy in the process of transmission, key technologies such as wind turbines, floating wind power platforms, submarine cables and power transmission have been launched one after another. Among them, HVDC technology has the characteristics of strong controllability, small footprint and efficient long-distance power transmission. It can effectively transport the power generated by offshore wind power to onshore substation, improve the reliability of power supply, reduce the loss of transmission lines, increase the transmission capacity of transmission lines and provide protection for offshore wind power and other large-scale access [11–15]. Therefore, HVDC technology has become one of the main means of cross-sea large-scale power transportation. It has become a new choice for power supply in mega-cities and isolated islands [16,17], and it is also one of the best choices for large-scale clean energy power generation to be connected and transmitted by high-voltage DC power grids in the future [18,19].

Compared with the HVDC system with its traditional two-level or three-level converter, a modular multilevel converter (MMC) has the characteristics of low switching loss, low harmonic loss and fast response to power grid fluctuations, which can realize the efficient, stable and controllable operation of an HVDC system [20–22]. However, due to the structure of MMC itself, the instantaneous voltages of the sub-modules of the upper and lower bridge arms are different, which leads to the internal circulation between the phases in the normal operation of the system. The existence of circulating current not only increases power consumption and energy loss and causes voltage fluctuation and voltage distortion but also shortens the service life of power electronic devices [23,24]. Well-specified circulating current suppressors play an important role in improving power quality, reducing loss and optimizing system performance in HVDC systems. The commonly used circulating current suppressor are divided into passive circulating current suppressors and active circulating current suppressors. Passive circulating current suppressors mainly rely on physical methods to adjust the elements to achieve the suppression of interphase circulation, but the flexibility and adaptability are general. The active circulating current suppressor dynamically suppresses the circulation through the software method; the response speed is fast, it avoids the problems caused by the passive circulating current suppressors and it is the main method to restrain the circulation [25,26]. Therefore, the design of a reasonable and effective circulation suppression strategy is of great significance to promote the progress of HVDC technology and the large-scale expansion of offshore wind power [27,28].

Traditional methods of circulating current suppression often employ Proportional Integral (PI) control [29,30]. However, the decoupling process involved in the control loop leads to increased workload, and the system stability and effectiveness in suppressing circulating currents are limited. Reference [31] introduces the Proportional Resonant (PR) controller, which eliminates coordinate transformations, as well as the coupling terms and feedforward compensation items affected by circuit parameters. In response to issues such as high parameter sensitivity, poor disturbance rejection and narrow bandwidth in PR controllers, literature [32] proposes a quasi-PR controller. However, it incorporates a notch filter to enhance control effectiveness and necessitates stability analysis of the system. Literature [33] combines a low-pass filter with Internal Model Control (IMC) to enhance the robustness of the system. Compared with the above methods, ADRC provides stronger robustness through active interference suppression, has relatively low requirements for dynamic characteristics of the system, and can quickly respond to changes and disturbances of the system [34]. ADRC circulation suppression strategy helps to improve the overall stability of the power system, which plays a key role in preventing system instability, oscillation or even collapse. Reference [35], utilizing the characteristic of LADRC that does not rely on a detailed model of the MMC circulating current, designs a circulating current suppression strategy that simplifies the mechanisms of circulating current control, demonstrating good robustness but constrained by bandwidth limitations. In the literature [36], a circulating current suppressor based on virtual impedance sliding mode control is proposed, and its transient performance is simulated. However, no explicit criteria are provided for evaluating the effectiveness of the circulating current suppressor.

This paper aims at the problem of negative sequence current circulation in long distance MMC-HVDC in marine environments. Based on the analysis of the working mechanism of MMC bridge arm circuit, an improved LADRC circulation suppression strategy of MMC based on offshore wind power is proposed in this paper. By changing the structure of the linear extended state observer (LESO), a new differential term is introduced into LESO. Augmented Order Decoupling LADRC (AD-LADRC) is designed to increase the bandwidth of the controller and decouple the disturbance term and tracking term of the controller. Through modeling, simulation and comparing the coupling relationship, the rapidity and stability of the ADRC are improved. The parameter setting process is simplified, and the system performance is optimized. The effectiveness of the improved LADRC circulating current suppressor is verified by a Matlab & Simulink simulation, and the circulation suppression effects of stable operation and voltage sudden change are analyzed and compared.

2. Advantages of HVDC Technology

HVDC provides strong technical support for the development and application of offshore wind farms. The offshore wind power HVDC project involved in this paper is shown in Figure 1 [37].

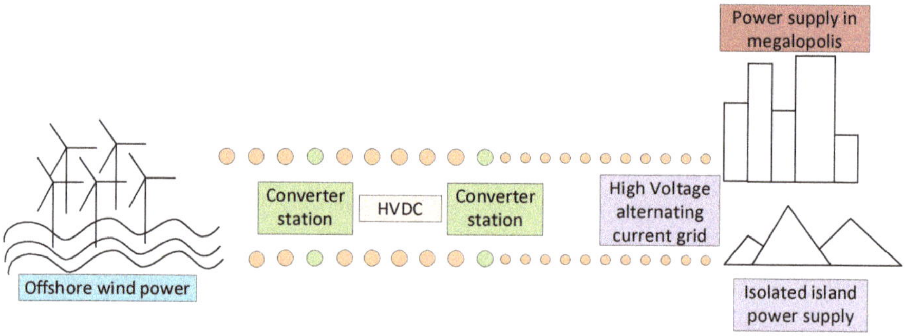

Figure 1. Overview of Offshore Wind Power HVDC Project.

As can be seen from Figure 1, the converter station is the key equipment of HVDC technology and an important hub connecting offshore wind farms and power loads. By controlling the turn-on and turn-off time of switching devices, the stepless conversion of AC and DC energy between different forms of power grids or between different voltage levels is realized; thus harmonics and losses are reduced.

An LADRC circulation suppression strategy of MMC-HVDC based on offshore wind power proposed in this paper has the following advantages:

(1) The harmonic level is low.

Compared with traditional DC transmission technology, the HVDC system with a modular multilevel converter generates voltage and current close to sinusoidal waveform, which has a significant effect on improving the quality and stability of power transmission in offshore wind farms.

(2) The stability of the system is enhanced.

By reducing the influence of external disturbances, the ADRC circulation suppression strategy helps to improve the overall stability of the offshore wind farm HVDC system. This plays a key role in preventing instability, oscillation or even collapse of the system.

(3) There is no reactive power compensation and no commutation failure.

HVDC technology uses turn−off devices to control the turn-on and turn-off time. This method does not rely on the AC side to provide commutation current and reverse voltage and effectively avoids the loss of a large amount of reactive power. In addition, it has nothing to do with the current direction and fundamentally solves the problem of commutation failure.

(4) It has the ability to supply electricity to isolated islands.

Because HVDC technology can realize commutation independently and can operate in passive inverter mode without external commutation voltage, the receiving end system can be a passive network.

(5) It is suitable for constructing a multi-terminal system.

The current of the HVDC system can flow in both directions, and the polarity of the DC voltage remains unchanged when the power flow is reversed. This makes it suitable for the construction of multi-terminal systems, in which the power flow between converter stations is more flexible, and it can provide a convenient connection form for the follow-up large-scale construction of new energy at sea and the combination of multi-energy situations.

3. Analysis of Working Mechanism of MMC Bridge Arm Circuit

3.1. MMC Topology

The single-phase equivalent circuit model of the Modular Multilevel Converter is depicted in Figure 2 [38].

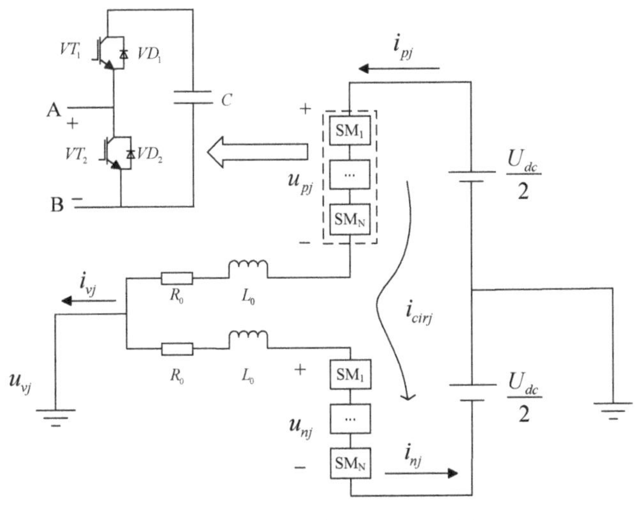

Figure 2. MMC single-phase equivalent circuit model.

It comprises upper and lower bridge arms, each formed by equivalent resistors, inductors and N sub-modules. The sub-module capacitor voltages are denoted as u_{pj} and u_{nj}; the bridge arm currents are labeled as i_{pj} and i_{nj}; the internal circulating current is denoted as i_{cirj}. In this context, L_0 represents the inductance of the bridge arm, and R_0 denotes the resistance of the bridge arm. On the AC side, the single-phase voltage is represented as u_{vj}, and the current is denoted as i_{vj}. On the DC side, the total voltage is denoted as U_{dc}. Each sub-module includes two IGBT switches, labeled as VT_1 and VT_2, Each sub-module includes two anti-parallel diodes, labeled as VD_1 and VD_2, and a capacitor, labeled as C.

3.2. Analysis of the Principle of MMC Circulation

Obtained from the MMC single-phase equivalent circuit (1):

$$\begin{cases} i_{cirj} = \frac{i_{pj}+i_{nj}}{2} \\ e^j = \frac{u_{nj}-u_{pj}}{2} \end{cases} \tag{1}$$

In the equation, e^j, represents the internal electromotive force of the three-phase system (abc).

Based on Kirchhoff's laws, the expression (2) for the reference values of the capacitor voltages in the upper and lower bridge arms can be derived:

$$\begin{cases} u_{pj}{}^* = \frac{U_{dc}}{2} - e_j - u_{diffj} \\ u_{nj}{}^* = \frac{U_{dc}}{2} - e_j - u_{diffj} \end{cases} \tag{2}$$

In the equation, u_{diffj} represents the circulating voltage drop across a single bridge arm.

Define the modulation ratios for MMC output voltage and output current (3):

$$\begin{cases} \mathrm{k} = \frac{U_m}{U_{dc}/2} \\ m = \frac{I_m/2}{I_{dc}/3} \end{cases} \tag{3}$$

In the equation, U_m represents the amplitude of the AC side voltage, and I_m denotes the amplitude of the AC side current.

The voltage and current equations for the upper and lower bridge arms are as shown in Equation (4):

$$\begin{cases} u_{pj} = \frac{1}{2}U_{dc}(1 - k\sin\omega t) \\ u_{nj} = \frac{1}{2}U_{dc}(1 + k\sin\omega t) \\ i_{pj} = \frac{I_{dc}}{3}[1 + m\sin(\omega t + \varphi)] \\ i_{nj} = \frac{I_{dc}}{3}[1 - m\sin(\omega t + \varphi)] \end{cases} \tag{4}$$

In the equation, φ represents the initial phase angle of the AC side phase j current.

From Equation (4), the instantaneous power of a single-phase upper or lower bridge arm can be obtained, and by integrating, the total energy of a single-phase upper or lower bridge arm can be calculated. Neglecting the DC component, the AC component is given by Equation (5):

$$W_{PM\_AC}(t) = \frac{mkU_{dc}I_{dc}}{12\omega_0}\sin(2\omega_0 + \varphi) \tag{5}$$

From Equation (5), it can be observed that the bridge arm current not only has the fundamental frequency component but also exhibits circulating currents with a negative sequence nature, primarily at twice the fundamental frequency.

Therefore, by incorporating the second harmonic component into Equation (4) and substituting it into Equation (1), the expression for the three-phase circulating current can be obtained:

$$\begin{cases} i_{diffa} = \frac{I_{dc}}{3} + I_{2f}\sin(2\omega t + \varphi) \\ i_{diffb} = \frac{I_{dc}}{3} + I_{2f}\sin(2\omega t + \varphi + \frac{2\pi}{3}) \\ i_{diffc} = \frac{I_{dc}}{3} + I_{2f}\sin(2\omega t + \varphi - \frac{2\pi}{3}) \end{cases} \tag{6}$$

In the equation, i_{diffa} represents the circulating current in phase A, i_{diffb} for phase B and i_{diffc} for phase C.

4. MMC Circulation Suppression Strategy of Improved LADRC

4.1. Design of Circulating Current Suppressor Based on LADRC

4.1.1. Traditional LADRC Controller Structure

The core concept of traditional LADRC is to achieve precise control through accurate estimation and suppression of both internal and external disturbances in the system. It does not require precise modeling of the system and emphasizes estimation and compensation of disturbances, thereby achieving robust control performance. The structure of a first-order linear active disturbance rejection controller is illustrated in Figure 3.

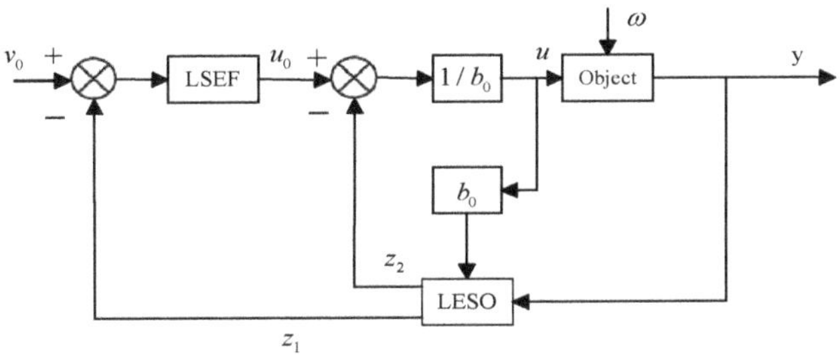

Figure 3. Second-order linear rejection controller block diagram.

LSEF refers to the linear state error feedback rate. Generally speaking, LSEF in LADRC can be in the form of PD controller, which can realize part of the functions of TD to improve the overall control performance of the system. LESO is a kind of observer used to estimate the state of the system, which can estimate the extended state of the system in real time, including internal dynamics and external disturbances.

LESO can effectively compare the actual state of the system with the expected state in the controller, thus helping the controller to accurately calculate the required compensation control signal.

A first-order linear system can be represented as

$$\dot{y} = -a_1 y + \omega + bu \tag{7}$$

In the equation, u is the input value, y is the output value, b is the gain, a_1 is the control parameter and ω represents the external disturbance.

The generalized disturbance equation can be expressed as

$$f(y, \dot{y}, \omega) = -a_1 y + \omega + (b - b_0)u \tag{8}$$

In the equation, b_0 is an estimate of the system b.

For the aforementioned equation, by selecting state variables as follows, the equation can be further derived:

$$\begin{cases} \dot{x}_1 = x_2 + b_0 u \\ \dot{x}_2 = h \\ y = x_1 \end{cases} \tag{9}$$

The expression for the LESO is

$$\begin{cases} \dot{z}_1 = z_2 - \beta_1(z_1 - y) + b_0 u \\ \dot{z}_2 = -\beta_2(z_1 - y) \end{cases} \tag{10}$$

In the equation, z_1 is the tracking signal of y, and z_2 tracks the total disturbance f.

By selecting appropriate observer gains, real-time tracking of system variables can be achieved, and the control rate can be designed as

$$u = \frac{u_0 - z_2}{b_0} \tag{11}$$

In a first-order LADRC, a proportional controller is commonly used instead of LESF and the equation is expressed as

$$u_0 = k_p(v_0 - z_1) \tag{12}$$

The controlled object can be set as

$$y = \frac{1}{s}(f + b_0 u) \tag{13}$$

$$\begin{cases} k_p = \omega_c \\ \beta_1 = 2\omega_0 \\ \beta_2 = {\omega_0}^2 \end{cases} \tag{14}$$

In the equation, v_0 represents the given controlled signal, k_p is the controller coefficient ω_c is the controller bandwidth and ω_0 is the observer bandwidth.

$$G_1(s) = \frac{\beta_2}{s^2 + \beta_1 s + \beta_2} \tag{15}$$

The closed-loop transfer function of a first-order feedback system is given by

$$y = \frac{k_p}{s + k_p} v + \frac{s^2 + (\beta_1 + k_p)s}{(s + k_p) + (s^2 + \beta_1 s + \beta_2)} f \tag{16}$$

The LADRC controller system's disturbance term and tracking term are both related to k_p, demonstrating a coupling relationship between the two. This coupling makes it challenging to adjust parameters. Additionally, from Equation (15), it can be inferred that while increasing the observer bandwidth ω_0 is beneficial for enhancing the disturbance observation capability of traditional LESO, the presence of observation noise limits the practical adjustment of the bandwidth in engineering applications.

4.1.2. Design of Circulating Current Suppressor Based on Traditional LADRC

Treating each harmonic component in the circulating current as an external disturbance, compensation is achieved through LESO. Subsequently, a circulating current suppression strategy based on LADRC is designed, and the suppression process is illustrated in Figure 4 Firstly, according to Equation (6), obtain the three-phase circulating current i_{diffj}, and through Park transformation, derive circulating currents i_{diffd} and i_{diffq}. Similarly, compare them with their respective reference values i_{diffd_ref} and i_{diffq_ref}. When the reference value is set to zero, it can effectively suppress the second harmonic in the circulating current; through the LADRC controller, the circulating current drops U_{diffd_ref}, U_{diffq_ref} are obtained and, finally, the three-phase circulating current drops are obtained through the inverse Park transformation.

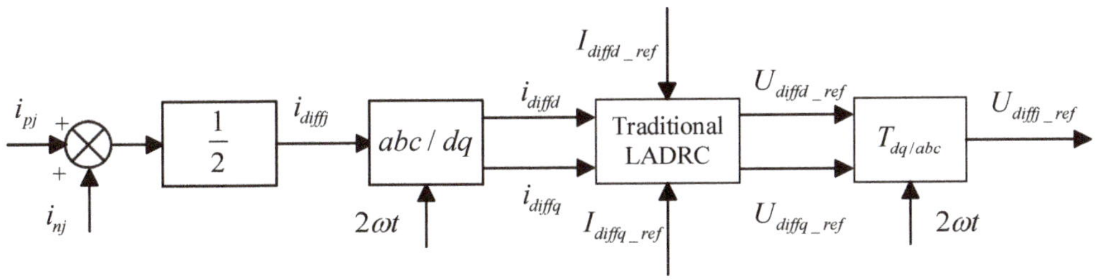

Figure 4. Diagram of the circulation suppression process.

4.1.3. MMC Control System Based on LADRC Circulating Current Suppressor

In this paper, double closed loop feedforward decoupling control is adopted in the main control part of MMC, and the control structure diagram is shown in Figure 5 [39,40]. In the figure, according to the two PI controllers, the output reference voltages u_{d_ref} and u_{q_ref}. These are then converted into three-phase voltage reference values V_{jref} through Park inverse transformation. U_{diffj_ref} represents the circulatory voltage drop obtained through the circulation suppression control strategy. Finally, the corresponding PWM pulse signals for controlling the sub-module switches are generated through the nearest level approximation. "*" represents the reference value of the corresponding quantity.

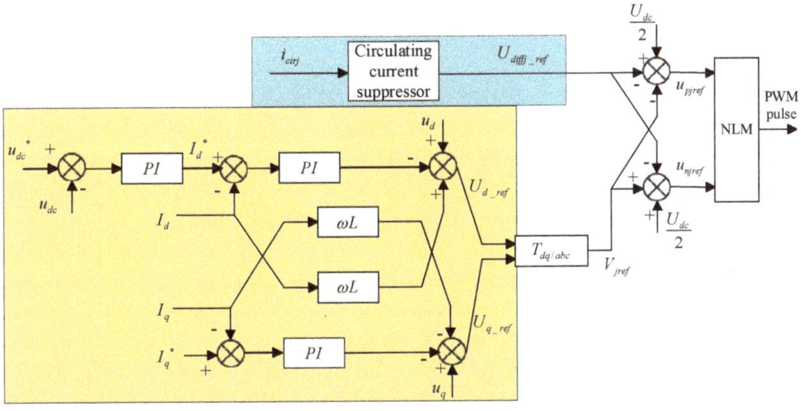

Figure 5. MMC structure control chart.

4.2. Design of Circulating Current Suppressor Based on AD−LADRC

4.2.1. Improving the Structural Design of LESO

In order to achieve faster and more effective circulation suppression, this paper introduces an additional state variable z_3 on the basis of the traditional second-order LESO. This variable is utilized to observe the changing trend of the total disturbance. Simultaneously, improvements are made to compensate for the total disturbance, enhancing the structure of the extended state observer for the decoupling of disturbance and tracking terms. The improvement in the LADRC is illustrated in Figure 6. The expression for the improved LESO is as follows:

$$\begin{cases} \dot{z}_1 = z_2 + b_0 u \\ \dot{z}_2 = z_3 - \beta_1(z_1 - y) \\ \dot{z}_3 = -(\beta_2 + \beta_3)(z_1 - y) \end{cases} \tag{17}$$

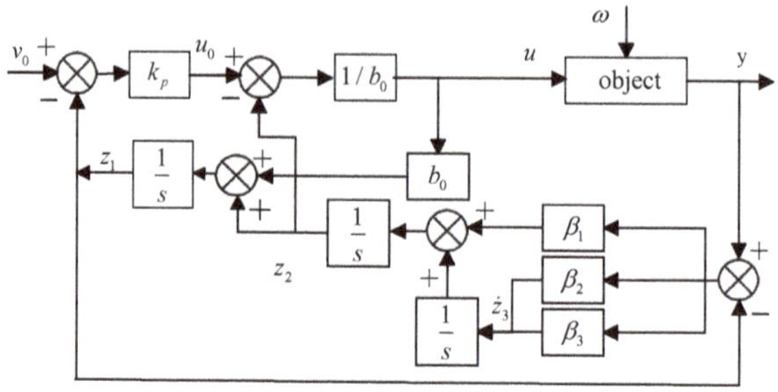

Figure 6. AD–LADRC structure diagram.

Applying the inverse Laplace transform to Equation (17) yields

$$\begin{cases} z_1 = \frac{\beta_1 s^2+\beta_2 s+\beta_3}{s^3+\beta_1 s^2+\beta_2 s+\beta_3} y + \frac{b_0 s^2}{s^3+\beta_1 s^2+\beta_2 s+\beta_3} u \\ z_2 = \frac{(\beta_1 s^2+\beta_2 s+\beta_3)s}{s^3+\beta_1 s^2+\beta_2 s+\beta_3} y - \frac{b_0(\beta_1 s^2+\beta_2 s+\beta_3)}{s^3+\beta_1 s^2+\beta_2 s+\beta_3} u \\ z_3 = \frac{(\beta_1 s^2+\beta_2 s+\beta_3)s^2}{s^3+\beta_1 s^2+\beta_2 s+\beta_3} y - \frac{b_0 s(\beta_1 s^2+\beta_2 s+\beta_3)}{s^3+\beta_1 s^2+\beta_2 s+\beta_3} u \end{cases} \quad (18)$$

The disturbance observation transfer function of the improved LESO is

$$G_2(s) = \frac{z_2}{f} = \frac{\beta_1 s^2 + \beta_2 s + \beta_3}{s^3 + \beta_1 s^2 + \beta_2 s + \beta_3} \quad (19)$$

In the formula, the gains are

$$\begin{cases} \beta_1 = 3\omega_0 \\ \beta_2 = 3\omega_0{}^2 \\ \beta_3 = \omega_0{}^3 \end{cases} \quad (20)$$

Based on Equations (11)–(13) and (17), the closed–loop transfer function of AD LADRC is as follows:

$$y = \frac{k_p}{s + k_p} v + \frac{s^2}{s^3 + \beta_1 s^2 + \beta_2 s + \beta_3} f \quad (21)$$

From Equations (19) and (21), it can be observed that compared to the closed–loop transfer function of the traditional LADRC control system, the AD-LADRC introduces an increased LESO bandwidth. The disturbance term is only influenced by β_1, β_2 and β_3, while the tracking term is only related to k_p. This achieves the decoupling of disturbance and tracking terms. From Figure 7, it is evident that the system bandwidth of the improved LESO has significantly increased, and there is a substantial reduction in the phase lag in the mid–frequency range.

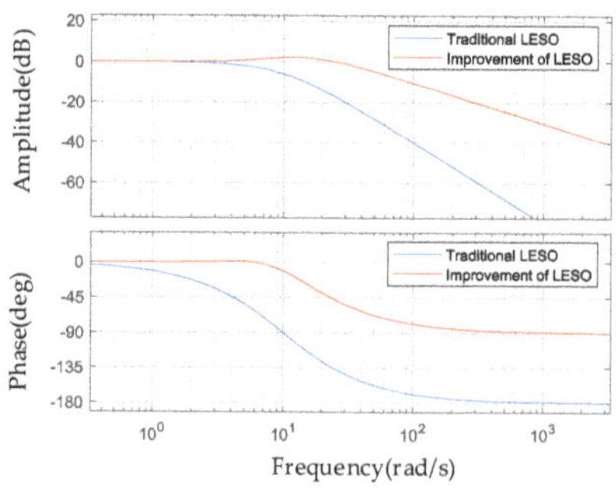

Figure 7. Disturbance observation transfer function characteristic curve.

4.2.2. Analysis of Suppressor Performance

(1) Frequency Domain Characteristics of ω_0 Changes

From Equation (17), the transfer functions for observation noise δ_n and input disturbance δ_c are derived as follows:

$$\frac{z_1}{\delta_n} = \frac{\beta_1 s^2 + \beta_2 s + \beta_3}{s^3 + \beta_1 s^2 + \beta_2 s + \beta_3} \tag{22}$$

$$\frac{z_1}{\delta_c} = \frac{b_0 s^2}{s^3 + \beta_1 s^2 + \beta_2 s + \beta_3} \tag{23}$$

When b_0 is fixed and ω_0 is set to 10, 20, 30, 40 and 50, respectively, the frequency domain characteristic curves are obtained as shown in Figure 8a, Observation Noise. From the graph, it can be seen that as ω_0 increases step by step, the response speed of the improved suppressor is enhanced. This reduces the high-frequency noise gain on its output side, effectively suppressing the equivalent measurement noise on the input side.

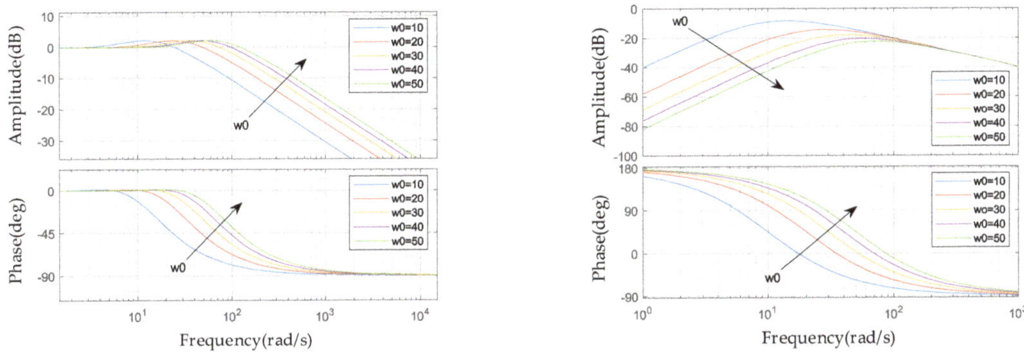

(**a**) Observation noise (**b**) Input disturbance

Figure 8. Frequency domain characteristic curves at different bandwidths.

According to Equation (23), when ω_0 is set to 10, 20, 30, 40 and 50, the frequency domain characteristic curves of this transfer function are obtained in Figure 8b, Input Disturbance. It can be observed from the graph that increasing ω_0 can reduce the gain of the disturbance term, enhance the system's disturbance rejection performance and have a significant impact on mid- to low-frequency disturbances.

(2) Frequency Domain Characteristic Curves of LADRC Before and After Improvement

Comparison of the frequency domain characteristic curves for LADRC before and after improvement is shown in Figure 9. The improved LADRC exhibits superior performance in terms of both speed and robustness compared to the traditional LADRC, making it better equipped to handle external disturbances.

When the disturbance is a unit step signal, the system output response can be derived from Equation (21) as follows:

$$Y(s) = \frac{s}{s^3 + 3\omega_0 s^2 + 3\omega_0 s + \omega_0{}^3} \tag{24}$$

Applying the Laplace inverse transform to Equation (24) and taking the limit, we can obtain $\lim_{t\to\infty} y(t) = 0$. Moreover, as the observer bandwidth ω_0 increases, the rate of decrease in $y(t)$ also increases. This indicates that the LADRC with the improved LESO has strong disturbance rejection capabilities.

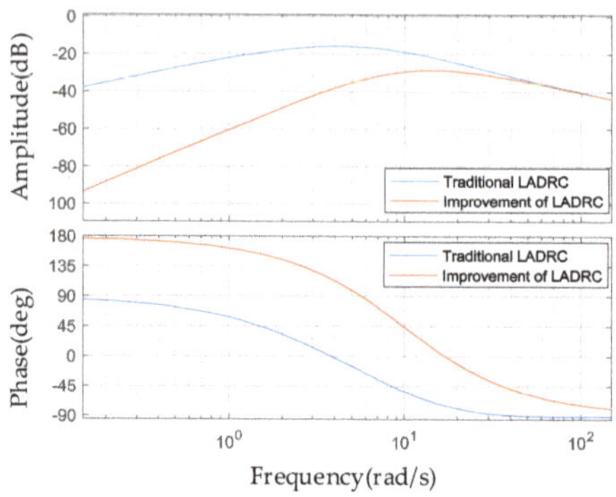

Figure 9. Comparison of frequency domain characteristic curves.

5. Simulation Analysis

In order to analyze the feasibility of the improved LADRC circulating current suppressor described in this paper, an 18-level modular multilevel converter simulation model is built based on Matlab (9.12.0.1884302 (R2022a)) & Simulink simulation software. The simulation parameters are as shown in Table 1, and controller parameters are as presented in Table 2. The circulation suppression effects of three kinds of circulating current suppressors, bridge arm current and sub-module capacitance voltage under steady state and DC side voltage sudden change are simulated and analyzed, respectively.

Table 1. Simulation parameters.

Parameter Names	Numerical Values
Number of MMC Arm Sub-Modules	18
Sub-Module Capacitance Value/mF	1.88
Arm Inductance Value/mH	5
Initial Value of Sub-Module Capacitor Voltage/V	1400
Frequency/Hz	50
DC Side Voltage/kV	25.2

Table 2. Controller parameters.

ω_0	ω_c	b_0
350	1050	2000

5.1. Steady-State Operating Condition

5.1.1. Circulation Suppression Effect

Three kinds of circulating current suppressors are put into the HVDC system, namely PI circulating current suppressor, LADRC circulating current suppressor and AD-LADRC circulating current suppressor. The circulation changes of the MMC bridge arm under the action of three kinds of controllers are simulated and compared, and the results are shown in Figure 10.

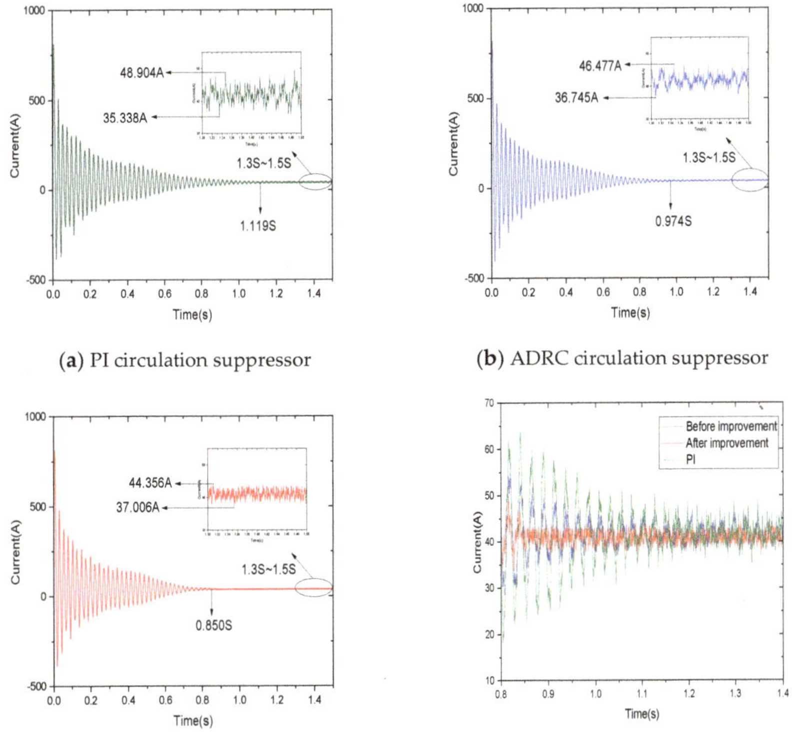

(**a**) PI circulation suppressor (**b**) ADRC circulation suppressor

(**c**) AD–ADRC circulation suppressor (**d**) 0.8 Seven 1.4 S Circulation comparison

Figure 10. Simulation diagram of circulation suppressor.

Figure 10a–c shows the circulation waveforms of the PI circulation suppressor, the LADRC circulation suppressor and the AD-LADRC circulation suppressor in turn. Figure 10d is a comparison of the 0.8 S~1.4 S waveforms selected from the above three figures in order to clearly observe that the suppression effect of the improved controller is better. Observe Figure 10a: when the circulation reaches a steady state at 1.119S, select and enlarge the 1.3 S~1.5 S waveform; the upper peak of the circulation is 48.904 A, the lower peak is 35.338 A and the peak difference is 13.566 A. Observe Figure 10b: when the circulation reaches a steady state at 0.974 S, select and enlarge the 1.3 S~1.5 S waveform; the upper peak is 46.477 A, the lower peak is 38.745 A and the peak difference is 9.732 A. Observe Figure 10c: when the circulation reaches a steady state at 0.850 S, select and enlarge the 1.3 S~1.5 S waveform; the upper peak is 44.356 A, the lower peak is 37.006 A and the peak difference is 7.350 A. In terms of stabilization time, the improved circulation suppressor is 24.04% and 12.73% shorter than that of PI and LADRC, respectively. In terms of peak deviation, the improved circulation suppressor is reduced by 45.82% and 24.48% compared with PI and LADRC, respectively. This shows that during the steady-state operation of the system, the improved circulation suppressor has been effectively improved in terms of speed and stability.

5.1.2. Arm Current and Sub-Module Capacitor Voltage

The simulation analysis of bridge arm current and sub-module capacitance voltage is an important means to evaluate the performance of HVDC system, and the waveform reflects the suppression effect of the circulating current suppressor. The simulation waveforms of the A-phase bridge arm current and sub-module capacitor voltage using the AD−LADRC circulatory suppressor are shown in Figures 11 and 12.

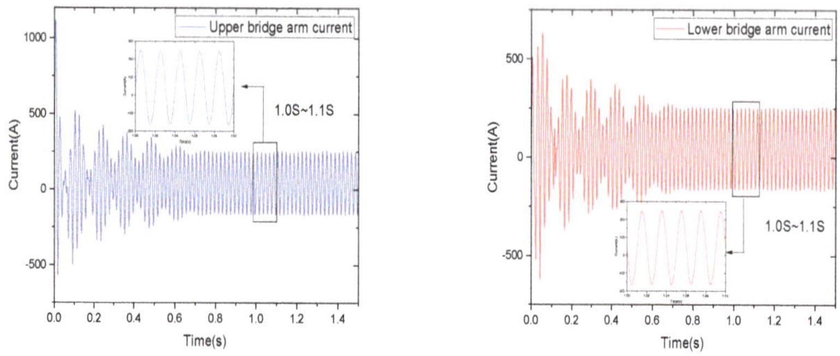

(**a**) Upper bridge arm current (**b**) Lower bridge arm current

Figure 11. A−phase bridge arm current.

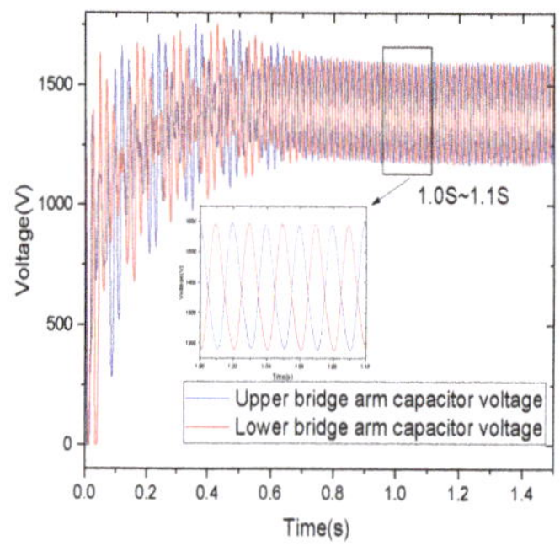

Figure 12. A-phase sub-module capacitor voltage.

The second harmonic component is significantly suppressed. By analyzing the waveform from 1.0 S to 1.1 S, it is observed that the bridge arm current waveform tends to be sinusoidal with fluctuations ranging between −165 A and 250 A. Furthermore, the fluctuation amplitude and waveform distortion rate of the capacitor voltage are significantly reduced, with a fluctuation range of 1150 V to 1600 V.

In addition, in order to directly reflect the influence of suppressor on the system, the bridge arm current and capacitor voltage when the suppressor is used or not are compared as shown in Figure 13. The above bridge arm current is taken as an example. Figure 13a shows the A phase bridge arm current using the suppressor, and Figure 13b shows the A phase bridge arm current without the suppressor. It can be seen from the figure that the waveform of the bridge arm current is obviously distorted due to the harmonic generated by the circulation current. The use of the suppressor can improve the influence of the circulation current and make the waveform closer to the sine wave. The peak current difference of bridge arm with the suppressor is 4.933 A and that without the suppressor is 29.305 A.

The capacitance voltage waveforms of sub-modules with and without suppressors are depicted in Figures 13c and 13d, respectively. It is evident from the figures that the difference in capacitance voltage peak between sub-modules with suppressors is 1.956 A, whereas for those without suppressors it is 23.719 A. Consequently, the utilization of suppressors leads to a significant reduction in both the range of capacitance voltage fluctuation and the waveform distortion rate.

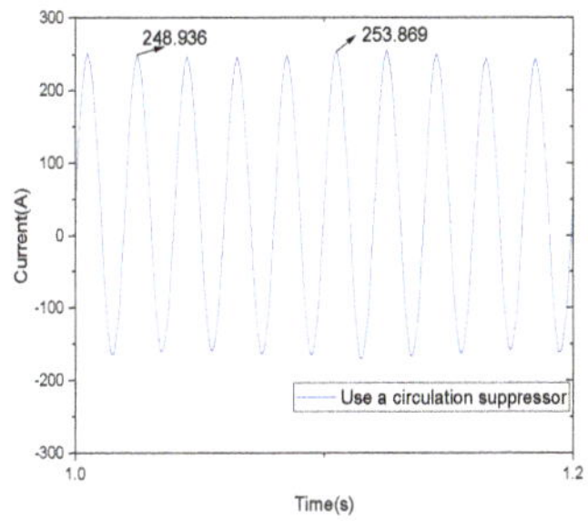

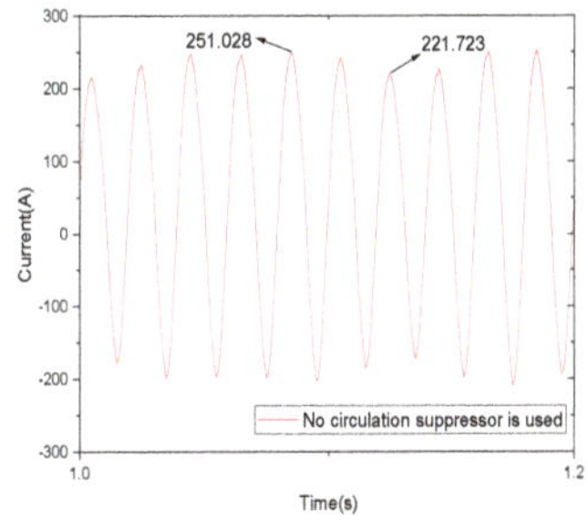

(**a**) Upper bridge arm current

(**b**) Upper bridge arm current

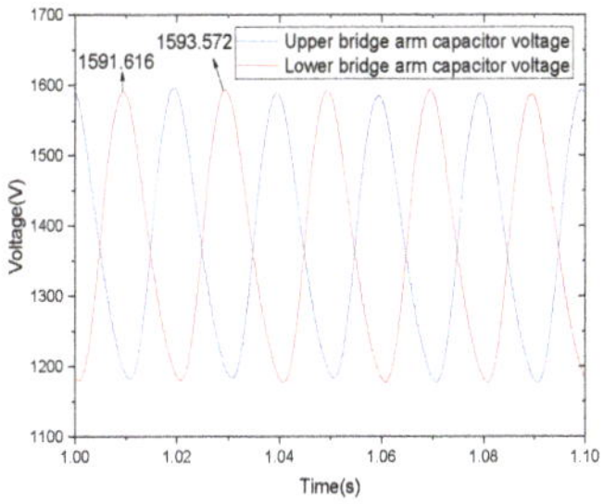

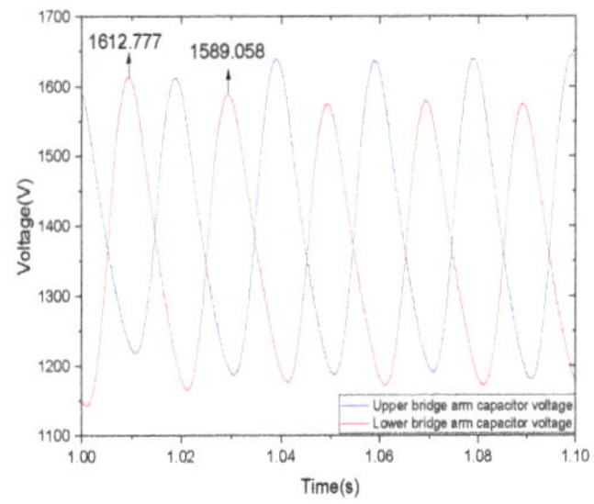

(**c**) capacitor voltage

(**d**) capacitor voltage

Figure 13. The effect of whether or not a suppressor is used on the arm current and capacitance voltage.

5.1.3. Bridge Arm Current THD

The harmonic analysis of the bridge arm current is helpful to understand the harmonic content of the system, and a good controller should be able to restrain harmonics effectively and ensure the stable operation of the system. Through the FFT analysis, it can further see the suppression effect of the suppressor on the second harmonic component of the bridge arm current. Taking t = 1.3S and 20 cycles as an example, as shown in Figure 14, when no circulation suppressor is used, the bridge arm current THD is 9.74%;when using the PI circulation suppressor, the bridge arm current THD is 1.82%; when using the LADRC circulation suppressor, the bridge arm current THD is 1.71%; and when using the AD-LADRC circulation suppressor, the bridge arm current THD is 1.63%, all of which are less than the 3% to 5% required for different grid−connection standards. Compared with PI and LADRC, the THD value of the AD-LADRC circulation suppressor proposed in this paper is reduced by 10.44% and 4.68%, respectively.

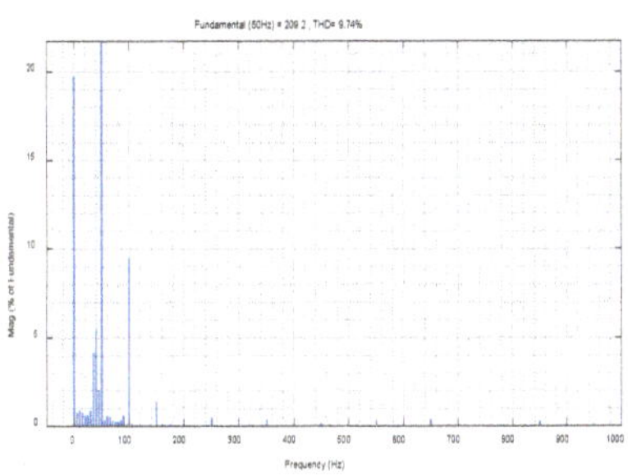

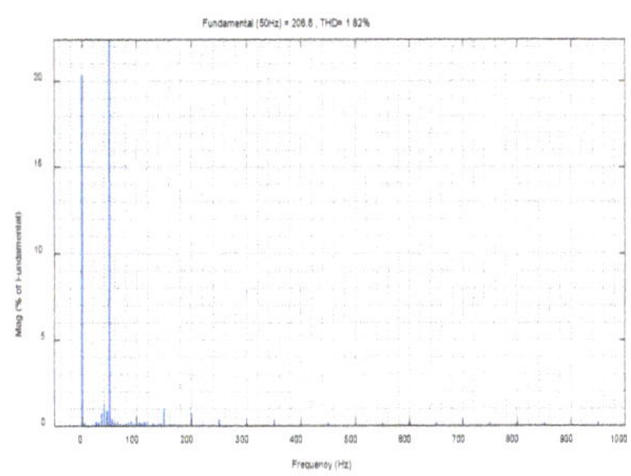

(**a**) THD value without suppressor

(**b**) PI–THD

Figure 14. *Cont.*

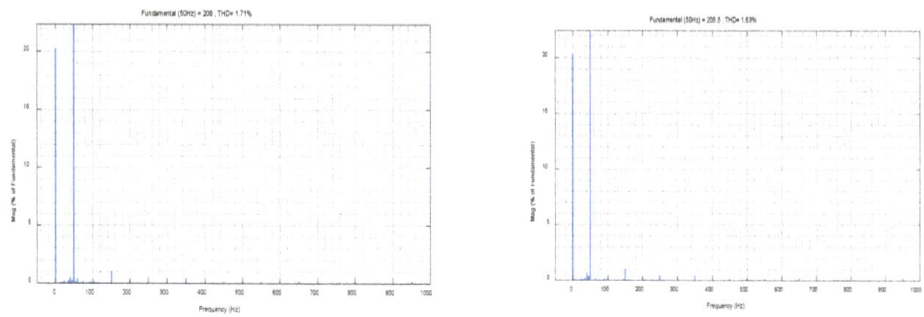

(**c**) ADRC–THD (**d**) AD-LADRC–THD

Figure 14. Bridge arm current THD.

5.2. DC Voltage Transient

By introducing DC side voltage sudden change, the dynamic response of the HVDC system in the face of voltage sudden change can be simulated, and the dynamic performance of the circulating current suppressor can be tested. Adjust the DC side voltage to 18 KV at 1.5 s, as shown in Figure 15.

As can be seen from Figure 15a, the system undergoes a voltage disturbance at 1.5S and the current suppressor continues to function properly. The power waveform remains relatively stable, with active power tracking the reference well and reaching a steady state. Reactive power is essentially maintained at zero, indicating that the system is in an active grid-connected state.

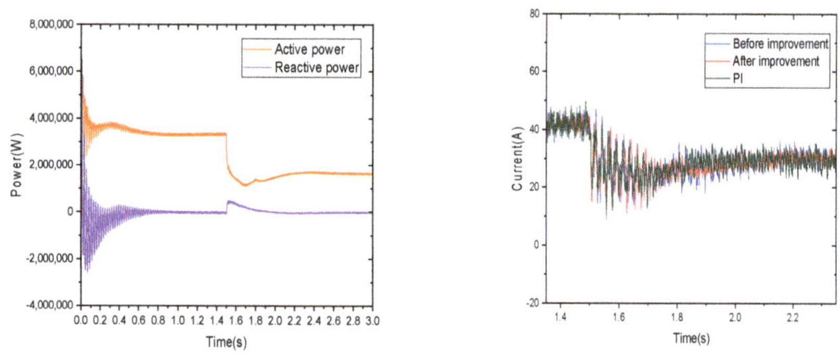

(**a**)Power change (**b**)Circulation comparison

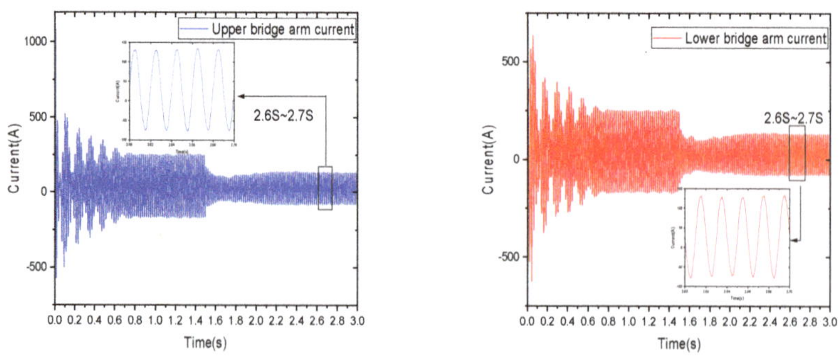

(**c**)Upper bridge arm current (**d**)Lower bridge arm current

Figure 15. *Cont.*

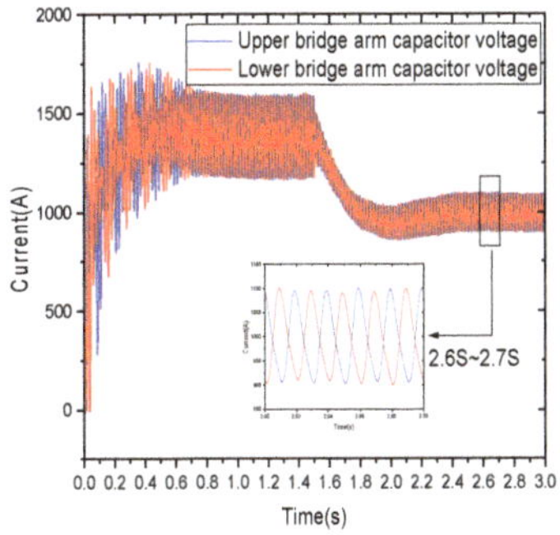

(**e**)Capacitive voltage

Figure 15. Voltage mutation simulation comparison.

The 1.35 S~2.35 S waveform is selected as shown in Figure 15b; the performance of the improved circulating current suppressor is better than the traditional controller. The former has a smaller oscillation amplitude and requires less time to reach stability. Moreover, observing the A-phase bridge arm current in Figure 15c,d and sub-module capacitance voltage in Figure 15e at this time, it can be corrected in time in the face of sudden voltage change, and the waveform is flat. Select 2.6S~2.7S waveforms, respectively; the bridge arm current fluctuates in the range of −75~135 A, and the A-phase sub-module capacitor voltage fluctuates in the range of 900~1100 V. Performing FFT analysis on the bridge arm current at this time, the improvement is 2.29%, the improvement before is 2.74% and PI is 2.89%. This indicates that the improved circulating current controller has excellent anti-interference capability and suppression effects.

6. Conclusions

In this paper, by combining MMC-HVDC transmission technology with an offshore wind farm, the mechanism of negative sequence current circulation dominated by second harmonic generation in MMC is analyzed. Based on the traditional LADRC circulation suppression strategy, an improved negative sequence current circulation suppression method is proposed. This method can improve the dynamic response ability and anti-interference ability of the system, and the fast performance and anti-disturbance performance are better than the traditional circulation suppression strategy. Through computer simulation, the effects of circulation suppressors of PI, LADRC and AD-LADRC under different working conditions of stable operation and voltage sudden change are compared and analyzed. The results show that

(1) The improved circulating current suppressor can effectively improve the waveform distortion effect of the bridge arm current. In terms of stabilization time, the improved circulating current suppressor is 24.04% and 12.73% shorter than PI and LADRC, respectively. In terms of peak deviation, the improved circulation suppressor is reduced by 45.82% and 24.48% compared with PI and LADRC, respectively.

(2) The second harmonic component is obviously suppressed, and the influence of the circulation can be improved after using the suppressor, so that the waveform is closer to the sine wave.

(3) The fluctuation amplitude of the capacitor voltage and the rate of waveform distortion decreased obviously. The peak difference of the capacitor voltage of the sub-module with suppressor was 1.956 A and that of the sub-module without suppressor was 23.719 A.

(4) When the circulating current suppressor is not used, the bridge arm current THD is 9.74%. When using a PI circulating current suppressor, the bridge arm current THD is 1.82%. When using a LADRC circulating current suppressor, the bridge arm

current THD is 1.71%. When using an AD-LADRC circulating current suppressor, the bridge arm current THD is 1.63%. The THD is controlled within the range of 3% THD and the improved controller is reduced by 10.44% and 4.68% compared with PI and LADRC, respectively.

(5) The increased-order decoupling auto−disturbance rejection suppressor proposed in this paper has stronger rapidity, adaptability and robustness, which is helpful to improve the grid-connected ability of offshore wind farms and provide some theoretical and application support for the development and large-scale utilization of offshore wind power.

Author Contributions: Conceptualization, X.X. and D.W.; methodology, X.X., X.Z. and L.T.; software, D.W.; validation, X.X. and D.W.; formal analysis, X.Z.; investigation, D.W. and L.T.; resources, X.X.; datacuration, D.W.; writing—original draft preparation, D.W.; writing—review and editing, X.X. and X.Z.; visualization, D.W.; supervision, X.X.; project administration, X.X.; funding acquisition, X.X. All authors have read and agreed to the published version of the manuscript.

Funding: This research was funded by the Tianjin Science and Technology Program: Deep Learning based Wind Power Prediction and Self-Anti Disturbance Control Methods and Engineering Applications, grant number 23YDTPJC00530, and the APC was funded by the Tianjin Science and Technology Program: Deep Learning-based Wind Power Prediction and Self-Anti Disturbance Control Methods and Engineering Applications.

Institutional Review Board Statement: Not applicable.

Informed Consent Statement: Not applicable.

Data Availability Statement: Data are contained within the article.

Conflicts of Interest: The authors declare no conflicts.

References

1. Zeng, M.; Cai, Y.; Shen, K. Optimal configuration of new energy grid connected energy storage capacity from the perspective of dual carbon. *Int. J. Energy Technol. Policy* **2023**, *18*, 326–342. [CrossRef]
2. Watson, C.S.; Somerfield, J.P.; Lemasson, J.A. The global impact of offshore wind farms on ecosystem services. *Ocean Coast Manag.* **2024**, *249*, 107023. [CrossRef]
3. Faraggiana, E.; Ghigo, A.; Sirigu, M.; Petracca, E.; Giorgi, G.; Mattiazzo, G.; Bracco, G. Optimal floating offshore wind farms for Mediterranean islands. *Renew. Energy* **2024**, *221*, 119785. [CrossRef]
4. Han, S.; Rui, H.; Hugo, M. Power quality monitoring in electric grid integrating offshore wind energy: A review. *Renew. Sustain Energy Rev.* **2024**, *191*, 114094.
5. Joanna, S.; Mariusz, C.; Joanna, P. Reliability of Renewable Power Generation using the Example of Offshore Wind Farms. *Folia Oeconomica Stetin.* **2023**, *23*, 228–245.
6. Samsó, R.; Crespin, J.; Olivares, G.A. Examining the Potential of Marine Renewable Energy: A Net Energy Perspective. *Sustainability* **2023**, *15*, 8050. [CrossRef]
7. Olczak, P.; Surma, T. Energy Productivity Potential of Offshore Wind in Poland and Cooperation with Onshore Wind Farm. *Appl Sci.* **2023**, *13*, 4258. [CrossRef]
8. Milad, S.; Mateo, R.; Alejandro, H.; Rodrigo, A. A Review of Offshore Renewable Energy in South America: Current Status and Future Perspectives. *Sustainability* **2023**, *15*, 1740.
9. Mohammad, B.; Turaj, A.; Deniz, S.V. Floating Offshore Wind Turbines: Current Status and Future Prospects. *Energies* **2022**, *16*, 2
10. David, G.; Jensen Paul, D. Chasing after the wind? Green economy strategies, path creation and transitions in the offshore wind industry. *Reg. Stud.* **2022**, *56*, 1671–1682.
11. Zhong, P.; Rong, Y.; Tai, L. Optimization Design of Voltage Level of Flexible DC Transmission with Offshore Wind Power Based on Genetic Algorithm. *J. Phys. Conf. Ser.* **2023**, *2527*, 012066.
12. Mujahid, E.; Pillai, A.C.; Longbin, T. Implications of wave–current interaction on the dynamic responses of a floating offshore wind turbine. *Ocean Eng.* **2024**, *292*, 116571.
13. González, G.W.; Montoya, D.O.; Rodríguez, T.L.C. Optimal Integration of Flexible Alternating Current Transmission Systems in Electrical Distribution Grids Using a Mixed-Integer Convex Model. *Algorithms* **2023**, *16*, 420. [CrossRef]
14. Peng, L.; Fu, Z.; Xiao, T. An Improved Dual Second-Order Generalized Integrator Phased-Locked Loop Strategy for an Inverter of Flexible High-Voltage Direct Current Transmission Systems under Nonideal Grid Conditions. *Processes* **2023**, *11*, 2634. [CrossRef]
15. Jiaqi, W.; Zhichao, F.; Xin, L.; Daoyuan, M. A PLL-free control strategy for flexible DC transmission systems. *Energy Rep.* **2022** *8*, 1413–1420.

16. Zhu, S.; Liu, K.; Qin, L.; Qing, W.; Yuye, L. Analysis and suppression of DC oscillation caused by DC capacitors in VSC-based offshore island power supply system. *IEEJ Trans. Electr. Electron. Eng.* **2019**, *14*, 545–555. [CrossRef]
17. Mengting, C.; Peiqiang, S.; Guipeng, C.; Fengyan, F.; Xinlin, Q. Multiple criteria analysis for electrifying off-grid island using renewable energy microgrid or submarine cable. *Electr. Power Syst. Res.* **2023**, *224*, 109728.
18. Weili, Z.; Tianning, Z. Application of Flexible DC Transmission Technology in Power Grid under Large-scale Development of New Energy. *IOP Conf. Ser. Earth Environ. Sci.* **2020**, *440*, 032035.
19. Chenhao, L.; Kuan, L.; Changhui, M.; Pengfei, Z.; Qi, T.; Yingtao, S.; Xin, W. Flexible control strategy for HVDC transmission system adapted to intermittent new energy delivery. *Glob. Energy Interconnect.* **2021**, *4*, 425–433.
20. Xingyang, H.; Kun, C.; Ting, W.; Zengrui, H. Analysis of typical fault characteristics of modular multilevel converter for flexible HVDC transmission. *J. Phys. Conf. Ser.* **2022**, *2369*, 012067.
21. Pan, R.; Yang, Y.; Yang, J.; Liu, D. Enhanced grid forming control for MMC-HVDC with DC power and voltage regulation. *Electr. Power Syst. Res.* **2024**, *229*, 110166. [CrossRef]
22. Pan, R.; Liu, D.; Yang, Y.; Yang, J. Network based impedance analysis of grid forming based MMC-HVDC with wind farm integration. *Electr. Power Syst. Res.* **2024**, *229*, 110120. [CrossRef]
23. Xiaolei, W.; Quan, Z.; Jianying, Z. Control Strategy of Circulating Current Suppression for Modular Multilevel Converter. *J. Phys. Conf. Ser.* **2023**, *2564*, 012005.
24. Chun, W.; Wenxu, Y.; Wenyuan, W.; Hongyu, N.; Jie, C. The Suppression of Modular Multi-Level Converter Circulation Based on the PIR Virtual Impedance Strategy. *World Electr. Veh. J.* **2023**, *14*, 17.
25. Manchala, R.N.; J, S.; Mandi, P.R. Circulating Current Suppression Control in Surrogate Network of MMC- HVDC System. *Int. J. Recent Technol. Eng.* **2020**, *8*, 29–34.
26. Debdeep, S.; Tanmoy, B.; Saurav, D. A Reduced Switching Frequency Sorting Algorithm for Modular Multilevel Converter with Circulating Current Suppression Feature. *IEEE Trans. Power Electron.* **2019**, *34*, 10480–10491.
27. Jinyu, W.; Jun, L.; Chengfu, W.; Chengfu, W.; Xiaoming, D. Circulating Current Suppression for MMC-HVDC under Unbalanced Grid Conditions. *IEEE Trans. Ind. Appl.* **2017**, *53*, 3250–3259.
28. Mingguang, Z.; Yao, S.; Huzhong, S.; Richang, G. MMC-HVDC circulating current suppression method based on improved proportional resonance control. *Energy Rep.* **2020**, *6*, 863–871.
29. Qingrui, T.; Zheng, X.; Minyuan, G.; Xiang, Z.; Jing, Z. Design of circulating current suppression controller for modular multilevel converter. *Power Syst. Autom.* **2010**, *34*, 57–61.
30. Xiahui, Z.; Minxiao, H.; Jinggang, Y.; Xiangkun, M.; Zijian, Q. Analysis of the influence mechanism of DC side voltage fluctuation on the circulating current of MMC and its suppression method. *Power Syst. Autom.* **2021**, *45*, 122–131.
31. Semih, I.; Mohammed, A.; Subhashish, B. An Optimized Circulating Current Control Method Based on PR and PI Controller for MMC Applications. *IEEE Trans. Ind. Appl.* **2021**, *57*, 5074–5085.
32. Zhouzhou, L. MMC-HVDC control and circulation suppression strategy based on quasi-PR controller. *J. Power Syst. Autom.* **2016**, *28*, 70–75.
33. Xianzheng, L.; Xingcheng, W.; Kai, Z. IMC based circulating current controller for MMC-HVDC. In Proceedings of the 2017 36th Chinese Control Conference (CCC), Dalian, China, 26–28 July 2017; pp. 617–622.
34. Jingqing, H. Auto disturbance rejection controller and its application. *Control. Decis.* **1998**, *13*, 19–23.
35. Fang, Z.; Guangyao, Z.; Yan, C. MMC Circulation suppression Strategy based on Linear Auto disturbance rejection Control. *J. Electr. Power Syst. Its Autom.* **2018**, *30*, 71–78.
36. Yang, X.; Li, Z.; Zheng, Q.T.; Zheng, T. Virtual Impedance Sliding Mode Control-Based MMC Circulating Current Suppressing Strategy. *IEEE Access* **2019**, *7*, 26229–26240. [CrossRef]
37. Farooq, A. Influence of Unified Power Flow Controller on Flexible Alternating Current Transmission System Devices in 500 kV Transmission Line. *J. Electr. Electron. Eng.* **2018**, *6*, 12–19. [CrossRef]
38. Zhang, G.; Song, J.; Li, C.; Gu, X. A Circulating Current Suppression Strategy for MMC Based on the 2*N*+1 PWM Approach. *World Electr. Veh. J.* **2023**, *14*, 106. [CrossRef]
39. Kun, W.; Kaipei, L.; Zhixuan, Z.; Wei, L.; Liang, Q. Capacitor voltage equalization strategy of modular multilevel converter based on fast sorting algorithm. *Electr. Meas. Instrum.* **2018**, *55*, 1–7.
40. Xitang, T.; Hongmei, Z.; Qinyue, Z.; Jiangbin, C. Capacitor voltage fluctuation analysis and equalization control of MMC module. *Power Electron. Technol.* **2016**, *50*, 1–4.

Journal of Marine Science and Engineering

Article

Performance Analysis of a Floating Wind–Wave Power Generation Platform Based on the Frequency Domain Model

Mingsheng Chen [1,2,3], Jiang Deng [2], Yi Yang [4], Hao Zhou [5], Tao Tao [4], Shi Liu [4], Liang Sun [2,*] and Lin Hua [6]

[1] Key Laboratory of High Performance Ship Technology (Wuhan University of Technology), Ministry of Education, Wuhan 430063, China; mschen@whut.edu.cn
[2] School of Naval Architecture, Ocean and Energy Power Engineering, Wuhan University of Technology, Wuhan 430063, China; 272022@whut.edu.cn
[3] Sanya Science and Education Innovation Park of Wuhan University of Technology, Sanya 572019, China
[4] China Southern Power Grid Technology Co., Ltd., Guangzhou 510080, China; yyxt007@sina.cn (Y.Y.); taotao@gd.csg.cn (T.T.); 13925041516@139.com (S.L.)
[5] China Ship Scientific Research Center, Wuxi 214000, China; zhouhao@cssrc.com.cn
[6] Hubei Key Laboratory of Advanced Technology for Automotive Components, Wuhan University of Technology, Wuhan 430070, China; hualin@whut.edu.cn
* Correspondence: liang.sun@whut.edu.cn; Tel.: +86-13986243560

Abstract: Integrating wave energy converters (WECs) onto floating offshore wind turbine platforms has emerged as a recent focal point of research aiming to achieve synergistic marine energy utilization and enhance the spatial efficiency of renewable energy. The power performance of WECs relies on hydrodynamic interactions with the floating platform. However, the coupled dynamic response and power generation mechanism remain unclear. This study establishes a multi-body model for the constrained motion of floating-platform and point-absorber WECs in the frequency domain. The power performance of WECs under different arrangements is compared and optimized. The performance of different hydrodynamic models in the South China Sea is analyzed further. The results indicate that exceptional peak performance is achieved when a single point absorber is placed on the floating platform. However, its performance under the full spectrum of wave frequencies in real sea conditions is suboptimal. Conversely, as the number of point absorbers on the floating platform increases, the performance of the hybrid system becomes more stable in real sea conditions. Furthermore, the array arrangement of point absorbers on the floating platform leads to multiple peaks in their power performance, and in selected array arrangements, the average power generation at specific frequencies is significantly superior to that of a single point absorber on the floating platform

Keywords: floating wind–wave power generation platform; frequency domain; constrained motion wave energy converter; optimization

Citation: Chen, M.; Deng, J.; Yang, Y.; Zhou, H.; Tao, T.; Liu, S.; Sun, L.; Hua, L. Performance Analysis of a Floating Wind–Wave Power Generation Platform Based on the Frequency Domain Model. *J. Mar. Sci. Eng.* **2024**, *12*, 206. https://doi.org/10.3390/jmse12020206

Academic Editor: Markel Penalba

Received: 6 December 2023
Revised: 2 January 2024
Accepted: 5 January 2024
Published: 23 January 2024

1. Introduction

In recent years, the development of floating offshore wind turbines has progressed rapidly as a crucial trend. A significant number of commercial and demonstrative projects have been constructed, such as Hywind Scotland wind farm (30 MW) in UK [1], Hywind Tampen wind farm (88 MW) in Norway [2], Floatgen (2 MW) in France [3], the Kincardine offshore wind farm (50.4 MW) in Britain [4], and "Haiyou Guanlan" demonstration project recently completed in China [5]. However, there are some limitations in the utilization of resources, construction cost, and energy stability of floating offshore wind turbines, and these limitations restrict their further development. Integrating WECs into a floating platform can address the variability in wind energy, allowing for the synergistic utilization of both energy sources and enhancing power generation stability and reliability [6]. Additionally, WECs can restrain the motion of the platform to a certain extent and improve the safety and stability of the overall structure [7].

Numerous scholars have undertaken conceptual designs and research for wind–wave power generation platforms. Based on different forms of floating foundations, wind–wave power generation platforms can be roughly categorized into four types. The first type is spar-type foundations with WECs, with typical representatives such as the "Spar-Torus Combination" (STC) [8–13] that combines a spar-type floating wind turbine with a torus point-absorber WEC, as shown in Figure 1a; the "Wind WEC" [14], which combines a spar-type floating wind turbine with the "Wavestar" WEC; and the "Hywind-Wavebob-NACA Combination" (HWNC) [15], which combines a spar-type floating wind turbine with wave energy and tidal energy converters. The second type is semi-submersible foundations with WECs. Typical representatives are the "Semi-submersible Flap Combination" (SFC) [16–20], which combines a semi-submersible floating wind turbine with three elliptical flap-type WECs, as shown in Figure 1b; "Wind-to-power" (W2P) [21], which combines a semi-submersible floating wind turbine with retractable blades and three-degrees-of-freedom (3-DOF) WECs; and the "Sea Flower" [22], which features gyro-stabilized devices on semi-submersible floating wind turbines with multiple WEC installations. The remaining two types are barge-type foundations with WECs and Tension Leg Platform (TLP) foundations with WECs, respectively. These are typically represented by the "D-Hyp" [23], which combines a barge floating platform with a semi-circular base with an oscillating water column WEC, which generates power using pitch motion, as shown in Figure 1c; the "C-Hyp" [24], which combines a barge floating platform with a WEC with a circular base; the "TWindWave" [25], which combines a TLP with four point-absorber WECs, as shown in Figure 1d; and the " TLP-WT-WEC-Combination" (TWWC) [26], which combines a TLP with a heave-type WEC.

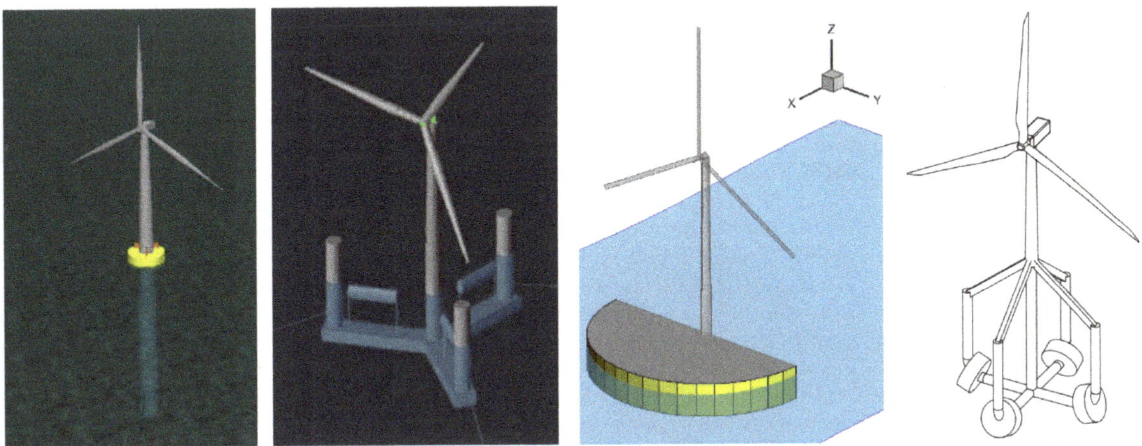

(**a**) SFC (**b**) STC (**c**) D-Hyp (**d**) TWindWave

Figure 1. Four concepts of floating wind–wave power generation platforms.

Among the above-mentioned wind–wave power generation platforms, the combination of a semi-submersible floating platform and a WEC is the most widely studied. The primary reasons for this widespread focus likely include the adaptability and high stability of a semi-submersible foundation, its ease of construction and maintenance, and its suitability for accommodating multiple WECs. In these studies, Luan et al. [16] carried out time-domain numerical simulation for an SFC, and linear rotational damping was used to simulate the influence of a Power Take-Off (PTO) system. The results show that the PTO damping coefficient and the mass of the WEC have great influences on the power generation of the WEC. Based on the concept of an SFC proposed by Luan [16], Michailides et al. [17] studied the influence of the WEC on the motion of the floating platform. They considered two different flap plate layouts and compared them with the case of a single floating wind turbine. The results show that power generation can be effectively improved by combining the flap WEC with the semi-submersible platform without significantly affecting the critical response of the semi-submersible platform. Gao et al. [18] carried out numerical simulation and experimental tests on an STC and SFC. The numerical and experimental results obtained under the combined action of wind and waves, considering the wind and wave conditions in operation and survival conditions, were compared for selected responses (such as motion, PTO force, and power generation efficiency). The results show that compared with the STC, the SFC had lower motion response under extreme conditions. Michailides et al. [19] conducted an experimental study on the survivability of the SFC and compared it with the numerical results. The results show that the numerical simulation overestimated some motion of the combined power generation system, which proves its good survivability in extreme sea conditions. Michailides et al. [20] conducted a study on the performance of an SFC under operating conditions through numerical analysis and physical model tests. The results indicate that when wind load is applied, the response of the platform in roll and pitch motion is greater under regular waves, while the power generation of the WEC is improved under irregular waves.

In addition to the extensively studied concepts of the STC and SFC, Chen et al. [21] conducted research on the W2P, showing the best wind coefficient of retractable blades can be achieved by adjusting the blade diameter and rotation speed, and a WEC with three degrees of freedom can effectively absorb wave energy. They integrated them and tested the power generation efficiency of the combined power generation platform under different arrangements. Fenu et al. [22] conducted a numerical simulation on the "Sea Flower" in adverse sea conditions, and the results indicate that the gyro device could significantly improve the pitch stability of the platform. Wang et al. [27] proposed combining a semi-submersible floating wind turbine with WECs. The study analyzed the hydrodynamic response of a combined structure through numerical simulation and analysis, focusing on the interaction between wind energy and wave energy. The difference between the scattering force and radiation force was compared with and without consideration of the hydrodynamic coupling effect. The results indicate that the combination of wind and wave energy alters the diffraction force and radiation force on the structure, ultimately affecting its hydrodynamic response. Gaspar et al. [28] analyzed the feasibility of using WECs to assist in the hydrodynamic compensation of floating platforms. The experimental results demonstrate that WECs can aid the hydrodynamic compensation system under the tested sea conditions and even expand the operating sea conditions. Additionally, it was observed that WECs located on the downwind and upwind sides of a platform have distinct functions in the dynamic compensation of the platform. Hallak et al. [29] combined DeepCwind with conical point-absorber WECs to investigate the linear hydrodynamic interaction between the platform and WECs. They explored the influence of changes in WEC diameter on the overall motion. Li et al. [30] studied the dynamic response of the system and investigated the influence of wind–wave coupling effects. They used numerical simulation and analysis to study the interaction between wind energy and wave energy in the combined system. The results show that the wind–wave coupling effect has a substantial influence on the dynamic response of the hybrid system. Si et al. [31] proposed the combination of a semi-submersible floating platform with three point absorbers. They conducted a preliminary feasibility study on the concept by performing verified numerical simulations of aero-hydro-servo-mooring dynamic integration. Shi et al. [32] presented a dimensionless optimization approach for shaping a hollow cylindrical WEC tailored for a wind–wave power generation platform. The optimal configuration was achieved using the differential evolution (DE) algorithm. Wang et al. [33] combined a semi-submersible floating wind turbine with a torus-type WEC. The results indicated that the use of a concave-type WEC led to improved dynamic response and increased power generation for the integrated system. The study examined four distinct geometries for WECs. Tian et al. [34] also investigated the impact of the number of WECs on motion, based on the model of Wang et al. [33]. The finding demonstrated that the combination concept with three WECs had the highest power generation efficiency. However, the authors only conducted time-domain numerical simulations for a limited number of WECs, without extending the model to address a frequency-domain model for an arbitrary number of WECs. In a comprehensive analysis, the primary challenges of a floating wind–wave power generation platform lie in determining the optimal combination of floating wind turbine platforms and WECs to achieve maximum performance. The aim is to find a balance between the high power generation efficiency of the hybrid system under typical sea states and its robust survivability in extreme sea states. Furthermore, it is crucial to explain how WECs reduce the motion response of floating platforms, allowing for peak-shaving and valley-filling electricity generation from the upper wind turbines.

Recently, Chen et al. [35] conducted a study on a wind–wave power generation platform consisting of an OC4 semi-submersible floating platform and a single point-absorber WEC positioned at the platform's center. Zhang et al. [36] expanded on this by using computational fluid dynamics (CFD) to implement viscous damping corrections for the point absorbers. Additionally, extensive analysis was conducted on the hydrodynamic characteristics of the hybrid system. The study also investigated the impact of different sizes of floaters on wave energy conversion efficiency. However, if the point absorber is not centrally located, the motion of the platform interacts with the heave motion of the point absorber, affecting its response and power generation characteristics. Furthermore, the configuration of multiple point absorbers can also affect the performance of the hybrid system [37–41]. Hence, it is essential to investigate the coupled dynamics of this hybrid system. In this study, the work of Zhang et al. [36] has been reconsidered, and the effects of wind loads have been ignored. The research of Chen et al. [42] and Zou et al. [43] indicated that wind loads and the hydrodynamic interactions of a multi-body system had minimal impact on the performance of point absorbers in a floating wind–wave power generation platform. The primary focus was on investigating the variations in the power performance of point absorbers under non-centralized array configurations. The novelty of our research lies in the following aspects: While considering the placement of non-centralized point absorbers on the floating platform, a multi-body constrained motion model for the platform with an arbitrary number of point absorbers is established in the frequency domain. The parameters of the PTO system are optimized for different arrangements of WECs, revealing the best performance for various point-absorber configurations. Thirdly, this study calculates the expected power output of various models in the South China Sea, revealing the performance of different configurations under real sea conditions.

This article is arranged as follows: Section 2 outlines the derivation process of the multi-body constrained motion model in the frequency domain and provides a methodology for assessing the power generation of point absorbers as well as the computation procedure for expected power in real sea conditions. Section 3 introduces the key parameters of the wind–wave power generation platform and relevant structural features, presenting a comparative analysis of hydrodynamic coefficients in free-floating states among different hydrodynamic models. Section 4 presents the validation of the constructed frequency-domain model for constrained motion, which is followed by a parametric analysis of PTO system parameters for different hydrodynamic models, with performance evaluations conducted in the South China Sea. Finally, Section 5 summarizes the principal research findings.

2. Theory Background

2.1. Frequency-Domain Model for Multiple Floating Bodies with Constraints

For a single point absorber, considering the effect of linearized fluid viscosity and a PTO system, the frequency-domain motion equation in the heave DOF can be expressed as follows [44]:

$$\left[-\omega^2(M_{33}+A_{33}(\omega))-i\omega\left[B_{33}(\omega)+B_{pto}+B_v\right]+(C_{33}+K_{pto})\right]\widehat{x}_3(i\omega)=\widehat{f}_3^{exc}(i\omega) \quad (1)$$

where M_{33}, C_{33}, and $\widehat{f}_3^{exc}(i\omega)$ represent the mass, hydrostatic stiffness, and wave exciting force of the point absorber. $A_{33}(\omega)$ and $B_{33}(\omega)$ represent the added mass and radiation damping of the point absorber. K_{pto}, B_{pto}, and B_v represent the stiffness and damping coefficients of the PTO and the viscous coefficient matrix. $\widehat{x}_3(i\omega)$ is the heave motion of the point absorber. The force applied to the PTO $\widehat{f}_{pto}$ can be expressed as follows:

$$\widehat{f}_{pto}(i\omega) = -i\omega B_{pto}\widehat{x}_3(i\omega) - K_{pto}\widehat{x}_3(i\omega) \tag{2}$$

Based on the frequency-domain model, considering the non-central placement of the point absorber, as shown in Figure 2, the two-body coupling model comprises sever degrees of freedom. However, it is the two DOFs associated with the roll and pitch of the floating platform that exert a substantial coupling influence on the heave motion of the point absorber. Considering the DOFs of heave, roll, and pitch for the platform, as well as the heave DOF for the point absorber, and based on the small-angle assumption, a transformation matrix **E** is defined to describe this coupling term:

$$\mathbf{E} = \begin{bmatrix} 1 & (COG2_y - COG1_y) & -(COG2_x - COG1_x) & -1 \end{bmatrix}_{1\times 4} \tag{3}$$

where $COG1$ and $COG2$ represent the position of the center of gravity for the platform and point absorber. Subscripts x and y denote the specific values of the center of gravity positions in the x- and y-directions. Therefore, the relative displacement between the platform and the point absorber $\widehat{x}_{Relative}$ can be expressed as follows:

$$\widehat{x}_{Relative}(i\omega) = \mathbf{E}\cdot\begin{bmatrix} \widehat{x}_3(i\omega) \\ \widehat{x}_4(i\omega) \\ \widehat{x}_5(i\omega) \\ \widehat{x}_9(i\omega) \end{bmatrix} \tag{4}$$

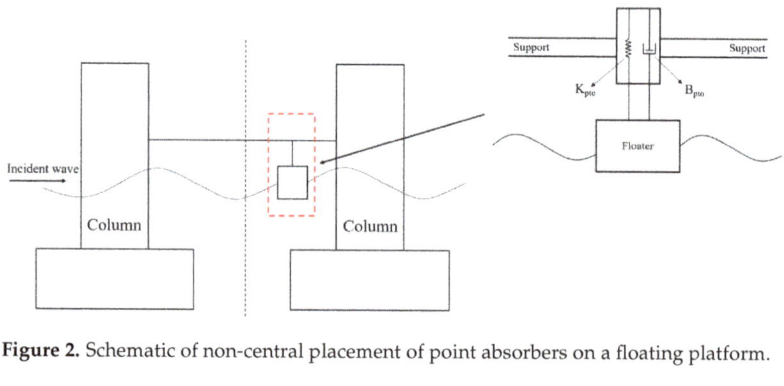

Figure 2. Schematic of non-central placement of point absorbers on a floating platform.

The forces acting on the PTO can be determined as follows:

$$\begin{bmatrix} \widehat{f}_{pto3}(i\omega) \\ \widehat{M}_4(i\omega) \\ \widehat{M}_5(i\omega) \\ \widehat{f}_{pto9}(i\omega) \end{bmatrix} = -i\omega\mathbf{E}^T B_{pto}\widehat{x}_{Relative}(i\omega) - \mathbf{E}^T K_{pto}\widehat{x}_{Relative}(i\omega) \tag{5}$$

where $\widehat{f}_{pto3}$ and $\widehat{f}_{pto9}$ represent the PTO forces exerted on both the floating platform and the point absorber; $\widehat{M}_4$ and $\widehat{M}_5$ represent the moments exerted on the floating platform in the roll and pitch DOFs.

A coupled heave DOF frequency-domain hydrodynamic model under heave, pitch, and roll motions for the wind–wave power generation platform can be established as follows:

$$\left\{ -\omega^2 \begin{bmatrix} M_{33}+A_{33}(\omega) & A_{34}(\omega) & A_{35}(\omega) & A_{39}(\omega) \\ A_{43}(\omega) & M_{44}+A_{44}(\omega) & A_{45}(\omega) & A_{49}(\omega) \\ A_{53}(\omega) & A_{54}(\omega) & M_{55}+A_{55}(\omega) & A_{59}(\omega) \\ A_{93}(\omega) & A_{94}(\omega) & A_{95}(\omega) & M_{99}+A_{99}(\omega) \end{bmatrix} - \right.$$
$$i\omega \left\{ \begin{bmatrix} B_{33}(\omega)+B_{v3} & B_{34}(\omega) & B_{35}(\omega) & B_{39}(\omega) \\ B_{43}(\omega) & B_{44}(\omega)+B_{v4} & B_{45}(\omega) & B_{49}(\omega) \\ B_{53}(\omega) & B_{54}(\omega) & B_{55}(\omega)+B_{v5} & B_{59}(\omega) \\ B_{93}(\omega) & B_{94}(\omega) & B_{95}(\omega) & B_{99}(\omega)+B_{v9} \end{bmatrix} + \mathbf{E}^T \cdot B_{pto} \cdot \mathbf{E} \right\} + \tag{6}$$
$$\left. \begin{bmatrix} C_{33} & & & \\ & C_{44} & & \\ & & C_{55} & \\ & & & C_{99} \end{bmatrix} + \mathbf{E}^T \cdot K_{pto} \cdot \mathbf{E} \right\} \begin{bmatrix} \widehat{x}_3(i\omega) \\ \widehat{x}_4(i\omega) \\ \widehat{x}_5(i\omega) \\ \widehat{x}_9(i\omega) \end{bmatrix} = \begin{bmatrix} \widehat{f}_3^{exc}(i\omega) \\ \widehat{f}_4^{exc}(i\omega) \\ \widehat{f}_5^{exc}(i\omega) \\ \widehat{f}_9^{exc}(i\omega) \end{bmatrix}$$

If we apply this model to a semi-submersible platform with N point absorbers, the frequency-domain hydrodynamic model for heave motion must consider (3 + N) degrees of freedom. The first three degrees of freedom are allocated to the platform in the heave, roll, and pitch DOFs, while the next N degrees of freedom are allocated to the point absorbers in the heave DOF. The mass matrix **M**, hydrostatic stiffness matrix **C**, viscous damping matrix **Bv**, damping coefficient matrix of the PTO **Bpto**, and stiffness matrix of the PTO **Kpto** are all diagonal and can be expressed as follows:

$$\mathbf{M} = diag(M_{11}, M_{22}, M_{33}, M_{44}, \cdots, M_{(N+3)(N+3)}) \tag{7}$$

$$\mathbf{C} = diag(C_{11}, C_{22}, C_{33}, C_{44}, \cdots, C_{(N+3)(N+3)}) \tag{8}$$

$$\mathbf{B_v} = diag(B_{v1}, B_{v2}, B_{v3}, B_{v4}, \cdots, B_{v(N+3)}) \tag{9}$$

$$\mathbf{B_{pto}} = diag(B_{pto1}, B_{pto2}, B_{pto3}, \cdots, B_{ptoN}) \tag{10}$$

$$\mathbf{K_{pto}} = diag(K_{pto1}, K_{pto2}, K_{pto3}, \cdots, K_{ptoN}) \tag{11}$$

The added mass matrix $\mathbf{A}(\omega)$ and radiation damping matrix $\mathbf{B}(\omega)$ with coupling terms can be expressed as follows:

$$\mathbf{A}(\omega) = \begin{bmatrix} A_{11}(\omega) & A_{12}(\omega) & A_{13}(\omega) & A_{14}(\omega) & \cdots & A_{1(N+3)}(\omega) \\ A_{21}(\omega) & A_{22}(\omega) & A_{23}(\omega) & A_{24}(\omega) & \cdots & A_{2(N+3)}(\omega) \\ A_{31}(\omega) & A_{32}(\omega) & A_{33}(\omega) & A_{34}(\omega) & \cdots & A_{3(N+3)}(\omega) \\ A_{41}(\omega) & A_{42}(\omega) & A_{43}(\omega) & A_{44}(\omega) & \cdots & A_{4(N+3)}(\omega) \\ \vdots & \vdots & \vdots & \vdots & \ddots & \vdots \\ A_{(N+3)1}(\omega) & A_{(N+3)2}(\omega) & A_{(N+3)3}(\omega) & A_{(N+3)4}(\omega) & \cdots & A_{(N+3)(N+3)}(\omega) \end{bmatrix} \tag{12}$$

$$B(\omega) = \begin{bmatrix} B_{11}(\omega) & B_{12}(\omega) & B_{13}(\omega) & B_{14}(\omega) & \cdots & B_{1(N+3)}(\omega) \\ B_{21}(\omega) & B_{22}(\omega) & B_{23}(\omega) & B_{24}(\omega) & \cdots & B_{2(N+3)}(\omega) \\ B_{31}(\omega) & B_{32}(\omega) & B_{33}(\omega) & B_{34}(\omega) & \cdots & B_{3(N+3)}(\omega) \\ B_{41}(\omega) & B_{42}(\omega) & B_{43}(\omega) & B_{44}(\omega) & \cdots & B_{4(N+3)}(\omega) \\ \vdots & \vdots & \vdots & \vdots & \ddots & \vdots \\ B_{(N+3)1}(\omega) & B_{(N+3)2}(\omega) & B_{(N+3)3}(\omega) & B_{(N+3)4}(\omega) & \cdots & B_{(N+3)(N+3)}(\omega) \end{bmatrix} \tag{13}$$

where the viscous coefficient in the heave DOF for the point absorber was determined through a free decay test conducted using STAR-CCM+ [36]. The viscous damping coefficients B_{v1}, B_{v2}, and B_{v3} for the floating platform are all set to 8% of their critical damping values [45,46]. Additionally, the constraint matrix **E** can be rewritten as follows:

$$\mathbf{E} = \begin{bmatrix} 1 & (COG2_y - COG1_y) & -(COG2_x - COG1_x) & -1 & 0 & 0 & \cdots & 0 \\ 1 & (COG3_y - COG1_y) & -(COG3_x - COG1_x) & 0 & -1 & 0 & \cdots & 0 \\ 1 & (COG4_y - COG1_y) & -(COG4_x - COG1_x) & 0 & 0 & -1 & \cdots & 0 \\ \vdots & \vdots & \vdots & \vdots & \vdots & \vdots & \ddots & \vdots \\ 1 & (COG(N+1)_y - COG1_y) & -(COG(N+1)_x - COG1_x) & 0 & 0 & 0 & \cdots & -1 \end{bmatrix}_{N\times(N+3)} \quad (14)$$

where $COG(N+1)$ represents the center of gravity of the Nth point absorber.

2.2. Assessment of Point-Absorber Power Generation

Based on the multi-body frequency-domain hydrodynamic coupling model for multiple point absorbers, the average power generation of the Nth point absorber $P_{ave(N)}$ can be determined as follows [47]:

$$P_{ave(N)}(i\omega) = \frac{1}{2} B_{pto}^N \omega^2 \left| \widehat{x}_3(i\omega) - \widehat{x}_{(6N+3)}(i\omega) \right|^2 \quad (15)$$

The q factor represents the impact of wave interactions on power absorption for the WEC arrays and can be expressed as follows [48]:

$$q(\omega) = \frac{P_{ave(1)}(\omega) + P_{ave(2)}(\omega) + \cdots + P_{ave(N)}(\omega)}{NP_{isolated}(\omega)} \quad (16)$$

If $q < 1$, it indicates that the average power generation of each WEC in the array is less than the power generation of an isolated WEC. Therefore, it can be considered that wave interactions have an adverse effect on the energy absorption of WECs in the wave field. Conversely, if $q > 1$, this effect is considered advantageous.

Additionally, the Capture Width Ratio (CWR) is a crucial metric for assessing the efficiency of a WEC in capturing wave energy. It can be expressed as follows [49]:

$$CWR = \frac{P}{DP_w} \quad (17)$$

where P represents the output power of the point absorber. D is the capture width, which refers to the diameter of the point absorber. P_w is the incident power of the wave per unit width and can be expressed as follows:

$$P_w = \frac{1}{16} \rho g H^2 \frac{\omega}{k_0} \left(1 + \frac{2k_0 d}{\sinh 2k_0 d} \right) \quad (18)$$

where k_0 represents the wave number, d is the water depth, and H is the wave height.

To estimate the power generation of the point absorber in real sea conditions, the use of the Power Spectral Density (PSD) spectrum to describe the wave energy density and obtain the probability distribution of waves at different frequencies is proposed. This will be combined with information about the wave conditions at the testing site and the average power generation obtained from Equation (15) to obtain the expected power in the target marine area. The widely used JONSWAP spectrum, with its wave spectral density function $S(\omega)$, can be expressed as follows [50]:

$$S(\omega) = \frac{5}{16} \cdot H_s^2 \omega_p^4 \cdot \omega^{-5} \exp\left(-\frac{5\omega_p^4}{4\omega^4} \right) \cdot A_\gamma \cdot \gamma^a \quad (19)$$

where H_s represents the significant wave height, ω_p is the peak frequency, γ is the peak factor, $A_\gamma = 1 - 0.287\ln(\gamma)$ is a normalizing factor, and the coefficient a can be expressed as follows:

$$a = \exp[-\frac{(\omega - \omega_p)^2}{2\sigma^2\omega_p^2}] \tag{20}$$

When $\omega \leq \omega_p, \sigma = 0.07$, and when $\omega > \omega_p, \sigma = 0.09$.

Therefore, the expected power of the point absorber $W_{Expected}$ in the target marine area can be calculated as follows [51]:

$$W_{Expected} = \sum\left(F(\omega_p, H_s) \cdot \int \frac{P_{ave}(\omega) \cdot S(\omega)}{\int_0^\infty S(\omega) \cdot d\omega} d\omega\right) \tag{21}$$

where $F(\omega_p, H_s)$ is the two-dimensional probability distribution related to the peak frequency and significant wave height, which is based on the wave conditions at the testing site.

3. Layout of Wind–Wave Platform and Hydrodynamic Analysis

3.1. Layout of Floating Wind–Wave Power Generation Platform

The floating wind–wave power generation platform consists of an OC4 semi-submersible floating wind turbine [52] and point absorbers with identical sizes, as shown in Figure 3. The point absorbers and the floating platform only experience vertical relative motion. The PTO system is installed on top of the platform support, utilizing the relative motion for power generation. Tables 1 and 2 provide detailed dimensions of the platform and point absorbers.

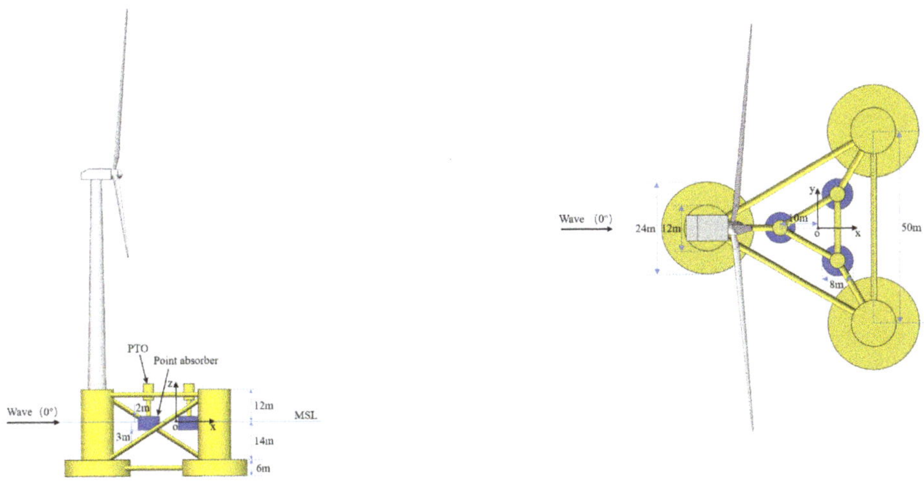

(a) Side view (b) Top view

Figure 3. The wind–wave power generation platform comprising the combination of the OC4 platform and point absorbers.

Table 1. Dimensions of OC4 platform.

Item	Value	Unit
Diameter of base columns	24	m
Diameter of upper columns	12	m
Length of base columns	6	m
Length of upper columns	26	m
Column center to center	50	m
Operating draft	20	m
Bracing diameter	1.6	m
Distance between COG and SWL	13.46	m
Displacement tonnage	1.3473×10^4	t

Table 2. Dimensions of point absorbers.

Item	Value	Unit
Radius	4	m
Length	5	m
Operating draft	3	m
Distance between COG and SWL	1	m
Displacement tonnage	1.5457×10^2	t

3.2. Multi-Body Hydrodynamic Analysis

Four models for different layouts were established in hydrodynamic analysis: single point absorber (SPA), single point absorber combined with a floating wind turbine platform (SPA-WP), three point absorbers (TPA), and three point absorbers combined with a floating wind turbine platform (TPA-WP), as shown in Figure 4. The wave exciting forces and hydrodynamic coefficients were determined using commercial software ANSYS-AQWA version 2023, and the simulation flow chart is shown in Figure 5. Waves with a frequency range of $\omega = 0.03$ rad/s $-$ 3.00 rad/s were selected, and an increment of 0.03 rad/s was used to ensure the accuracy of frequency-domain calculations. The results of the added mass and radiation damping for the point absorbers with different layouts are shown in Figure 6. Overall, the added mass of the point absorbers in different layouts tends to remain constant at high wave frequencies. When a floating platform exists (SPA-WP and TPA-WP in Figure 6), the added mass of the point absorbers increases slightly at a low wave frequency, and this effect is further amplified when three point absorbers are arranged. Additionally, regardless of whether a floating platform or an array arrangement is being considered, the radiation damping of the point absorbers tends to approach zero at low and high wave frequencies, with varying degrees of peak increase. Due to the presence of the platform, multiple peaks in radiation damping become significant, with the SPA-WP model demonstrating the most significant impact. When comparing the surface contour of waves for different layouts under an incident wave heading of 0 degrees, as illustrated in Figure 7, it becomes apparent that the floating platform induces a pronounced wave-focusing effect. This effect leads to a substantial increase in wave crest amplitudes within the interior of the platform. Further investigation into the changes in the Response Amplitude Operator (RAO) for the motion of each model is shown in Figure 8. It is observed that considering a floating platform significantly increases the peak RAO of the point absorbers, and a secondary peak of similar magnitude appears due to hydrodynamic interference. Additionally, when considering a point-absorber array, the peak RAO of two of the point absorbers is enhanced, while the peak of the other one decreases. Therefore, the impact of array arrangement on the performance of the power generation platform requires further study.

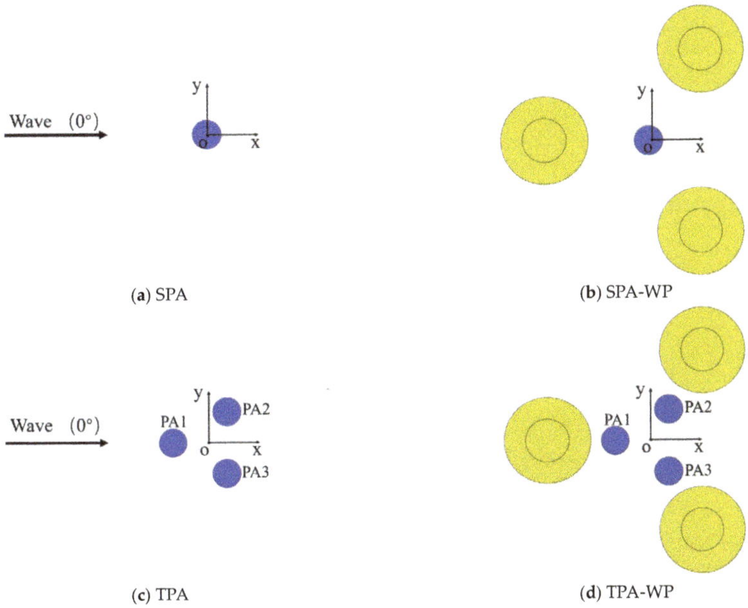

Figure 4. Four different layouts of the point absorbers.

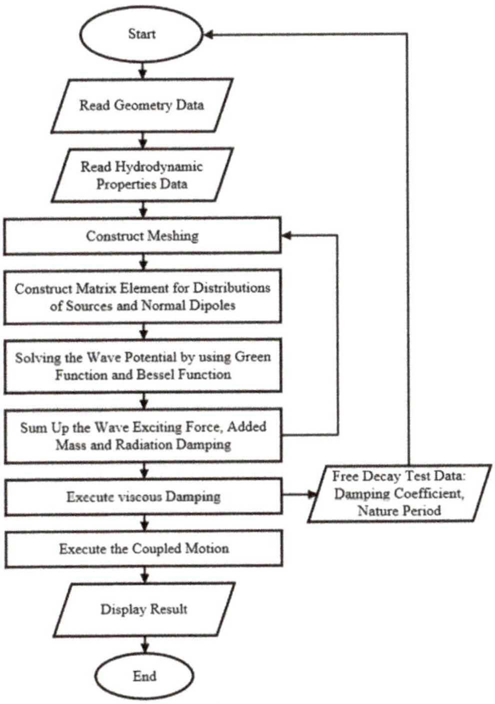

Figure 5. Simulation flow chart for ANSYS-AQWA.

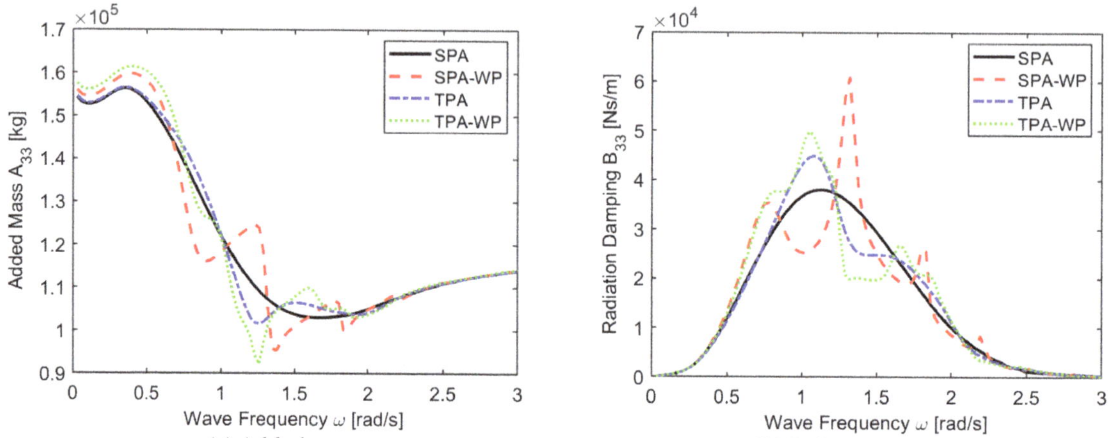

(**a**) Added mass (**b**) Radiation damping

Figure 6. Comparison of the hydrodynamic coefficients of the point absorber models.

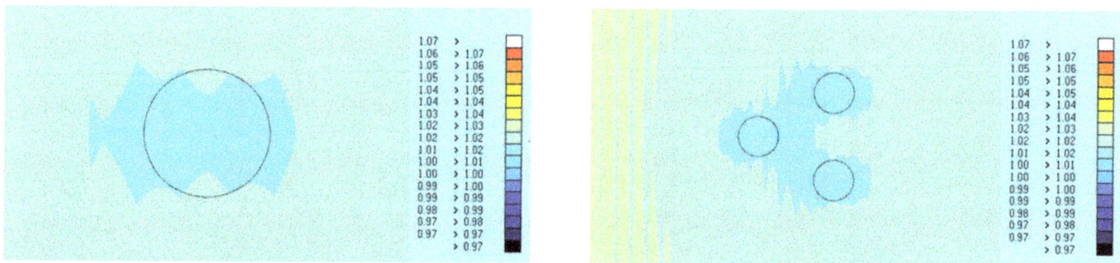

(**a**) SPA (**b**) TPA

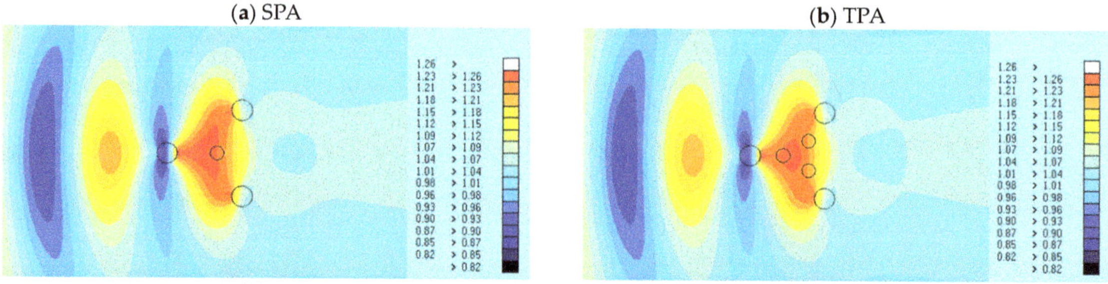

(**c**) SPA-WP (**d**) TPA-WP

Figure 7. Surface contours under different layouts ($H = 2$ m, $\omega = 0.66$ rad/s).

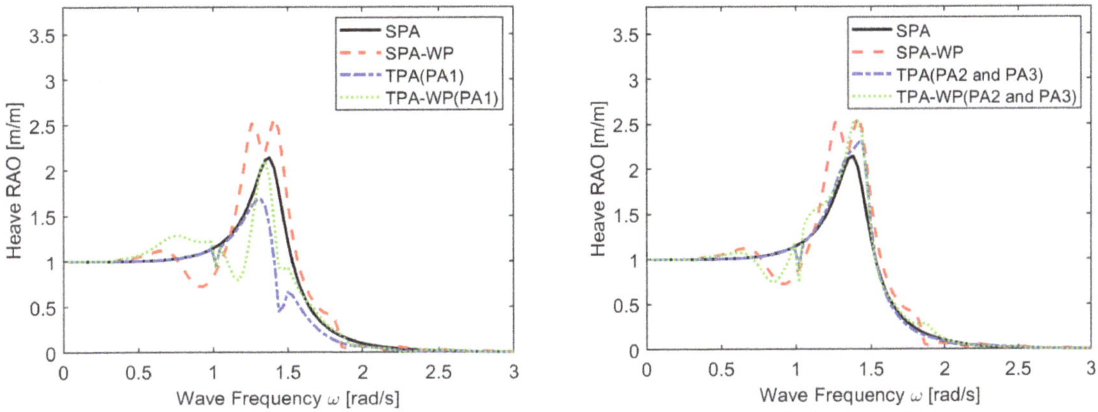

(**a**) PA1 (**b**) PA2 and PA3

Figure 8. Comparison of RAO for the point absorbers.

4. Optimization of Power Generation in Frequency Domain

4.1. Verification of Motions of Multiple Floating Bodies with Constraints

To validate the accuracy of the developed frequency-domain model, a linear PTO was added to the hybrid system. Taking the TPA-WP model as an example, a PTO stiffness of 5000 N/m and a PTO damping coefficient of 45,000 Ns/m were selected. The research of Zou et al. [43] shows that the numerical simulation results of ANSYS-AQWA for a multi-body system have good agreement with the experimental data when in open water. Thus, a time-domain model for constrained motion was established using ANSYS-AQWA, and the Fender module was utilized to simulate the PTO system. An infinite distance Hinged Joint module was employed to represent the vertically constrained motion between the floating platform and the point absorber, as illustrated in Figure 9a, in ANSYS-AQWA. The heave motion of the point absorbers and platform under regular waves from a direction of $-180°$ with a height of 2.0 m and frequency of 0.6 rad/s was calculated as shown in Figure 9b. The time histories of heave motion were statistically analyzed to derive response amplitudes. These amplitudes were then compared with the RAO directly computed through the frequency-domain model, as depicted in Figure 10. It can be observed that excellent agreements have been achieved between the results from present frequency-domain calculations and those obtained from the time domain using ANSYS-AQWA.

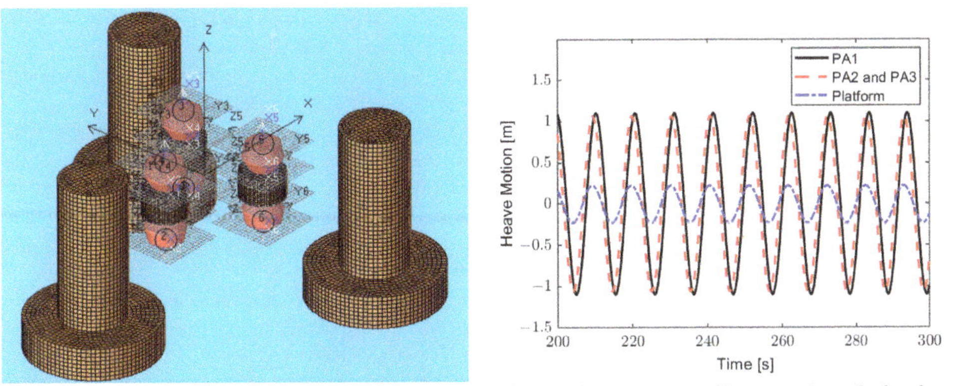

(**a**) Schematic of the TPA-WP model in ANSYS-AQWA (**b**) Time history curves of heave motion calculated using the ANSYS-AQWA time-domain model

Figure 9. Time-domain simulation in ANSYS-AQWA.

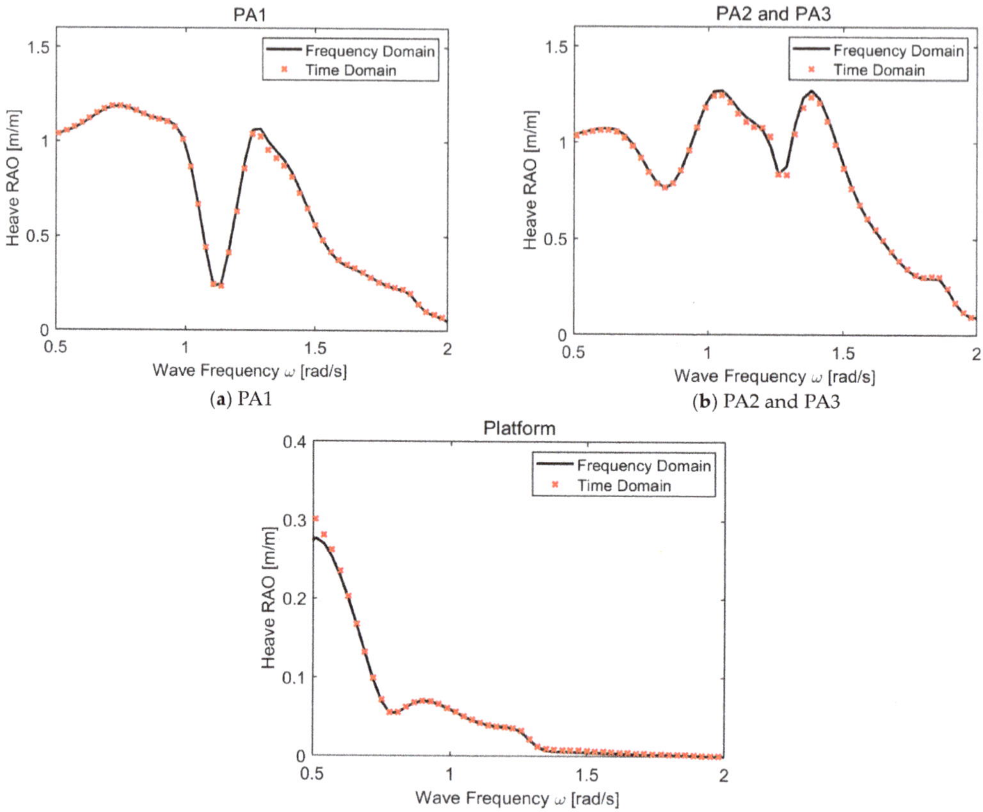

(**a**) PA1 (**b**) PA2 and PA3 (**c**) Platform

Figure 10. Comparison of RAO between present frequency-domain model and time-domain model in ANSYS-AQWA.

4.2. Optimization of PTO

The influence of the PTO on the performance of the point absorber was investigated With a given PTO damping coefficient of 45,000 Ns/m, the contour diagrams in Figure 11 show the CWR of the point absorber under different PTO stiffness values and wave frequencies for a $-180°$ head wave. In the simulated scenarios, the range of PTO stiffness was from 0 to 5×10^5 N/m. The maximum CWR values in Figure 11 were statistically analyzed, as shown in Table 3. It can be observed that the CWR of the point absorber decreases as the PTO stiffness increases in all four models. The optimal PTO stiffness values are found within the range of 0–10^4 N/m, with the maximum CWR occurring at a stiffness value of 0, corresponding to wave frequencies near the natural frequency of the point absorber. Furthermore, comparing Figures 11b and 11f, it is noted that when a floating platform is considered, the CWR exhibits a secondary peak near a wave frequency of 1.7 rad/s. The stiffness values corresponding to this secondary peak are different from those at the primary peak. Figure 12 illustrates the relative motion RAO comparison between the two models at the frequency corresponding to the secondary peak. It is observed that the RAO at the secondary peak frequency has increased, and the spectral width is wider compared to the case with a stiffness value of 0. This reflects that considering the floating platform alters the system's motion characteristics and alters the original power generation mode of the point absorber. Therefore, to determine the optimal PTO parameters, a PTO stiffness of 10^{-6} N/m is selected for further research.

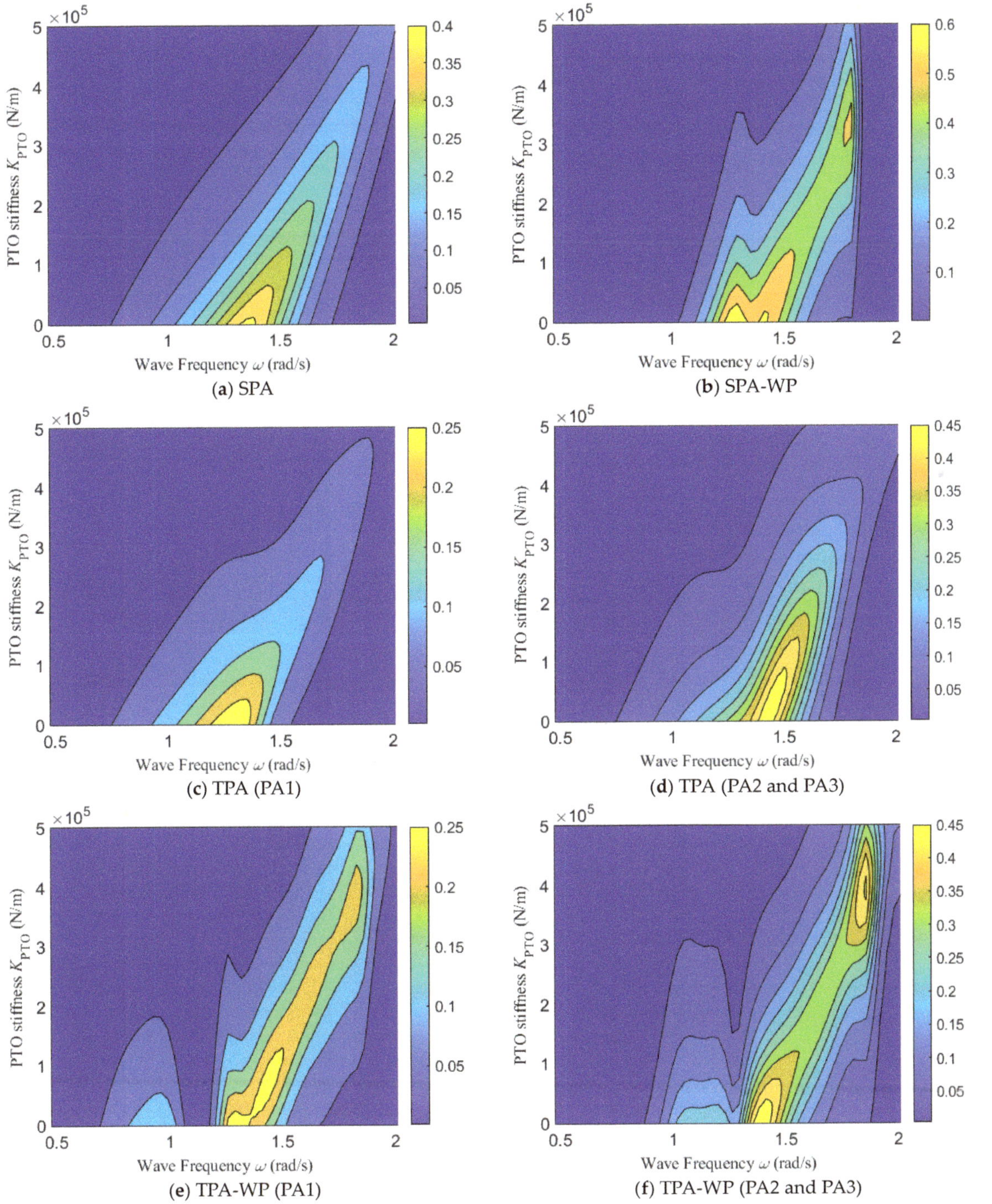

Figure 11. Contour diagram of the CWR of the point absorber for varying PTO stiffness values and wave periods under different models.

Table 3. The maximum CWR for each point absorber with optimal PTO stiffness.

Model	Number of Point Absorbers	Maximum CWR	Optimal PTO Stiffness (N/m)	Optimal Wave Frequency (rad/s)
SPA		0.4122	0	1.35
SPA-WP		0.6873	0	1.29
TPA	PA1	0.2993	0	1.29
	PA2 and PA3	0.4831	0	1.44
TPA-WP	PA1	0.2887	0	1.29
	PA2 and PA3	0.4969	0	1.38

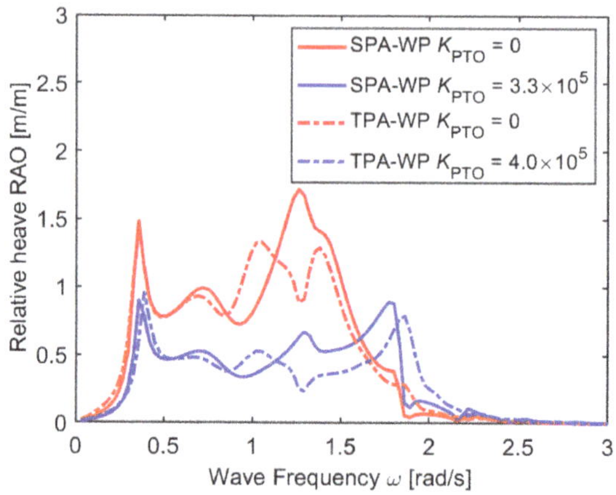

Figure 12. Comparison of relative RAO between the peak and secondary peak of CWR.

Further research on the influence of the PTO damping coefficient is presented. The contour diagrams in Figure 13 depict the CWR of the point absorber for different PTO damping coefficient values and wave frequencies. The statistical results for the maximum CWR are shown in Table 4. It is evident that considering the floating platform leads to a significant increase in CWR, regardless of the number of point absorbers. When a single point absorber is placed on the floating platform, the CWR increases 82%. Furthermore when three point absorbers are placed on the floating platform, a more pronounced bimodal pattern in the CWR emerges. One peak is close to the natural frequency of the point absorber, indicating the dominance of the individual point absorbers in relative motion The maximum CWR tends to reach another peak, with the corresponding wave frequency shifting towards the natural frequency of the platform. Additionally, the optimal PTO damping coefficient increases 372.3% and 386.7% separately. Comparing the CWR under the optimal damping coefficient for each model, as shown in Figure 14, it is evident that the SPA-WP model outperforms the SPA model overall. Under the TPA-WP model, the peak frequency shifts significantly towards lower wave frequency, and with increasing wave frequency, the CWR decreases more gradually compared to the other models. This suggests that placing point absorbers in an array configuration on a floating foundation can alter its working conditions to adapt to a wider range of wave frequencies.

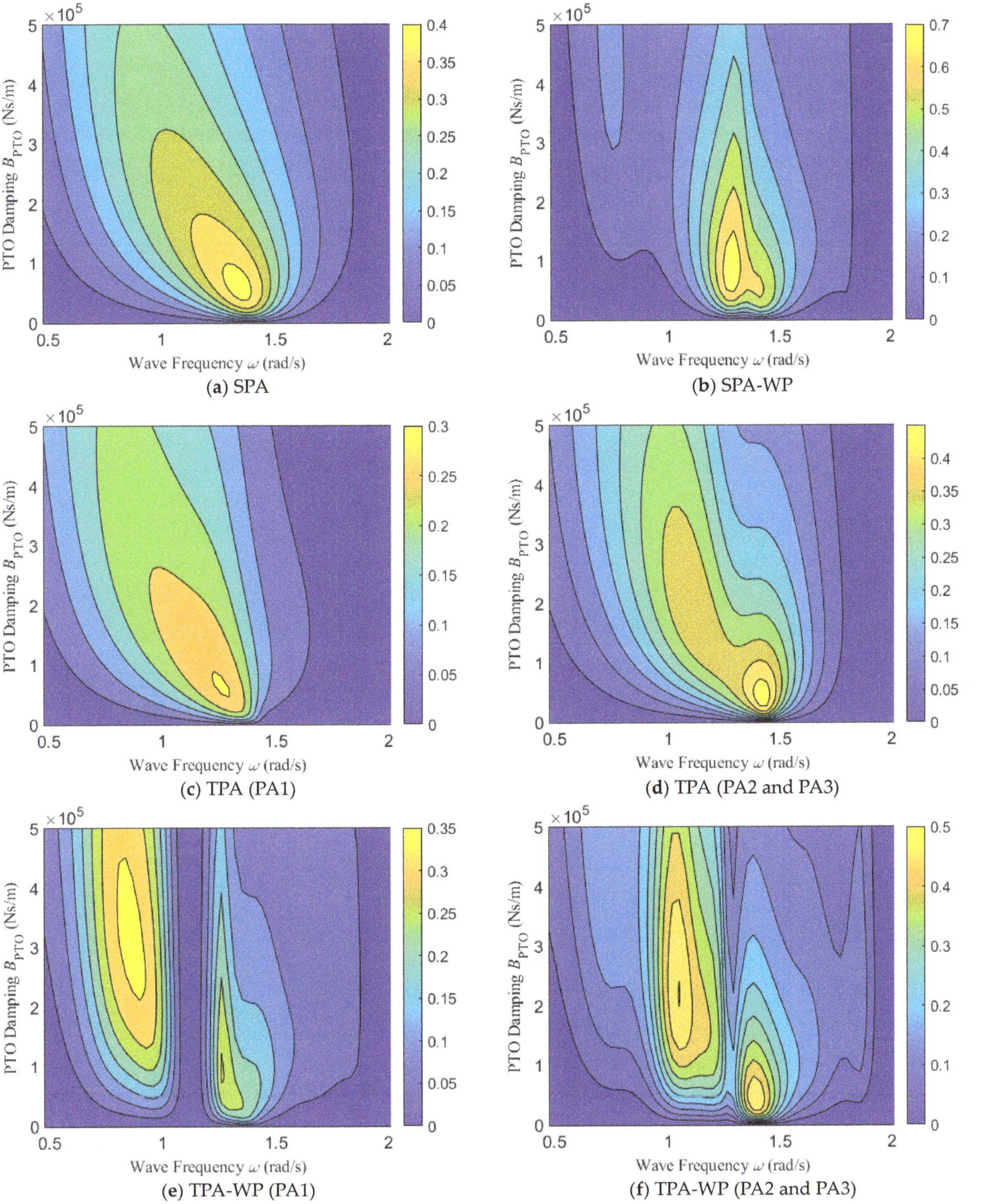

Figure 13. Contour diagram of the CWR of the point absorber for varying PTO damping coefficients and wave periods under different models.

Table 4. The maximum CWR for each point absorber with optimal PTO damping coefficient.

Model	Number of Point Absorbers	Maximum CWR	Optimal PTO Damping Coefficient (Ns/m)	Optimal Wave Frequency (rad/s)
SPA		0.4193	5.80×10^4	1.35
SPA-WP		0.7633	8.50×10^4	1.29
TPA	PA1	0.3044	6.50×10^4	1.26
	PA2 and PA3	0.4787	4.50×10^4	1.41
TPA-WP	PA1	0.3634	3.07×10^5	0.87
	PA2 and PA3	0.5017	2.19×10^5	1.05

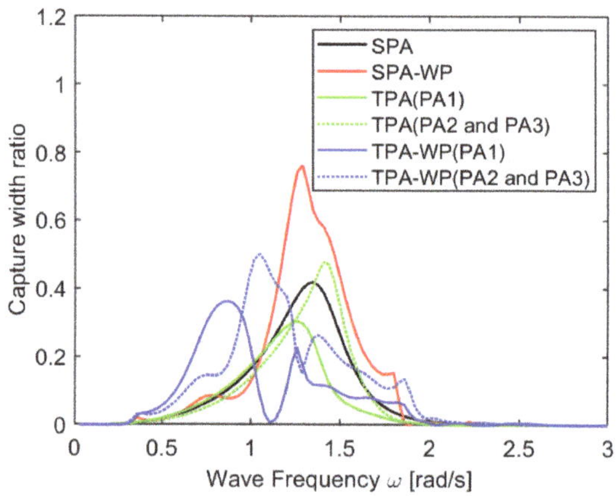

Figure 14. Comparison of CWR under the optimal PTO damping coefficient for different models.

To investigate the impact of array placement on the power generation performance of point absorbers placed within a floating platform, based on the TPA-WP model, two placement schemes were considered: the second placement scheme for three point absorbers combined with a floating wind turbine platform (TPA-WP2) and six point absorbers combined with a floating wind turbine platform (SIXPA-WP), as shown in Figure 15. Based on the previous analysis of PTO parameters, the contour diagrams in Figure 16 depict the total power generation of point absorbers in each model as a function of the PTO damping coefficient and wave frequency for a $-180°$ head wave. The maximum total power generation for each model is recorded in Table 5. It is evident that the total power generation of point absorbers with a floating platform is significantly greater than without a floating platform. Additionally, in both array placement schemes, the number of peak power generation points increases as the number of point absorbers increases. This demonstrates that array placement enables point absorbers to adapt to a wider range of wave frequencies, expanding the operational range of the hybrid system.

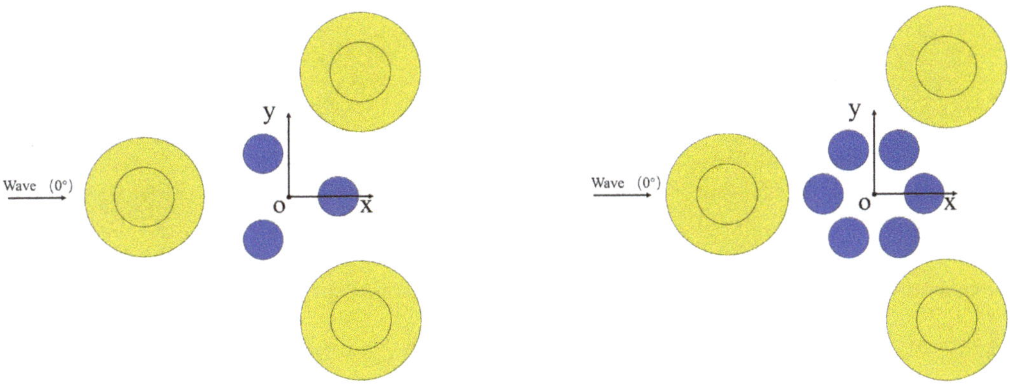

(a) TPA-WP2 (b) SIXPA-WP

Figure 15. The other two layouts for the point-absorber array.

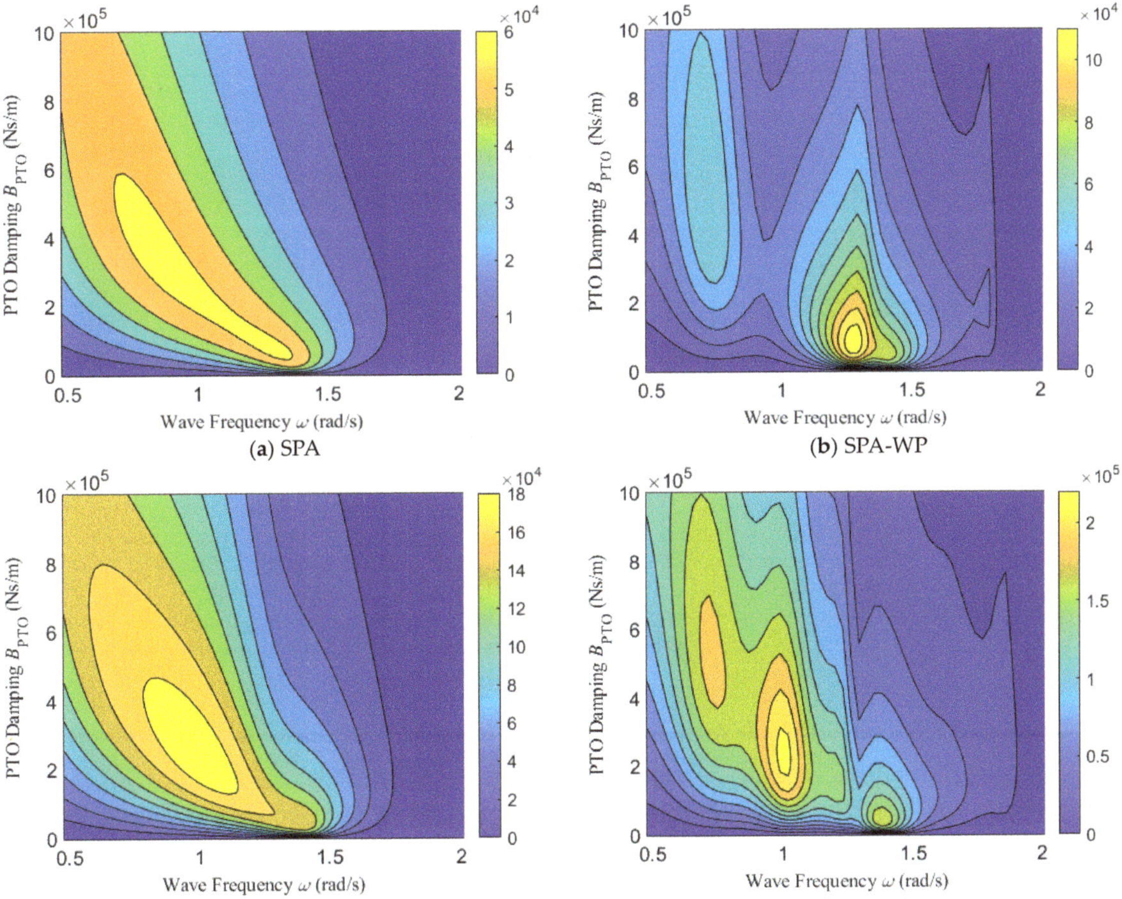

(a) SPA (b) SPA-WP

(c) TPA (d) TPA-WP

Figure 16. *Cont.*

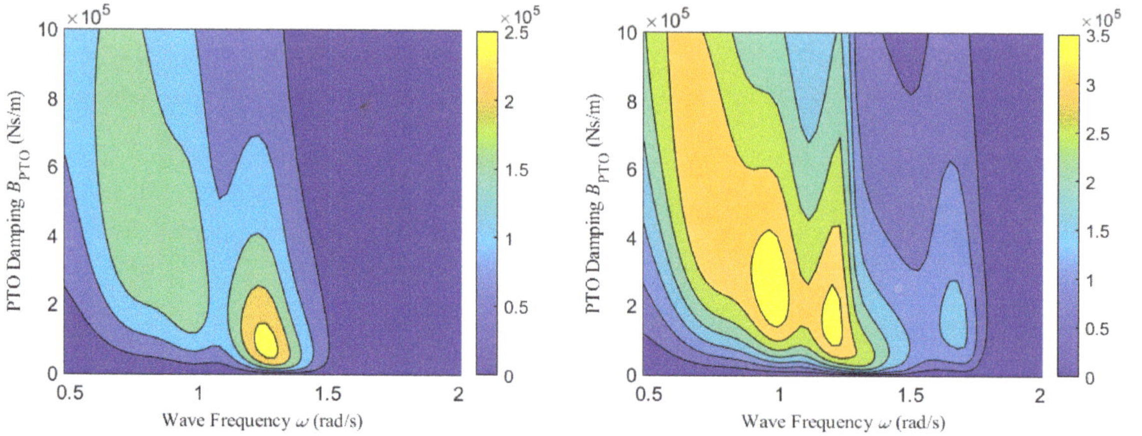

(**e**) TPA-WP2 (**f**) SIXPA-WP

Figure 16. Contour diagram of the total power generation of the point absorber for varying PTO damping coefficients and wave periods under different models.

Table 5. The maximum total power generation under different models.

Model	Maximum Total Power Generation (W)	Optimal PTO Damping Coefficient (Ns/m)	Optimal Wave Frequency (rad/s)
SPA	6.2417×10^4	7.60×10^4	1.29
SPA-WP	1.1666×10^5	8.50×10^4	1.29
TPA	1.9231×10^5	2.51×10^5	0.99
TPA-WP	2.2623×10^5	2.39×10^5	0.99
TPA-WP2	2.7278×10^5	8.40×10^4	1.26
SIXPA-WP	3.8422×10^5	2.38×10^5	0.99

To investigate the influence of wave angles on power generation, considering the model's geometric symmetry, only wave angles of $-180°$, $-165°$, $-150°$, $-135°$, and $-120°$ are considered. Figure 17 shows the comparison of the maximum total power generation for each model. It can be observed that when the floating platform is not considered, the maximum power generation of the three-point-absorber array is roughly three times that of a single point absorber, and it is not significantly affected by wave angles. However, when considering the floating platform, both single and multiple point absorbers show a significant increase in their maximum power generation. For the single point absorber on the floating platform, its maximum power generation at $-120°$ wave angle is even close to the total maximum power generation of the three-point-absorber array. Nevertheless, it is significantly influenced by the wave angle, with a nearly two-fold difference in maximum power generation observed between different wave angles When comparing the two layouts (TPA-WP and TPA-WP2) for three point absorbers on the floating platform, it was found that the first placement scheme has a slightly lower overall maximum power generation than the second one. However, it is less influenced by wave angles, demonstrating a more stable performance overall.

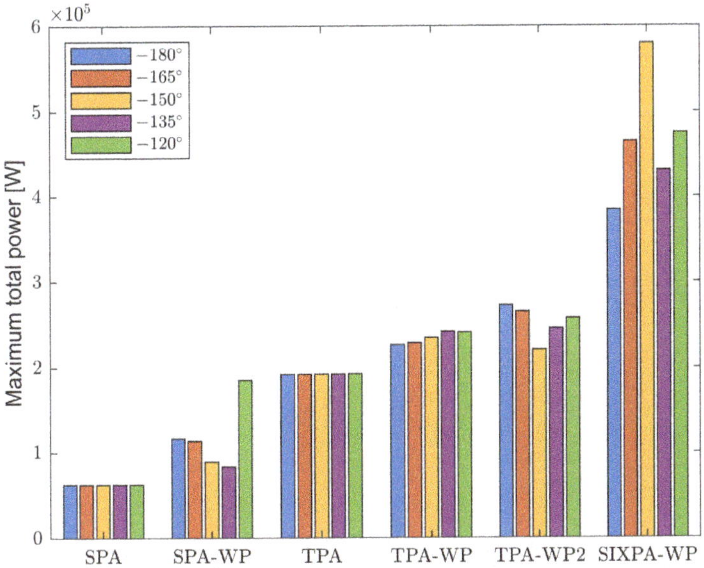

Figure 17. The variation in the maximum total power generation of point absorbers in different models.

Previous research has demonstrated that considering a floating platform can lead to an improvement in the peak performance of point absorbers. A study was conducted to investigate whether a floating platform leads to better average performance of point-absorber arrays compared to a single point absorber. A PTO damping of 2.3×10^5 Ns/m was selected, and two typical wave angles at $-180°$ and $-120°$ were compared in terms of the total power for each model, as shown in Figure 18. Subsequently, a comparison of the q factor for each model was conducted, as shown in Figure 19. The black dashed line represents a q factor of 1. It can be observed that at wave frequencies lower than 1 rad/s, the average performance of point absorbers in different arrays is close to that of a single point absorber. Additionally, the q factors for all three array arrangements are very similar. As the frequency increases, all three array arrangements display multiple peaks at the same frequency. The maximum peak of the q factor can reach up to 70 and occurs at high frequencies where the total power tends towards zero. This reflects that even though the motion responses of each model are small at high frequencies, the hydrodynamic resonance generated by the array of point absorbers on the floating platform significantly improves their average power performance at certain specific wave frequencies. Therefore, when conducting practical engineering design, it is advisable to select sea conditions that are close to these peak frequencies, which will also be the subject of future research.

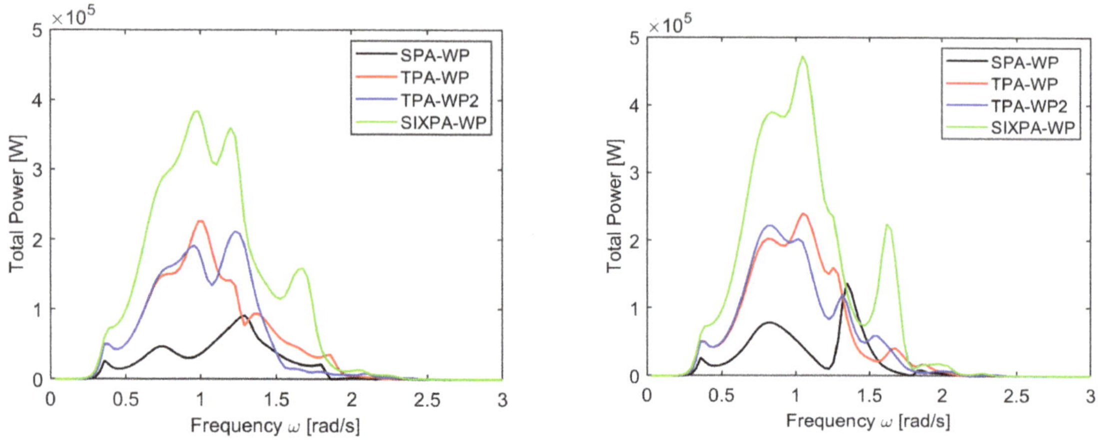

(**a**) −180° wave angle (**b**) −120° wave angle

Figure 18. The variation in total power with wave frequency for different models.

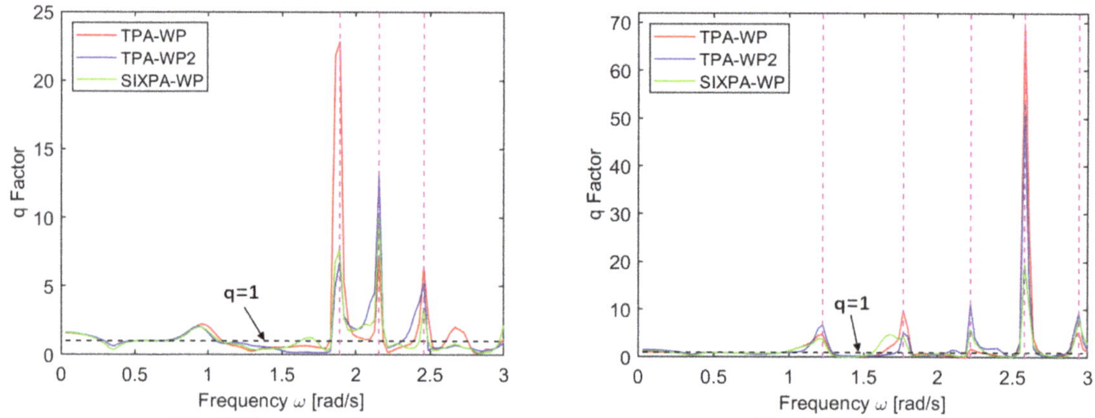

(**a**) −180° wave angle (**b**) −120° wave angle

Figure 19. The variation in q factor with wave frequency for different models.

4.3. The Expected Power in the South China Sea

Previous studies have concentrated on parameter optimization for point absorbers under regular wave conditions, which may not be universally applicable. To predict the performance of the model under real sea conditions, data from the South China Sea were selected. Figure 20 presents wave scatter data obtained from multiple locations in the northern part of the South China Sea, where water depths exceed 150 m [53]. The numbers in the first row represent the peak wave period Tp (s), while the numbers in the first column represent significant wave height Hs (m).

Hs \ Tp	2.5	3.5	4.5	5.5	6.5	7.5	8.5	9.5	10.5	11.5	12.5	13.5	14.5	15.5	16.5	17.5	18.5	19.5	20.5	Total
0.5	1.1	704.5	3306.8	6461.9	4611.1	1702.4	765.8	1972.4	1104.9	308.5	138.2	57.9	36.3	18.3	3.2	1.1	0.2	0.1	0.0	21195.0
1.5	0.0	12.9	927.8	7026.0	13642.5	13892.9	5605.7	1224.3	691.3	587.7	414.6	221.7	143.7	64.9	14.1	6.7	1.3	0.3	0.1	44478.5
2.5	0.0	0.0	0.8	24.4	1213.0	7748.9	8158.0	2228.2	321.5	129.3	111.2	95.1	78.1	38.1	7.8	3.3	0.5	0.2	0.0	20158.6
3.5	0.0	0.0	0.0	0.2	10.5	279.0	4053.5	3687.3	782.9	111.0	41.0	26.4	26.1	19.6	3.3	1.1	0.4	0.1	0.0	9042.6
4.5	0.0	0.0	0.0	0.0	0.1	4.8	144.7	1855.9	1370.0	198.3	30.3	13.5	8.8	5.5	1.3	0.8	0.3	0.1	0.0	3634.4
5.5	0.0	0.0	0.0	0.0	0.0	0.1	4.8	88.0	640.3	274.3	35.9	11.0	6.1	2.8	0.8	0.5	0.2	0.0	0.0	1064.8
6.5	0.0	0.0	0.0	0.0	0.0	0.0	0.3	8.6	57.6	165.5	35.3	8.6	4.6	1.9	0.6	0.4	0.1	0.0	0.0	283.5
7.5	0.0	0.0	0.0	0.0	0.0	0.0	0.0	1.1	11.3	30.9	24.4	5.5	2.8	1.2	0.4	0.2	0.0	0.0	0.0	77.9
8.5	0.0	0.0	0.0	0.0	0.0	0.0	0.0	0.0	2.4	10.9	14.7	3.6	1.5	1.0	0.3	0.1	0.0	0.0	0.0	34.4
9.5	0.0	0.0	0.0	0.0	0.0	0.0	0.0	0.0	0.2	3.6	6.9	3.7	0.9	0.7	0.1	0.0	0.0	0.0	0.0	16.0
10.5	0.0	0.0	0.0	0.0	0.0	0.0	0.0	0.0	0.0	0.5	3.0	2.8	0.8	0.6	0.0	0.0	0.0	0.0	0.0	7.9
11.5	0.0	0.0	0.0	0.0	0.0	0.0	0.0	0.0	0.0	0.0	0.7	1.7	0.8	0.5	0.0	0.0	0.0	0.0	0.0	3.8
12.5	0.0	0.0	0.0	0.0	0.0	0.0	0.0	0.0	0.0	0.0	0.1	0.6	0.6	0.3	0.0	0.0	0.0	0.0	0.0	1.7
13.5	0.0	0.0	0.0	0.0	0.0	0.0	0.0	0.0	0.0	0.0	0.0	0.1	0.4	0.2	0.0	0.0	0.0	0.0	0.0	0.8
14.5	0.0	0.0	0.0	0.0	0.0	0.0	0.0	0.0	0.0	0.0	0.0	0.0	0.2	0.1	0.0	0.0	0.0	0.0	0.0	0.3
Total	1.1	717.4	4235.4	13512.5	19477.2	23628.2	18732.7	11065.8	4982.5	1820.6	856.3	452.3	311.8	156.0	31.9	14.1	3.2	0.8	0.1	100000

Figure 20. Distribution of significant wave height and peak wave period.

According to Equation (19), the power density spectrum for each peak wave period and significant wave height was calculated, with a peak factor γ of 1.5 chosen based on the condition in the South China Sea. Figure 21 shows the PSD diagram for Tp = 7.5 s and Hs = 1.5 m. After the PSD was normalized and combined it with the total power generation results calculated in the frequency domain, two typical wave angles, $-180°$ and $-120°$, were selected. The variation in expected power with PTO damping for each model was obtained using equation 1, as shown in Figure 22. The dotted lines and asterisks represent the coordinates of the highest expected power. It can be observed that, unlike regular wave analysis in the frequency domain, under real sea conditions, the optimal PTO damping coefficient for each model is around 3.5×10^5 N/m, showing significant differences from the result in Table 1. Although the PA-WP model exhibits the best performance in terms of power generation, its expected power is only slightly improved compared to SPA and even occasionally falls below the performance of SPA. This suggests that while the SPA-WP solution may have better peak performance, its performance stability across all wave frequencies is relatively poor. Similarly, when comparing the two layout options for the three point absorbers (TPA-WP and TPA-WP2), it is evident that the second layout option yields significantly higher expected power than the first. The first layout option also exhibits performance close to that of TPA, indicating that it is more affected by the wave angle. Finally, the model with six point absorbers (SIXPA-WP) demonstrates more stable performance, suggesting that the placement scheme with six point absorbers is better suited for the sea conditions in the South China Sea.

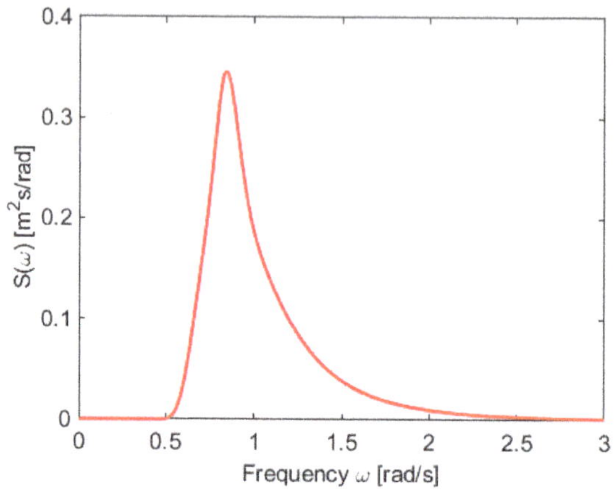

Figure 21. PSD based on JONSWAP spectrum.

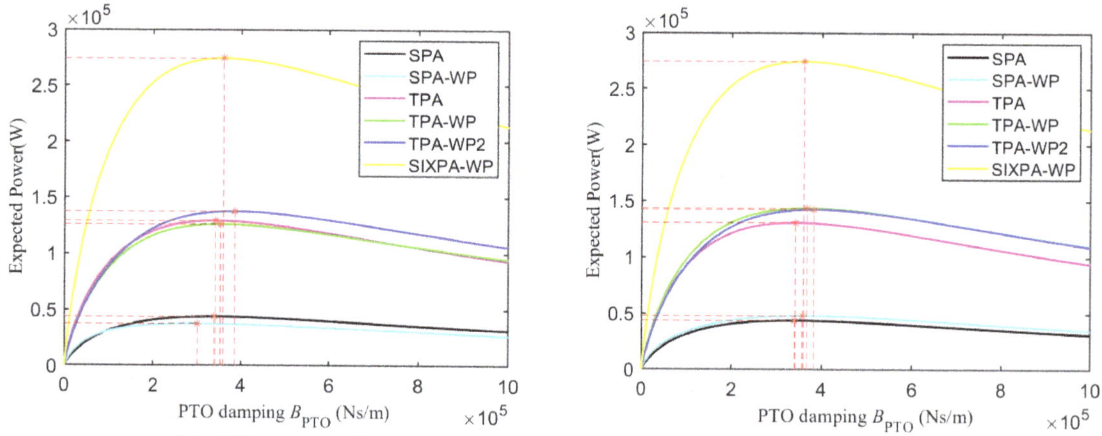

(**a**) −180° wave angle (**b**) −120° wave angle

Figure 22. The expected power for different models.

5. Conclusions

This paper discusses the motion and power generation mechanisms of a floating wind–wave power generation platform composed of multiple point absorbers and a semi-submersible floating platform. A comparative analysis was conducted to investigate the differences between different layouts in the frequency domain. The time-domain outcomes obtained through ANSYS-AQWA were utilized to validate the frequency-domain model of the multi-body constrained motion that had been established. Subsequently, a parametric analysis of PTO parameters was carried out, leading to the determination of optimal PTO parameters for each model, along with an analysis of the performance variations of point absorbers under different layouts. Finally, the expected power in the South China Sea was calculated for each model, revealing variations compared to the parameter analysis. From this research, several conclusions can be drawn:

(1) The hydrodynamic coefficients in the heave DOF of the point absorbers are significantly influenced by the floating platform. Regardless of the arrangement of point absorbers on the platform, both added mass and radiation damping exhibit varying degrees of increase. This leads to higher RAO peak values compared to the model

without a floating platform. The hydrodynamic coefficients of the floating platform, however, are minimally affected by the point absorbers.

(2) Regardless of the arrangement of point absorbers, the optimal PTO stiffness that maximizes the CWR tends towards zero. Only when the floating foundation is considered does the CWR exhibit a secondary peak, with the corresponding PTO stiffness of around 3×10^5–4×10^5 N/m. This suggests that increasing stiffness can change the resonance frequency of the hybrid system to some extent.

(3) When considering the optimal PTO damping coefficient, the CWR values of the point absorbers all exhibit a peak near their natural frequencies. When a three-point-absorber array is arranged on the floating platform, the peak shifts towards the natural frequency of the platform, and the optimal PTO stiffness significantly increases. It can be inferred that array arrangements can alter the working conditions and adaptability range of point absorbers.

(4) When considering different arrangements of point-absorber arrays on the floating platform, it is observed that when considering the floating platform, the point-absorber array's maximum total power generation is minimally affected by the wave angle but offers little improvement over individual point absorbers. When considering a floating platform, the maximum power generation of each model is improved compared to that for individual point absorbers. The arrangement of a single point absorber on the floating platform results in the highest increase in maximum power generation, but it is significantly affected by the angle of waves. On the other hand, both arrangements of three point absorbers exhibit more stable performance. Similarly, the arrangement of six point absorbers is also significantly influenced by the wave angle. Additionally, when considering whether the average performance of point-absorber arrays on a floating platform is superior to that of a single point absorber on the same platform, it was observed that due to the hydrodynamic resonance generated by the array arrangements, multiple peaks occurred in the q factor at the same frequency.

(5) The expected power performance of point absorbers in different arrangements in the South China Sea differs significantly from their performance in the maximum power analysis. While arranging a single point absorber on a floating foundation yields the best peak performance, its stability across all wave frequencies is poor, even dropping below that of a single point absorber. The more point absorbers are arranged in an array, the more stable their performance becomes, demonstrating better adaptability.

However, limitations still exist in this study. It is worth noting that this paper only presents a frequency-domain analysis of the wind–wave power generation platform and does not consider nonlinear factors such as end-stop [44] or mooring systems [53]. Future research will develop corresponding time-domain analysis programs for this model. Additionally, only limited numerical simulations were conducted; further investigation of short-crested waves [54] and validation with physical model tests will be included in future studies.

Author Contributions: Conceptualization, M.C. and J.D.; methodology, M.C. and L.S.; software, M.C., H.Z. and J.D.; validation, J.D., M.C. and Y.Y.; formal analysis, T.T.; investigation, Y.Y. and L.S.; resources, L.S. and M.C.; data curation, J.D.; writing—original draft preparation, J.D. and M.C.; writing—review and editing, L.S., H.Z. and T.T.; visualization, J.D.; supervision, S.L. and M.C.; project administration, Y.Y., S.L. and T.T.; funding acquisition, M.C., Y.Y. and L.H. All authors have read and agreed to the published version of the manuscript.

Funding: This research was funded by the National Natural Science Foundation of China, grant numbers 52171275 and 51961125103, and the Natural Science Foundation of Hainan Province, China, grant number 520MS072. The work presented in this paper is part of the project "Research and development of offshore floating multi-energy complementary comprehensive energy supply system" supported by China Southern Power Grid Technology Co., Ltd., grant No. NYJS2020KJ005-23. The financial support is greatly appreciated.

Institutional Review Board Statement: Not applicable.

Informed Consent Statement: Not applicable.

Data Availability Statement: Data are contained within the article.

Conflicts of Interest: Author Yi Yang, Tao Tao, and Shi Liu were employed by the company China Southern Power Grid Technology Co., Ltd. The remaining authors declare that the research was conducted in the absence of any commercial or financial relationships that could be construed as a potential conflict of interest.

Nomenclature

$\mathbf{A}(\omega)$	Added mass matrix [-]
$\mathbf{B}(\omega)$	Radiation damping matrix [-]
Bpto	Damping coefficient matrix of PTO [-]
Bv	Viscous damping matrix [-]
C	Hydrostatic stiffness matrix [-]
d	Water depth [m]
D	Capture width [m]
E	Constraint matrix [-]
$\widehat{f}_{pto}$	The force applied to the PTO [N]
H	Wave height [m]
H_s	Significant wave height [m]
Kpto	Stiffness matrix of PTO [-]
k_0	Wave number [-]
M	Mass matrix [-]
$P_{ave(N)}$	The average power generation of the Nth point absorber [W]
P	The output power of the point absorber [W]
P_w	Incident power of the wave per unit width [W]
$W_{Expected}$	The expected power [W]
ω_p	Peak frequency [rad/s]
$\widehat{x}_{Relative}$	The relative displacement between the platform and the point absorber [m]
γ	Peak factor [-]

Abbreviations

CFD	Computational fluid dynamics
CWR	Capture Width Ratio
DOF	Degree of freedom
PSD	Power Spectral Density
PTO	Power Take-Off
RAO	Response Amplitude Operator
SIXPA-WP	Six point absorbers combined with floating wind turbine platform
SPA	Single point absorber
SPA-WP	Single point absorber combined with floating wind turbine platform
TPA	Three point absorbers
TPA-WP	Three point absorbers combined with floating wind turbine platform
TPA-WP2	The second placement scheme for three point absorbers combined with floating wind turbine platform
WECs	Wave energy converters

References

1. Hill, J.S. Hywind Scotland, World's First Floating Wind Farm, Performing Better than Expected. Available online: https://cleantechnica.com/2018/02/16/hywind-scotland-worlds-first-floating-wind-farm-performing-better-expected/ (accessed on 1 September 2023).
2. Idcore. Equinor Officially Opens 88MW Hywind Tampen. Available online: https://renews.biz/87665/equinor-officially-opens-88mw-hywind-tampen/ (accessed on 1 September 2023).
3. Floatgen. The First Operational Results Confirm the Excellent Performance of Ideol's Floater. Available online: https://floatgen.eu/en/actualites/first-operational-results-confirm-excellent-performance-ideols-floater (accessed on 1 September 2023).
4. Cobra Group. KOWL: World's Largest Floating Windfarm Fully Operational. Available online: https://www.principlepower.com/news/kowl-worlds-largest-floating-windfarm-fully-operational (accessed on 1 September 2023).
5. Yang, X.; Gao, L. China's First Deep-Sea Floating Wind Power Platform "Haiyang Guanlan" Was Successfully Connected to the Grid and Put into Operation. Available online: http://finance.people.com.cn/n1/2023/0520/c1004-32690779.html (accessed on 1 September 2023).
6. Castro-Santos, L.; Martins, E.; Guedes Soares, C. Cost assessment methodology for combined wind and wave floating offshore renewable energy systems. *Renew. Energy* **2016**, *97*, 866–880. [CrossRef]
7. Slocum, A.; Kluger, J.; Mannai, S. Energy Harvesting and Storage System Stabilized Offshore Wind Turbines. In Proceedings of the 2019 Offshore Energy and Storage Summit (OSES), Brest, France, 10–12 July 2019; pp. 1–6.
8. Muliawan, M.J.; Karimirad, M.; Moan, T.; Gao, Z. STC (Spar-Torus Combination): A Combined Spar-Type Floating Wind Turbine and Large Point Absorber Floating Wave Energy Converter—Promising and Challenging. In Proceedings of the International Conference on Offshore Mechanics and Arctic Engineering, Rio de Janeiro, Brazil, 1–6 July 2012; American Society of Mechanical Engineers: New York, NY, USA, 2012; pp. 667–676.

9. Muliawan, M.J.; Karimirad, M.; Gao, Z.; Moan, T. Extreme responses of a combined spar-type floating wind turbine and floating wave energy converter (STC) system with survival modes. *Ocean Eng.* **2013**, *65*, 71–82. [CrossRef]
10. Wan, L.; Gao, Z.; Moan, T.; Lugni, C. Experimental and numerical comparisons of hydrodynamic responses for a combined wind and wave energy converter concept under operational conditions. *Renew. Energy* **2016**, *93*, 87–100. [CrossRef]
11. Wan, L.; Gao, Z.; Moan, T.; Lugni, C. Comparative experimental study of the survivability of a combined wind and wave energy converter in two testing facilities. *Ocean Eng.* **2016**, *111*, 82–94. [CrossRef]
12. Wan, L.; Ren, N.; Zhang, P. Numerical investigation on the dynamic responses of three integrated concepts of offshore wind and wave energy converter. *Ocean Eng.* **2020**, *217*, 107896. [CrossRef]
13. Mitra, A.; Sarkar, S.; Chakraborty, A.; Das, S. Sway vibration control of floating horizontal axis wind turbine by modified spar-torus combination. *Ocean Eng.* **2021**, *219*, 108232. [CrossRef]
14. Karimirad, M.; Koushan, K. WindWEC: Combining wind and wave energy inspired by hywind and wavestar. In Proceedings of the 2016 IEEE International Conference on Renewable Energy Research and Applications (ICRERA), Birmingham, UK, 20–23 November 2016; pp. 96–101.
15. Li, L.; Gao, Y.; Yuan, Z.; Day, S.; Hu, Z. Dynamic response and power production of a floating integrated wind, wave and tidal energy system. *Renew. Energy* **2018**, *116*, 412–422. [CrossRef]
16. Luan, C.; Michailides, C.; Gao, Z.; Moan, T. Modeling and Analysis of a 5 MW Semi-Submersible Wind Turbine Combined with Three Flap-Type Wave Energy Converters. In Proceedings of the International Conference on Offshore Mechanics and Arctic Engineering, San Francisco, CA, USA, 8–13 June 2014; American Society of Mechanical Engineers: New York, NY, USA, 2014.
17. Michailides, C.; Luan, C.; Gao, Z.; Moan, T. Effect of Flap Type Wave Energy Converters on the Response of a Semi-Submersible Wind Turbine in Operational Conditions. In Proceedings of the International Conference on Offshore Mechanics and Arctic Engineering, San Francisco, CA, USA, 8–13 June 2014; American Society of Mechanical Engineers: New York, NY, USA, 2014.
18. Gao, Z.; Moan, T.; Wan, L.; Michailides, C. Comparative numerical and experimental study of two combined wind and wave energy concepts. *J. Ocean Eng. Sci.* **2016**, *1*, 36–51. [CrossRef]
19. Michailides, C.; Gao, Z.; Moan, T. Experimental study of the functionality of a semisubmersible wind turbine combined with flap-type Wave Energy Converters. *Renew. Energy* **2016**, *93*, 675–690. [CrossRef]
20. Michailides, C.; Gao, Z.; Moan, T. Experimental and numerical study of the response of the offshore combined wind/wave energy concept SFC in extreme environmental conditions. *Mar. Struct.* **2016**, *50*, 35–54. [CrossRef]
21. Chen, W.; Gao, F.; Meng, X.; Chen, B.; Ren, A. W2P: A high-power integrated generation unit for offshore wind power and ocean wave energy. *Ocean Eng.* **2016**, *128*, 41–47. [CrossRef]
22. Fenu, B.; Attanasio, V.; Casalone, P.; Novo, R.; Cervelli, G.; Bonfanti, M.; Sirigu, S.A.; Bracco, G.; Mattiazzo, G. Analysis of a Gyroscopic-Stabilized Floating Offshore Hybrid Wind-Wave Platform. *J. Mar. Sci. Eng.* **2020**, *8*, 439. [CrossRef]
23. Soulard, T.; Babarit, A. Numerical Assessment of the Mean Power Production of a Combined Wind and Wave Energy Platform. In Proceedings of the International Conference on Offshore Mechanics and Arctic Engineering, Rio de Janeiro, Brazil, 1–6 July 2012; American Society of Mechanical Engineers: New York, NY, USA, 2012; pp. 413–423.
24. Soulard, T.; Babarit, A.; Borgarino, B.; Wyns, M.; Harismendy, M. C-HyP: A Combined Wind and Wave Energy Platform with Balanced Contributions. In Proceedings of the International Conference on Offshore Mechanics and Arctic Engineering, Nantes, France, 9–14 June 2013; American Society of Mechanical Engineers: New York, NY, USA, 2013.
25. Wright, C.; Pakrashi, V.; Murphy, J. Numerical Modelling of a Combined Tension Moored Wind and Wave Energy Convertor. In Proceedings of the European Wave and Tidal Energy Conference (EWTEC) Series 2017, Cork, Ireland, 27 August–1 September 2017.
26. Ren, N.; Ma, Z.; Shan, B.; Ning, D.; Ou, J. Experimental and numerical study of dynamic responses of a new combined TLP type floating wind turbine and a wave energy converter under operational conditions. *Renew. Energy* **2020**, *151*, 966–974. [CrossRef]
27. Wang, Y.; Zhang, L.; Michailides, C.; Wan, L.; Shi, W. Hydrodynamic Response of a Combined Wind–Wave Marine Energy Structure. *J. Mar. Sci. Eng.* **2020**, *8*, 253. [CrossRef]
28. Gaspar, J.F.; Kamarlouei, M.; Thiebaut, F.; Guedes Soares, C. Compensation of a hybrid platform dynamics using wave energy converters in different sea state conditions. *Renew. Energy* **2021**, *177*, 871–883. [CrossRef]
29. Ghafari, H.R.; Ghassemi, H.; He, G. Numerical study of the Wavestar wave energy converter with multi-point-absorber around DeepCwind semisubmersible floating platform. *Ocean Eng.* **2021**, *232*, 109177. [CrossRef]
30. Li, J.; Shi, W.; Zhang, L.; Michailides, C.; Li, X. Wind–Wave Coupling Effect on the Dynamic Response of a Combined Wind–Wave Energy Converter. *J. Mar. Sci. Eng.* **2021**, *9*, 1101. [CrossRef]
31. Si, Y.; Chen, Z.; Zeng, W.; Sun, J.; Zhang, D.; Ma, X.; Qian, P. The influence of power-take-off control on the dynamic response and power output of combined semi-submersible floating wind turbine and point-absorber wave energy converters. *Ocean Eng.* **2021**, *227*, 108835. [CrossRef]
32. Shi, W.; Li, J.; Michailides, C.; Chen, M.; Wang, S.; Li, X. Dynamic Load Effects and Power Performance of an Integrated Wind—Wave Energy System Utilizing an Optimum Torus Wave Energy Converter. *J. Mar. Sci. Eng.* **2022**, *10*, 1985. [CrossRef]
33. Wang, Y.; Shi, W.; Michailides, C.; Wan, L.; Kim, H.; Li, X. WEC shape effect on the motion response and power performance of a combined wind-wave energy converter. *Ocean Eng.* **2022**, *250*, 111038. [CrossRef]

34. Tian, W.; Wang, Y.; Shi, W.; Michailides, C.; Wan, L.; Chen, M. Numerical study of hydrodynamic responses for a combined concept of semisubmersible wind turbine and different layouts of a wave energy converter. *Ocean Eng.* **2023**, *272*, 113824. [CrossRef]
35. Chen, M.; Wang, R.; Xiao, P.; Zhu, L.; Li, F.; Sun, L. Numerical Analysis of a Floating Semi-Submersible Wind Turbine Integrated with a Point Absorber Wave Energy Convertor. In Proceedings of the 30th International Ocean and Polar Engineering Conference, Shanghai, China, 11–16 October 2020.
36. Zhang, X.; Li, B.; Hu, Z.; Deng, J.; Xiao, P.; Chen, M. Research on Size Optimization of Wave Energy Converters Based on a Floating Wind-Wave Combined Power Generation Platform. *Energies* **2022**, *15*, 8681. [CrossRef]
37. De Andrés, A.; Guanche, R.; Meneses, L.; Vidal, C.; Losada, I.J. Factors that influence array layout on wave energy farms. *Ocean Eng.* **2014**, *82*, 32–41. [CrossRef]
38. Mercadé Ruiz, P.; Nava, V.; Topper, M.B.; Ruiz Minguela, P.; Ferri, F.; Kofoed, J.P. Layout optimisation of wave energy converter arrays. *Energies* **2017**, *10*, 1262. [CrossRef]
39. Sun, P.; Hu, S.; He, H.; Zheng, S.; Chen, H.; Yang, S.; Ji, Z. Structural optimization on the oscillating-array-buoys for energy capturing enhancement of a novel floating wave energy converter system. *Energy Convers. Manag.* **2021**, *228*, 113693. [CrossRef]
40. Liu, Y.; Zheng, S.; Liang, H.; Cong, P. Wave interaction and energy absorption from arrays of complex-shaped point absorbers. *Phys. Fluids* **2022**, *34*, 097107. [CrossRef]
41. Zhong, Q.; Yeung, R.W. On optimal energy-extraction performance of arrays of wave-energy converters, with full consideration of wave and multi-body interactions. *Ocean Eng.* **2022**, *250*, 110863. [CrossRef]
42. Chen, M.; Xiao, P.; Zhou, H.; Li, C.B.; Zhang, X. Fully coupled analysis of an integrated floating wind-wave power generation platform in operational sea-states. *Front. Energy Res.* **2022**, *10*, 931057. [CrossRef]
43. Zou, M.; Chen, M.; Zhu, L.; Li, L.; Zhao, W. A constant parameter time domain model for dynamic modelling of multi-body system with strong hydrodynamic interactions. *Ocean Eng.* **2023**, *268*, 113376. [CrossRef]
44. Chen, M.; Xiao, P.; Zhang, Z.; Sun, L.; Li, F. Effects of the end-stop mechanism on the nonlinear dynamics and power generation of a point absorber in regular waves. *Ocean Eng.* **2021**, *242*, 110123. [CrossRef]
45. Hu, J.; Zhou, B.; Vogel, C.; Liu, P.; Willden, R.; Sun, K.; Zang, J.; Geng, J.; Jin, P.; Cui, L.; et al. Optimal design and performance analysis of a hybrid system combing a floating wind platform and wave energy converters. *Appl. Energy* **2020**, *269*, 114998. [CrossRef]
46. Zhou, B.; Hu, J.; Jin, P.; Sun, K.; Li, Y.; Ning, D. Power performance and motion response of a floating wind platform and multiple heaving wave energy converters hybrid system. *Energy* **2023**, *265*, 126314. [CrossRef]
47. Folley, M. *Numerical Modelling of Wave Energy Converters: State-of-the-Art Techniques for Single Devices and Arrays*; Academic Press: Cambridge, MA, USA, 2016.
48. Babarit, A. Review on the park effect in arrays of oscillating wave energy converters. *Renew. Energy* **2013**, *58*, 68–78. [CrossRef]
49. Eatock Taylor, R.; Taylor, P.; Stansby, P. A coupled hydrodynamic–structural model of the M4 wave energy converter. *J. Fluids Struct.* **2016**, *63*, 77–96. [CrossRef]
50. *GD04-2021*; Application Guide for Design and Evaluation of Offshore Structures Environmental Conditions. China Classification Society: Beijing, China, 2021.
51. Wen, Y.; Wang, W.; Liu, H.; Mao, L.; Mi, H.; Wang, W.; Zhang, G. A Shape Optimization Method of a Specified Point Absorber Wave Energy Converter for the South China Sea. *Energies* **2018**, *11*, 2645. [CrossRef]
52. Robertson, A.; Jonkman, J.; Masciola, M.; Song, H.; Goupee, A.; Coulling, A.; Luan, C. *Definition of the Semisubmersible Floating System for Phase II of OC4*; NREL/TP–5000–60601; National Renewable Energy Laboratory (NREL): Golden, CO, USA, 2014.
53. Liu, H.; Chen, M.; Han, Z.; Zhou, H.; Li, L. Feasibility Study of a Novel Open Ocean Aquaculture Ship Integrating with a Wind Turbine and an Internal Turret Mooring System. *J. Mar. Sci. Eng.* **2022**, *10*, 1729. [CrossRef]
54. Sun, L.; Zang, J.; Taylor, R.E.; Taylor, P.H.; Chen, M. Sensitivity Analysis of Wave Energy Converters in Short-Crested Waves. In Proceedings of the Thirteenth ISOPE Pacific/Asia Offshore Mechanics Symposium, Jeju, Republic of Korea, 14–17 October 2018.

Journal of
Marine Science and Engineering

MDPI

Article

Design and Performance Evaluation of an Enclosed Inertial Wave Energy Converter with a Nonlinear Stiffness Mechanism

ian Qin [1], Zhenquan Zhang [1], Xuening Song [1], Shuting Huang [1], Yanjun Liu [1,2,*,†] and Gang Xue [1,2,*,†]

1 Institute of Marine Science and Technology, Shandong University, Qingdao 266237, China; 201916191@mail.sdu.edu.cn (J.Q.); zh_zhenquan@mail.sdu.edu.cn (Z.Z.); 202336925@mail.sdu.edu.cn (X.S.); hst@sdu.edu.cn (S.H.)

2 School of Mechanical Engineering, Key Laboratory of High-Efficiency and Clean Mechanical Manufacture, Ministry of Education, National Demonstration Center for Experimental Mechanical Engineering Education, Shandong University, Jinan 250061, China

* Correspondence: lyj111@sdu.edu.cn (Y.L.); xuegang@sdu.edu.cn (G.X.); Tel.: +86-0532-58633267 (Y.L.)

† These authors contributed equally to this work.

Abstract: In order to enhance the power generation efficiency and reliability of wave energy converters (WECs), an enclosed inertial WEC with a magnetic nonlinear stiffness mechanism (nonlinear EIWEC) is proposed in this paper. A mathematical model of the nonlinear EIWEC was established based on the Cummins equation and the equivalent magnetic charge method, and the joint simulations were carried out using MATLAB/Simulink 2020 and AMESim 2020 softwares. The effect of the magnetic nonlinear stiffness mechanism (NSM) on the performance of the EIWEC system was investigated. The results show that the nonlinear negative stiffness property of NSM can significantly improve the motion response and output power of EIWEC under low-frequency waves. Compared to EIWEC without NSM (linear EIWEC), nonlinear EIWEC has a higher generation efficiency and wider frequency bandwidth. Additionally, the effects of linear spring, internal mass body, and hydraulic power take-off (PTO) system parameters on the energy conversion capability of the system were analyzed to provide a reference for the design of nonlinear EIWECs. In general, the proposed nonlinear EIWEC could provide good development potential for the scale utilization of wave energy resources.

Keywords: wave energy converter; nonlinear stiffness mechanism; hydraulic PTO system; wide bandwidth; AMESim

Citation: Qin, J.; Zhang, Z.; Song, X.; Huang, S.; Liu, Y.; Xue, G. Design and Performance Evaluation of an Enclosed Inertial Wave Energy Converter with a Nonlinear Stiffness Mechanism. *J. Mar. Sci. Eng.* **2024**, *12*, 191. https://doi.org/10.3390/jmse12010191

Academic Editor: Barbara Zanuttigh

Received: 16 December 2023
Revised: 15 January 2024
Accepted: 18 January 2024
Published: 20 January 2024

1. Introduction

With the rapid expansion of the global economy and population, the energy crisis and climate change have reached a heightened level of urgency [1]. Various measures have been implemented, including the United Nations Framework Convention on Climate Change [2] and the Paris Agreement [3], aimed at mitigating the impact of global warming and fostering sustainable development across economic, social, and environmental domains. There is now a widespread consensus among the international community to change the conventional fossil-fuel-based energy structure and actively advance the utilization of renewable energy resources [4,5].

The oceans encompass 71% of the Earth's surface and are abundant in energy resources [6]. Wave energy, one of the typical ocean energy resources, has advantages such as better availability, greater predictability, and higher energy density [7]. Wave energy can provide electricity support for island communities, ocean observation instruments, aquaculture activities [8,9], and hydrogen production and desalination [10], exhibiting vast application potential.

In recent years, many wave energy converters (WECs) have been proposed, such as oscillating buoy WECs, oscillating water column WECs, and overtopping WECs [11]. These diverse design concepts have contributed to effectively developing and utilizing wave energy resources. However, the diversity of concepts is also detrimental, in the sense that there is no "winning concept" at this moment. This hinders potential economies of scale and standardization gains, as well as diluting the investments made into wave energy. Moreover, in these concepts, almost all of the WEC's units, such as the power take-off (PTO) system, control system, and electrical system, are directly exposed to the marine environment. The entire WEC will be affected by all processes: biofouling, corrosion, extreme waves, etc. [12] These elements of the marine environment can make critical units (especially the PTO system) less reliable. These situations can increase the LCOE either through lower efficiency, higher downtimes or greater expenditures in operation and maintenance.

Enclosed inertial WECs (EIWECs) have attracted much attention from scholars because of their outstanding survivability and reliable working performance [13]. Prominent EIWECs include Searev [14], Pendulum WEC (PeWEC) [15], Inertial Sea WEC (ISWEC) [16], the Wello Oy Penguin [17], the E-Motions [18], the Seaturns [19] and the WITT [20], and their survivability has generally been verified under extreme wave scenarios. They feature the enclosure of every PTO component inside a rigid and watertight buoy. The encapsulation structure protects all the internal parts from potential risks of corrosion or flooding [21]. Under wave excitation, energy harvesting is achieved by relative motion (linear or rotational motion) between the buoy and the internal mass body [16,22,23]. Therefore, this study will focus on performance analysis and optimization of the EIWECs.

Improving the generation efficiency of the WECs plays an essential role in reducing the LCOE. Efficient power extraction from the ocean in EIWECs with linear stiffness springs (linear EIWECs) is achievable when the system operates near resonance [24]. However, wave energy is primarily distributed within a period range from 5 s to 15 s, exhibiting broadband and low-frequency characteristics [25]. It is difficult for linear WECs to adequately harvest energy from the wide frequency band of the ocean waves, thereby leaving most of the wave energy unutilized [26]. The natural frequency of the WEC is determined by the stiffness of the hydrostatic force, which is typically very high, resulting in a natural frequency much higher than the dominant frequencies in the spectrum of the incoming ocean waves [27]. For instance, for a point absorber with a diameter of 4 m, its natural frequency is about 1.78 rad/s, which means that the system can resonate at a wave period of 3.5 s [28]. This value is considerably lower than the common wave period. To accommodate the low-frequency characteristics of ocean waves, larger-sized buoys are necessary to achieve reduced intrinsic frequencies of WECs [29]. According to Kurniawan et al. [29], to resonate at 8 s (typical Wave Period), a point absorber would need a diameter of approximately 30 m. However, the large-sized floats may pose additional challenges, including wave-structure interactions, economies of scale and costs. Moreover, these larger-sized floats might be less hydrodynamically efficient than smaller bodies due to opposing forces over the body surface [30].

Numerous techniques have been put forth to enhance efficiency, including modifying the mass moment of inertia and actively controlling the PTO system [21,31–33]. These methods typically involve real-time parameter tuning based on estimating the wave excitation force and are challenging to implement in practice [34]. The prediction deviation of wave information might cause a significant decrease in control performance. Furthermore, these control strategies required extra control units, inevitably increasing installation and maintenance expenses.

In the field of mechanical vibration energy harvesting, nonlinearities and instabilities are typically introduced into the system to increase frequency bandwidth and the dynamic range of the excitation amplitude. Scapolan [35] and Giorgi et al. [36] introduced a time-varying damper, tackling the problem of the performance enhancement of a mechanical energy harvester, in terms of both frequency bandwidth and response amplitude, via time variations of the PTO damping coefficient. Some WECs adopted parametric resonance to improve the energy conversion performance of the system. To meet the configuration requirements for parametric resonance, the float is usually connected to the pendulum for inertial coupling in some schemes [37]. According to Giorgi [38], nonlinear hydrodynamics are also capable of inducing parametric resonance between different degrees of freedom of the float. In addition to the foregoing methods, the nonlinear stiffness approach, such as the bistable mechanism [39] and the collision mechanism [40], can also improve the frequency bandwidth and performance of the WECs. The bistable mechanism has two potential wells and a potential barrier, which can be obtained through the inclined structural configuration of springs or repulsive force between magnets [41,42]. For a bistable WEC system, there are three main motion patterns: periodic intrawell, interwell, and chaotic motion [43]. Among them, interwell motion can travel across the potential barrier and oscillate continuously between two potential wells, can generate a larger motion amplitude, and is considered to be a desirable high-energy orbit [28]. Recently, bistability has also been applied to EIWEC. Wang et al. [44] investigated the performance of a bistable snap-through PTO system operating inside a submerged EIWEC, highlighting the significance of the bistable mechanism in improving efficiency. Todalshaug et al. [45] proposed a bistable mechanism called "CorPower" and performed experiments on a 1/16-scale model. The results indicate that CorPower can provide phase control to the system, significantly increasing the power performance. The component was applied to the CorPower prototype [46]. Additionally, a novel X-shaped structure, offering an innovative bistable and quasi-zero stiffness, was applied in the EIWEC, thereby greatly enhancing the wave energy conversion capability [47]. These studies indicated that the utilization of nonlinear stiffness properties enables the passive tuning of the system. Nonlinear stiffness mechanisms are advantageous in adapting the system to wave frequency variations since the frequency bandwidth of the system is increased. However, this benefit is limited to large amplitude wave scenarios; when the wave height is small, the nonlinear stiffness may hinder the power extraction of EIWECs.

In our previous study, a novel magnetic nonlinear stiffness mechanism (NSM) was proposed and applied to a point absorber WEC (PAWEC) [28]. The NSM system exhibited a lower potential barrier, facilitating the buoyancy of PAWEC to generate high-energy interwell motions even in low amplitude waves. This characteristic significantly enhanced the energy harvesting efficiency and expanded the frequency bandwidth of the PAWEC. In this study, we extended the application of NSM in EIWEC (nonlinear EIWEC) and further assessed the influence of the nonlinear stiffness mechanism on the dynamics and power conversion performance of EIWEC.

Because of its great adaptability to low frequency waves and high efficiency, the hydraulic power take-off (PTO) system is a popular option for energy conversion in WECs [48]. Choi et al. [49] constructed and tested a 50 kW hydraulic PTO system, confirming the positive role of the accumulator in smoothing the power output. Gao et al. [50] considered the nonlinear effect of the hydraulic system and analyzed the influence of wave and hydraulic component parameters on motion response and performance. Liu et al. [51] employed the hydraulic PTO system to capture wave energy in a two-raft-type WEC. They discussed the variation in the optimal hydraulic PTO parameters with wave state. To improve the efficiency and stability of the rotating EIWEC, an adaptive hydraulic PTO system was designed and tested by Xue et al. [52]. The effects of the precharged pressure, delay time, and open state time of the accumulator on the average output power and power fluctuation index were experimentally analyzed.

The hydraulic PTO system typically consists of multiple hydraulic components, and variations in component parameters will result in complex energy conversion and force reaction characteristics [48]. To the authors' knowledge, in the field of wave energy harvesting, few studies focused on the analysis and performance evaluation of hydraulic PTO systems coupled with nonlinear stiffness mechanisms. In previous studies of WECs with nonlinear stiffening mechanisms [28,39,40,43,53], the PTO system is frequently modeled as a linear relationship between the damping coefficient and the velocity. In this paper, the hydraulic PTO system was further introduced into the nonlinear EIWEC to ensure that the analysis more closely resembles a real scenario. However, the complex dynamic behavior of the NSM will increase the design difficulty of the hydraulic PTO system. To address the issue, this paper investigated the effects of critical hydraulic PTO parameters on the performance of the nonlinear EIWEC system. The novelty of this paper is in two main aspects: (a) An enclosed inertial wave energy converter with a nonlinear stiffness mechanism was designed and proposed. The encapsulation structure can protect critical hydraulic PTO components from potential risks of corrosion or flooding. The nonlinear stiffness mechanism can increase the motion response and frequency bandwidth of the system, improving the operating performance of the EIWEC. (b) The hydraulic PTO system was further introduced into the nonlinear EIWEC. The results are expected to be a helpful reference for the hydraulic PTO system design of the nonlinear EIWEC.

The remainder of the paper is organized as follows. Section 2 describes the schematic design and models of the nonlinear EIWEC. In Section 3, the influence of nonlinear stiffness mechanisms and critical hydraulic PTO parameters on the motion response and performance of the system is discussed. Section 4 presents the main conclusions.

2. Schematic Design and Modeling

2.1. Schematic Design

The proposed enclosed inertial wave energy converter with a nonlinear stiffness mechanism (nonlinear EIWEC) is illustrated in Figure 1. As depicted in Figure 1a, the nonlinear EIWEC primarily consists of an energy harvester and a nonlinear stiffness mechanism (NSM), which are connected via a rod.

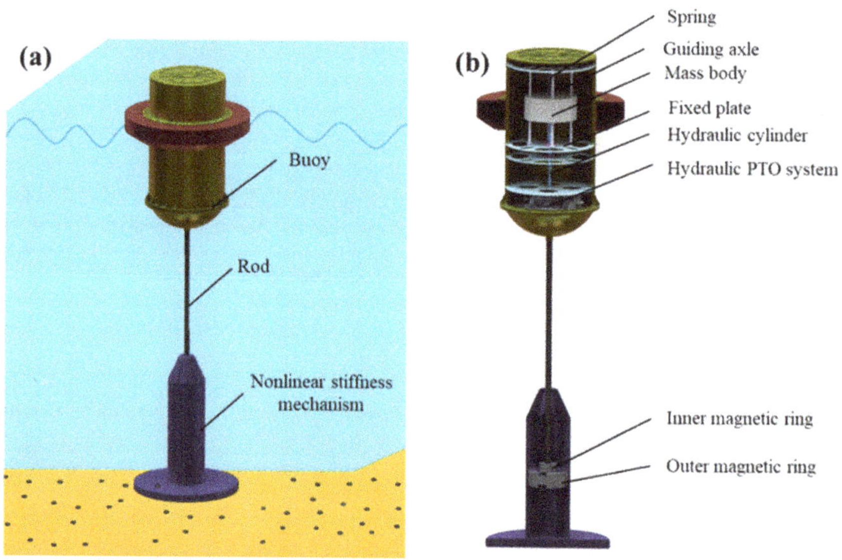

Figure 1. Nonlinear enclosed inertial wave energy converter (nonlinear EIWEC). (**a**) Working scenario, (**b**) internal structure.

Figure 1b shows a schematic diagram of the internal structure. The energy harvester comprises an inner mass body, a buoy, a linear spring, and a hydraulic PTO unit. These components are encapsulated inside a rigid and watertight buoy. The encapsulation structure can protect all the internal parts from potential risks of corrosion or flooding. The relative heaving motion between the buoy and the internal mass body can drive the hydraulic PTO system to generate electricity.

Figure 2 illustrates a working schematic of the nonlinear EIWEC. The buoy and inner mass body are connected by a linear spring that can transfer the wave energy captured by the buoy to the inner mass body. The cylinder and piston rod of the hydraulic cylinder are mounted on the inner mass body and the buoy, respectively. Because of the relative motion between the inner mass body and the buoy, the upper and lower chambers of the hydraulic cylinder produce a pressure difference, allowing the oil to flow into the hydraulic PTO system (Figure 2a). Four check valves rectify the oil flow in the hydraulic motor into one direction. An accumulator is utilized to absorb flow fluctuations within the pipeline and stabilize the output speed of the hydraulic motor. After that, the oil flows into the hydraulic motor through the throttle valve. The hydraulic motor rotates and drives the generator to produce electrical energy.

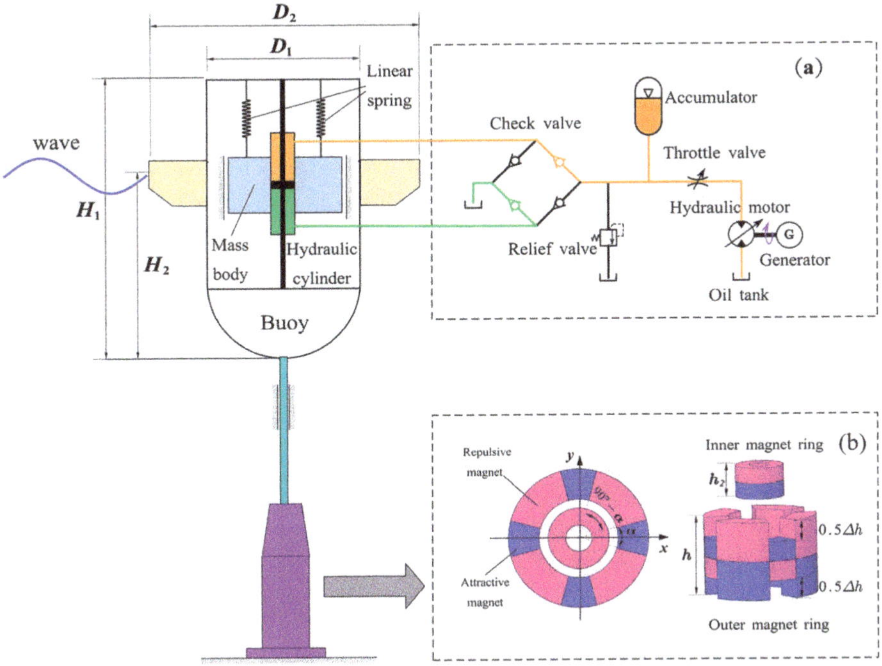

Figure 2. Working schematic of nonlinear EIWEC. (**a**) Hydraulic PTO unit, (**b**) nonlinear stiffness mechanism (NSM).

Figure 2b shows the three-dimensional structure of the NSM, and it comprises an inner and an outer magnetic ring. The inner magnetic ring is connected to the buoy via the rod, following the buoy to produce a heaving motion. The outer magnetic ring consists of two types of sector magnets with opposite magnetic poles and is mounted in the housing fixed to the seabed. The sector magnets with the same pole direction as the inner magnetic ring are 'repulsive magnets'. The remaining sector magnets are 'attractive magnets' [28]. The primary role of the NSM is to provide nonlinear stiffness force to the buoy to increase the motion amplitude and improve the energy conversion performance of the EIWEC.

2.2. Mathematical Modeling

According to Guo et al. [40], under wave excitation, the buoy of a wave energy converter moves mainly in the heave direction. To simplify the model, we only consider the heaving motion response in terms of the structural design of the EIWEC. The nonlinear stiffness mechanism is mounted on the seabed, connected to the float by a rigid linkage. As a result, the buoy of the EIWEC can only generate a motion response in the heave degree of freedom.

As shown in Figure 3, the nonlinear EIWEC consists of three stiffness terms: hydrostatic stiffness k_{hs}, linear spring stiffness k, and magnetic nonlinear stiffness k_m. When analyzing the effect of k_m on the system's performance, the effect of the other two stiffness terms should also be considered. The effects of k_m and k on the motion response of the EIWEC are evaluated in this section since k_{hs} is a constant term.

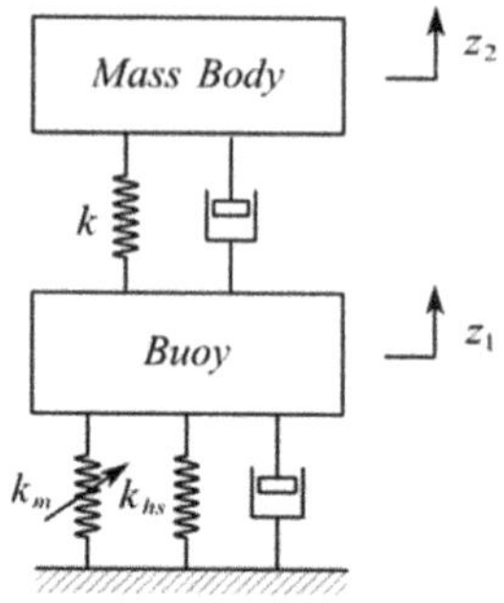

Figure 3. Vibration model of a nonlinear EIWEC.

The buoy is mainly subjected to wave excitation force F_e, radiation force F_r, hydrostatic restoring force F_{hs}, axial magnetic force of NSM F_m, and interaction force between the buoy and the mass body F_i. The dynamic model of the nonlinear EIWEC can be expressed as follows:

$$m_1\ddot{z}_1 = F_e + F_r + F_{hs} + F_m + F_i, \tag{1}$$

where m_1 denotes the mass of the buoy, and $\ddot{z}_1$ is the acceleration of the buoy in the heave direction.

The equations of motion of the internal mass body are as follows:

$$m_2\ddot{z}_2 = -F_i, \tag{2}$$

where m_2 denotes the mass of the mass body and $\ddot{z}_2$ is the heave acceleration of the mass body. The interaction force F_i depends on the relative motion between the buoy and the mass body and can be written as

$$F_i = k(z_2 - z_1) + F_{PTO}, \tag{3}$$

where k denotes the stiffness coefficient of the linear spring and F_{PTO} denotes the reaction force of the hydraulic PTO system. z_1 and z_2 represent the displacements of the buoy and the mass body, respectively. $z_3 = z_2 - z_1$, z_3 represents the relative displacement between the inner mass body and the buoy.

In addition, considering the limited internal space of the buoy, we use the initial elongation Δl instead of the spring stiffness k to avoid choosing some impractical stiffness.

$$\Delta l = \frac{m_2 g}{k}, \tag{4}$$

In regular waves, the wave excitation force F_e acting on a buoy can be expressed through the Haskind relation as [54]:

$$F_e = A\sqrt{\frac{2g^3\rho B_z(\omega)}{\omega^3}}\cos(\omega t + \varphi), \tag{5}$$

where A is the amplitude of the wave, ρ is the density of seawater, g is the acceleration of gravity, $B_z(\omega)$ is the radiation damping coefficient, ω is the frequency of the wave, and φ is the phase of the wave.

According to the Cummins equation [55], the wave radiation force F_r can be given as

$$F_r = -m_\infty \ddot{z}_1 - \int_0^t k_r(t-\tau)\dot{z}(\tau)d\tau, \tag{6}$$

where m_∞ represents the mass and added mass at infinite frequency of the buoy. $k_r(t)$ is the impulse response function (IRF) of the radiation force, which can be obtained by:

$$k_r = \frac{2}{\pi}\int_0^\infty B_z(\omega)\cos(\omega t)dt, \tag{7}$$

Wave excitation and radiation forces are modeled using linear hydrodynamic coefficients, including added mass and radiation damping coefficients. ANSYS-AQWA 2020 is a standard simulation package, and it can calculate these hydrodynamic parameters using a frequency domain potential flow boundary element method (BEM) solver. The BEM solutions of added mass and radiation coefficients are obtained by solving the Laplace equation for the velocity potential of the float, which assumes that the flow is inviscid, incompressible, and irrotational [56]. The added mass A_z and radiation damping B_z are shown in Figure 4.

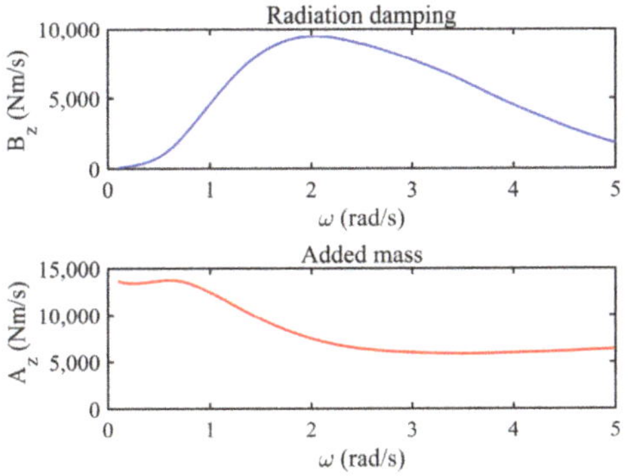

Figure 4. Added mass A_z and radiation damping B_z.

In order to improve the calculation speed of the radiation force, a state space model (SSM) was used instead of the convolution integral term $\int_0^t \tilde{k}_r(t-\tau)\dot{z}(\tau)d\tau$.

$$\begin{cases} \dot{\mathbf{X}}(t) = \mathbf{A}\mathbf{X}(t) + \mathbf{B}\dot{z}_1(t) \\ \int_0^t \tilde{k}_r(t-\tau)\dot{z}(\tau)d\tau = \mathbf{C}\mathbf{X}(t) \end{cases}, \tag{8}$$

where $\mathbf{X}(t)$ is the n-order state vector. $\mathbf{A}$, $\mathbf{B}$, and $\mathbf{C}$ are constant coefficient matrices of the SSM. $\tilde{k}_r(t)$ is the approximate value of $k_r(t)$. $\dot{z}_1(\tau)$ is the velocity of the buoy. $\mathbf{A}$, $\mathbf{B}$, and $\mathbf{C}$ are constant coefficient matrices of the state space model that can be determined using the realization theory approach.

$$\mathbf{A} = \begin{bmatrix} -5.539 & -51.02 & -176.5 & -646.6 & -991.7 & -978.8 \\ 1 & 0 & 0 & 0 & 0 & 0 \\ 0 & 1 & 0 & 0 & 0 & 0 \\ 0 & 0 & 1 & 0 & 0 & 0 \\ 0 & 0 & 0 & 1 & 0 & 0 \\ 0 & 0 & 0 & 0 & 1 & 0 \end{bmatrix}$$

$$\mathbf{B} = \begin{bmatrix} 1 & 0 & 0 & 0 & 0 & 0 \end{bmatrix}^T$$

$$\mathbf{C} = \begin{bmatrix} 17340 & 98150 & 745500 & 2352000 & 6298000 & -82460 \end{bmatrix}$$

In the time domain, the hydrostatic restoring force F_{hs} can be expressed as:

$$F_{hs} = -\frac{1}{4}\rho g \pi {D_2}^2 z_1(t), \tag{9}$$

where D_2 is the diameter of the float, and D_2 = 4 m in this study.

Figure 5 shows the equivalent magnetic charge model for coaxial inner and outer sector magnets. The axial magnetic force of the NSM F_m can be calculated based on the equivalent magnetic charge theory [28].

$$F_m = \frac{B_r^2}{\pi\mu_0}\int_0^{\alpha}\int_{\frac{\pi}{2}-\alpha}^{\frac{\pi}{2}}\int_{R_1}^{R_2}\int_{R_3}^{R_4}\left(\frac{{z_{13}}^{(r)}}{\left|\vec{r}_{13}^{\,(r)}\right|}+\frac{{z_{24}}^{(r)}}{\left|\vec{r}_{24}^{\,(r)}\right|}-\frac{{z_{23}}^{(r)}}{\left|\vec{r}_{23}^{\,(r)}\right|}-\frac{{z_{14}}^{(r)}}{\left|\vec{r}_{14}^{\,(r)}\right|}\right) r_1 r_2 dr_1 dr_2 d\theta d\delta - \frac{B_r^2}{\pi\mu_0}\int_0^{\alpha}\int_{\alpha}^{\frac{\pi}{2}}\int_{R_1}^{R_2}\int_{R_3}^{R_4}\left(\frac{{z_{13}}^{(a)}}{\left|\vec{r}_{13}^{\,(a)}\right|}+\frac{{z_{24}}^{(a)}}{\left|\vec{r}_{24}^{\,(a)}\right|}-\frac{{z_{23}}^{(a)}}{\left|\vec{r}_{23}^{\,(a)}\right|}-\frac{{z_{14}}^{(a)}}{\left|\vec{r}_{14}^{\,(a)}\right|}\right) r_1 r_2 dr_1 dr_2 d\theta d\delta, \tag{10}$$

where P_i is a point on the magnetic pole surface S_i of the inner sector magnet (i = 1,2) and P_j is a point on the magnetic pole surface S_j of the outer sector magnet (j = 3,4). $\vec{r}_{ij}$ is the position vector between points P_i and P_j, and z_{ij} is the axial component of $\vec{r}_{ij}$. The superscripts (r) and (a) denote repulsive and attractive magnets, respectively. B_r is the residual magnetic flux density. μ_0 is the vacuum permeability.

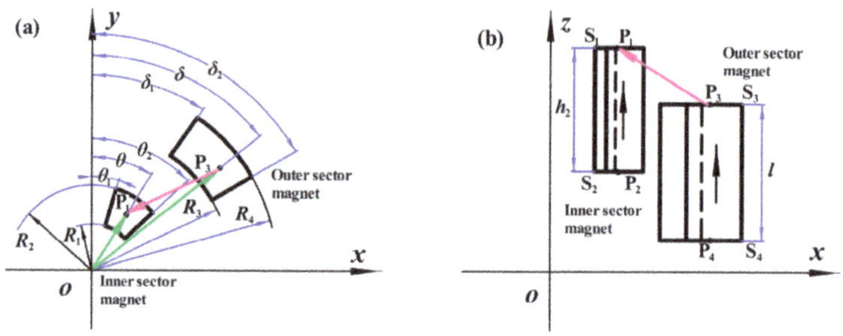

Figure 5. Equivalent magnetic charge model for coaxial inner and outer sector magnets. (**a**) Front view of the magnetic charge model, (**b**) Top view of the magnetic charge model.

The reaction force of the hydraulic PTO system acting on the buoy can be calculated by

$$F_{PTO} = (p_t - p_b)A_p, \tag{11}$$

$$A_p = \pi\left(R_p^2 - R_r^2\right), \tag{12}$$

where R_p and R_r are the radius of piston and rod, respectively. p_t and p_b are the oil pressure of the upper and bottom chamber of hydraulic cylinder, respectively. The value of p_t is approximately equal to the accumulator pressure p_{ac}, which can be given by [49]:

$$p_t = p_{ac}, \tag{13}$$

$$\frac{\dot{p}_{ac}}{p_{pre}} = -\kappa\frac{q_{ac}}{V_{g0}}, \tag{14}$$

where p_{pre} is the pre-charged gas pressure of the accumulator; V_{g0} is the initial gas pressure of the accumulator; κ is the gas adiabatic index; and q_{ac} denotes the flow rate of oil into the accumulator. q_{ac} can be obtained by

$$q_{ac} = q_t - q_v, \tag{15}$$

$$q_t = \left(z_1 - z_2\right)A_p, \tag{16}$$

where q_t denotes the flow rate of oil out of the upper chamber and q_v denotes the outlet flow rate of the throttle valve.

The throttle valve is used to regulate the oil flow into the hydraulic motor. In this study, the outlet flow rate of the throttle valve can be expressed by [49]:

$$q_v = C_d A_v \sqrt{\frac{2(p_M - p_{ac})}{\rho_0}}, \tag{17}$$

where C_d is the flow coefficient, A_v denotes the valve port through-flow area, p_M is the oil pressure at the inlet of the hydraulic motor, and ρ_0 is the density of the hydraulic oil.

Neglecting the compressibility and leakage of the hydraulic oil, the angular velocity ω_M of the hydraulic motor can be determined by:

$$q_M = D_M \omega_M, \tag{18}$$

where q_M is the flow rate into the hydraulic motor, $q_M = q_v$, and D_M is the displacement of the hydraulic motor.

The kinetic equations for the hydraulic motor and generator are given by:

$$J_t \dot{\omega}_M = D_M(p_M - p_{tank}) - T_G, \tag{19}$$

where J_t is the total rotational inertia of the hydraulic motor and generator, p_{tank} is the tank pressure taken as 0, and T_G is the resistive torque of the permanent magnet synchronous generator. When the electrical load is purely resistive, the value of T_G can be calculated by

$$T_G = \frac{E^2}{R\omega_G}, \tag{20}$$

where R is the load resistance; ω_G is the angular velocity of the generator, $\omega_G = \omega_M$; and E is the phase electromotive force of the generator, which can be calculated by

$$E = \frac{4.44 N k_e u n_G \phi_e}{60}, \tag{21}$$

where N is the number of series turns per phase winding; k_e is the electromotive force winding factor; u is the number of pole pairs of the generator armature windings; ϕ_e is the magnetic flux; and n_G is the generator rotational speed with $n_G = 2\pi\omega_G$.

The mean output power of the generator during period t to $t + T$ can be defined as

$$P_i = \frac{E^2}{R}, \tag{22}$$

$$P_m = \frac{1}{T}\int_t^{t+T} P_{Gi} dt, \tag{23}$$

where P_i is the instantaneous power of the generator, and P_m is the mean power of the generator.

To evaluate the motion response of the nonlinear EIWEC, the buoy, mass, and relative RAOs are defined as [40]:

$$RAO_1 = \frac{\max(z_1)}{A}, \tag{24}$$

$$RAO_2 = \frac{\max(z_2)}{A}, \tag{25}$$

$$RAO_3 = \frac{\max(z_3)}{A}, \quad (26)$$

where A is the wave amplitude.

2.3. Software Application and Configuration

The calculation framework of this study is shown in Figure 6. Wave excitation and radiation forces are modeled using linear hydrodynamic coefficients, including added mass and radiation damping coefficients, which can be obtained by ANSYS-AQWA. A joint simulation calculation between MATLAB/Simulink 2020 and AMESim 2020 is applied for the time domain analysis. The former provides the motion response signals to the latter and the latter transmits the computed PTO forces to the former.

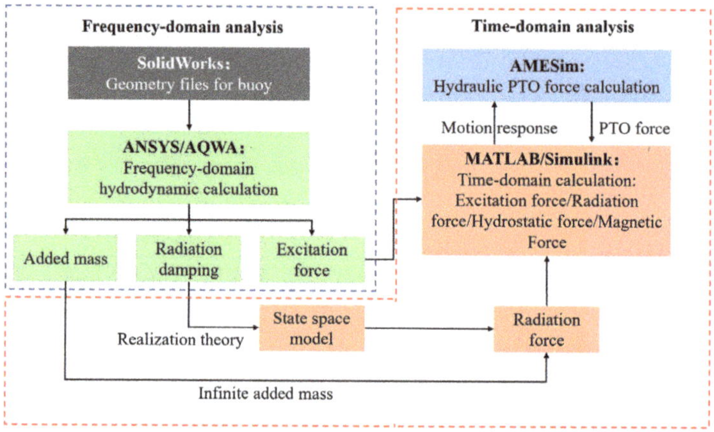

Figure 6. Calculation framework.

2.4. Validation of the Time-Domain Model

To validate the time-domain model of the WEC based on the state space model in this study, the results were compared with published works. The response amplitude operator (RAO) for the heave motion of the WEC system reported by Tampier and Grueter [57] was used for comparison. From Figure 7, the simulation results are in good agreement with the previously reported results, indicating that the proposed time-domain simulation model can accurately simulate the motion response.

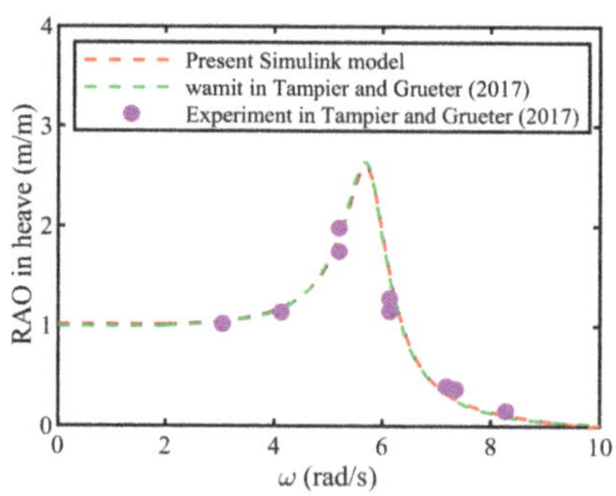

Figure 7. Comparison of RAO in heave obtained from the present time-domain model and simulation and experimental results reported by Tampier and Grueter (2017) [57].

3. Results and Discussion

3.1. Mechanical Property Analysis of NSM

3.1.1. Structural Parameters

The structural parameters of NSM are shown in Table 1. The outer magnetic ring comprises four attractive magnets and four repulsive magnets. The attractive and repulsive magnets have the same inner and outer diameters, and their height difference is Δh.

Table 1. Structural parameters of NSM.

Parameter	Symbol	Value
Inner diameter of inner magnetic rings (m)	R_1	0.2
External diameter of inner magnetic rings (m)	R_2	0.5
Inner diameter of outer magnetic rings (m)	R_3	0.6
External diameter of outer magnetic rings (m)	R_4	1.0
Center angle of attractive magnet (°)	α	18
Center angle of repulsive magnet (°)	$90° - \alpha$	72
Height difference between attractive and repulsive magnets (m)	Δh	0.08
Height of inner magnetic ring (m)	h_1	0.15
Permeability of vacuum (H/m)	μ_0	$4\pi \times 10^{-7}$
Residual magnetic flux density (T)	B_r	1.25

3.1.2. Magnetic Force and Stiffness of NSM

Figure 8a shows the magnetic force F_m as a function of displacement z_1, and the corresponding stiffness curves of the NSM are shown in Figure 8b. For comparison, the variation curves of hydrostatic force F_{hs} and hydrostatic stiffness k_{hs} are also plotted in Figure 8.

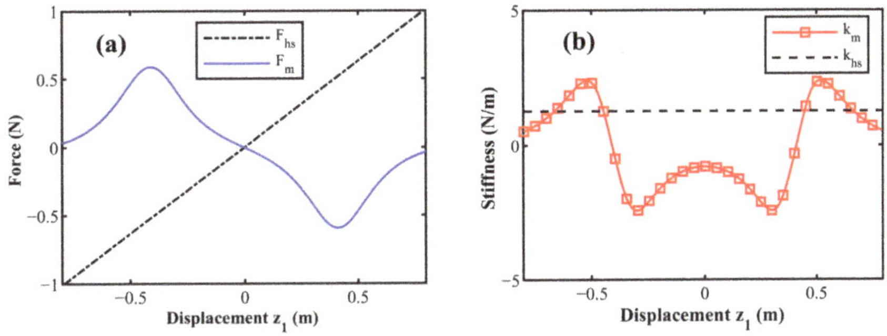

Figure 8. (**a**) Magnetic force of NSM, (**b**) stiffness of NSM.

From Figure 8a, because of the difference in magnetic field strength at various positions, the magnetic force and stiffness of the inner magnetic ring exhibit a nonlinear variation. At some positions, the NSM exhibits negative stiffness properties and has a good potential for improving the frequency bandwidth and energy conversion efficiency of energy harvesting systems. To investigate the effect of NSM on the EIWEC, we further analyzed the motion response and output power of the nonlinear EIWEC in Section 3.2.

3.2. Effect of Nonlinear Stiffness Mechanism (NSM)

3.2.1. Motion Response Analysis

Figure 9 illustrates the RAO as a function of frequency for linear and nonlinear EIWECs with different k. The simulation was executed for 300 s at each wave frequency (the calculation time step is 0.01 s.), and then the maximum motion response amplitude in the time range of 150 s to 300 s was then extracted as RAOs to eliminate transient effects in the initial stage. As depicted in Figure 9, the RAOs of the linear and nonlinear EIWECs are plotted by blue dashed and red dotted lines, respectively. Notably, the maximum stroke limit (end-stops) of the buoy for the device is ±4 m. All of the displacements in Figure 9 are within the range. When the wave frequency is less than 1 rad/s, the RAOs of the buoy and mass body of the linear EIWEC remain around 1, and the relative motion between the buoy and mass body is small. As the wave frequency increases, the motion response of the buoy and mass body first increases and then decreases. According to Figure 9c,f,i, the frequency bandwidth of the linear EIWEC is narrower compared with the nonlinear EIWEC. Accordingly, the linear EIWEC is not capable of adequately capturing energy from the wide frequency band of the ocean waves, leaving most of the wave energy unutilized.

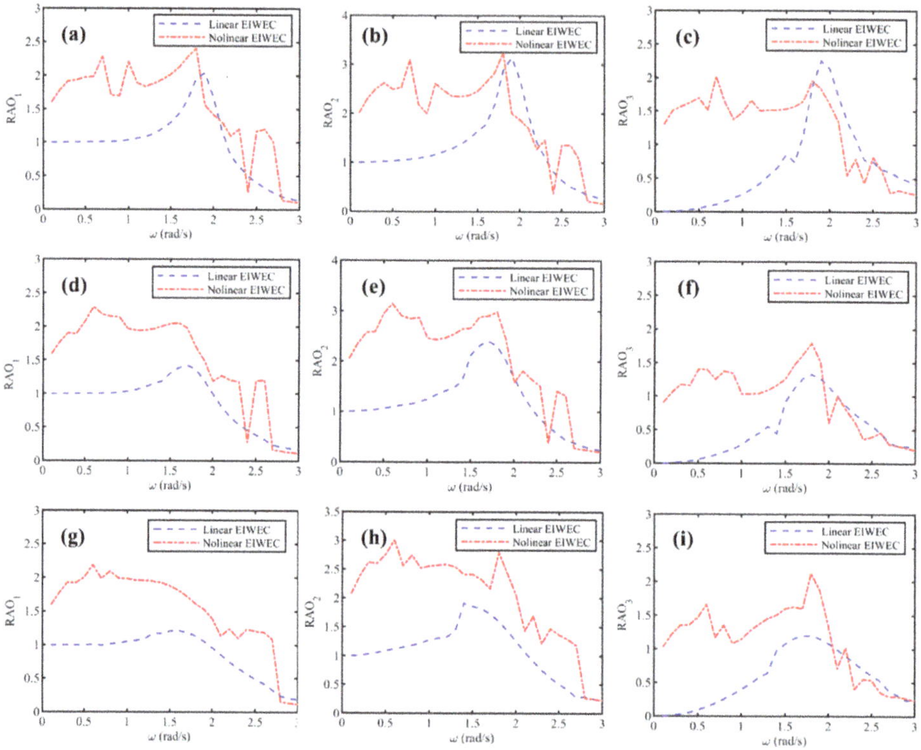

Figure 9. RAOs of the buoy, mass body, and relative motion. (**a–c**) $\Delta l = 1$ m; (**d–f**) $\Delta l = 2$ m; (**g–i**) $\Delta l = 3$ m. Simulation conditions: m_2 = 6670 kg, D_m = 80 cc/rev, R = 50 Ω, p_{pre} = 30 bar, and V_{g0} = 40 L.

For a linear EIWEC, it resonates around 1.7 rad/s. Wave frequencies below or above this value will cause a significant power reduction. According to Figure 8, the NSM has a negative stiffness property, which can reduce the system's natural frequency. Therefore, when the wave frequency is less than 1.7 rad/s, the nonlinear EIWEC is close to the resonant state, and the RAOs of relative motion are much higher than those of linear EIWEC. The large amplitude relative motion provides favorable conditions for energy transfer and conversion, contributing to the increase in output power of the EIWEC.

Meanwhile, the resonant frequency of the system is no longer a fixed value because of the influence of the nonlinear stiffness property, resulting in multiple resonance peaks in the RAOs curve of the nonlinear EIWEC. These characteristics effectively increase the frequency bandwidth of the EIWEC system, enabling EIWEC to be more robust in the face of the change in sea state. When the wave frequency is greater than 1.7 rad/s, the negative stiffness property of NSM causes frequency detuning of the EIWEC, and the low natural frequency of the nonlinear EIWEC cannot match the high-frequency incident wave. Therefore, the motion response of the nonlinear EIWEC decreases rapidly with strong fluctuations. The response reduction for the nonlinear EIWEC occurs at periods of about 3.7 s. These results indicate that the nonlinear EIWEC is not adept at capturing wave energy from high-frequency components, and its performance may even be lower than linear EIWEC under some frequency conditions. However, this period value is relatively low in actual waves and may not be of practical interest. According to Falnes et al. [25], the wave period range is typically between 5 s and 15 s. In this period range, the motion response of the nonlinear EIWEC is consistently at high values, which shows good development potential for extracting energy from actual waves.

Meanwhile, the RAO curves of the nonlinear EIWEC show irregular fluctuations as the frequency changes. The fluctuations may produce erratic, chaotic behavior and wear out the device more in the long term. To explain the reason for the fluctuations and avoid those sudden downward spike regions, we further analyzed the motion response of the nonlinear EIWEC at different frequencies in terms of nonlinear dynamics.

Figure 10 illustrates the phase trajectories of the motion response of the buoy and the internal mass body for nonlinear EIWECs with different spring stiffnesses when the wave frequencies are 2.3 rad/s, 2.4 rad/s, and 2.5 rad/s, respectively.

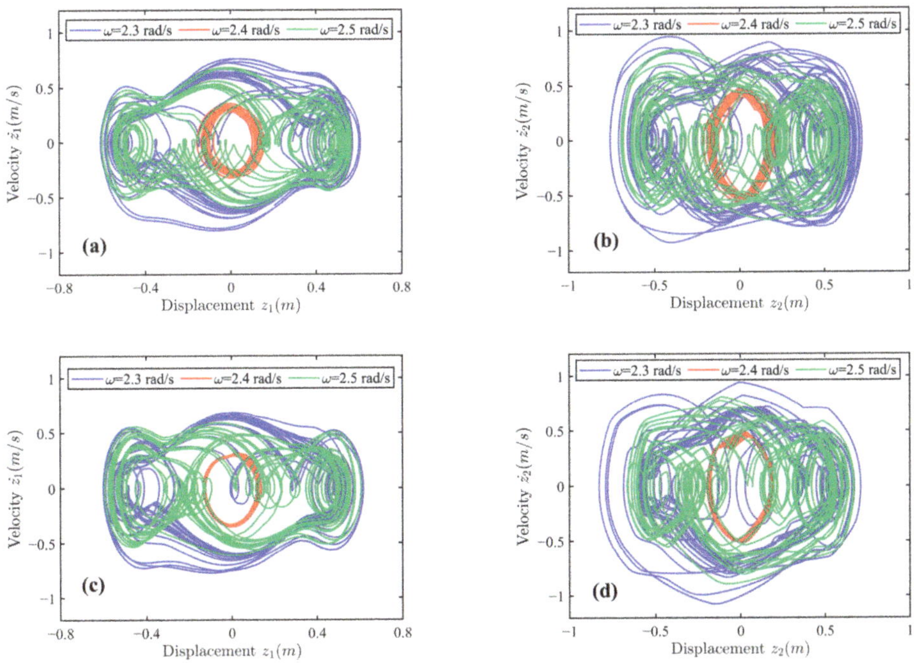

Figure 10. Phase trajectories of the motion response of the buoy and the internal mass body for nonlinear EIWECs with different spring stiffnesses when the wave frequencies are 2.3 rad/s, 2.4 rad/s, and 2.5 rad/s, respectively. (**a**,**b**) $\Delta l = 1$ m; (**c**,**d**) $\Delta l = 2$ m.

From Figure 10, when the wave frequencies are 2.3 rad/s and 2.5 rad/s, the buoy and internal mass body of the nonlinear EIWEC produce complex chaotic motions. The chaotic motion is irregular and contains inter-well and intra-well components, making the energy output strongly random and fluctuating. At a wave frequency of 2.4 rad/s, the buoy and the internal mass body are trapped in the middle potential well and can only produce intra-well motion, causing a sudden drop in the output power. In practice, a nonlinear stiffness mechanism might make the system more sensitive at high-frequency conditions (greater than the resonance frequency), and the EIWEC produces complex and diverse nonlinear motions, resulting in irregularities in the RAOs curves.

The stiffness of the nonlinear mechanism changes drastically around −0.5 m and 0.5 m. In order to investigate the effect of the change on the system, the phase trajectories of the buoys of linear and nonlinear EIWEC are plotted for different stiffness conditions when the wave height is 1 m and the frequency is 1 rad/s, as illustrated in Figure 11.

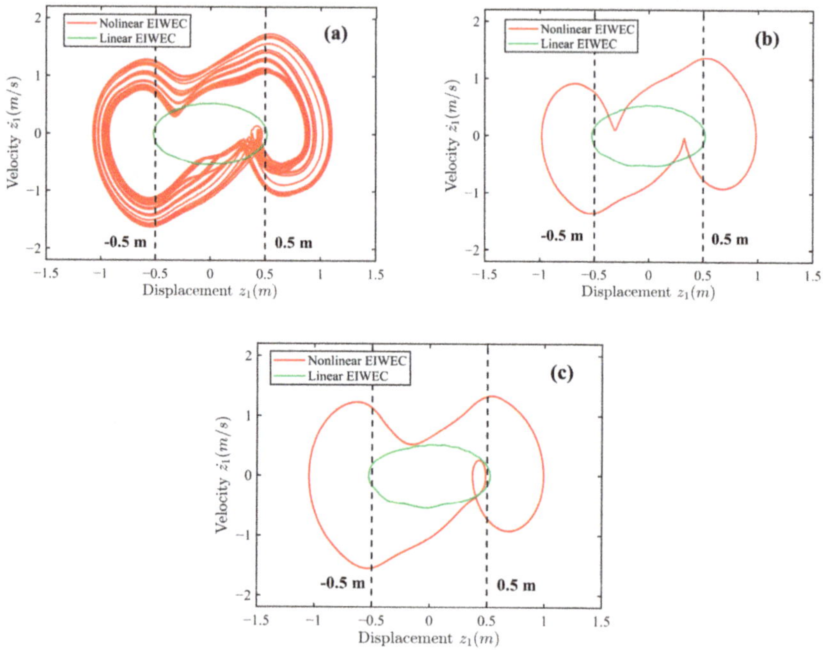

Figure 11. Phase trajectories of the buoys of linear and nonlinear EIWEC for different stiffness conditions when the wave height is 1 m and the frequency is 1 rad/s. (**a**) $\Delta l = 1$ m; (**b**) $\Delta l = 2$ m; (**c**) $\Delta l = 3$ m.

For the nonlinear EIWEC, the velocity of the buoy changes drastically at these two positions, which could cause fluctuations in the energy output. Interestingly, these stiffness variations seem to bring remarkable gains to the system. For the linear EIWEC, the motion response of the buoy is limited to these two threshold intervals. Due to the stiffness variations, the motion response of the float increases rapidly for the nonlinear EIWEC, which favors the power output capability of the system.

3.2.2. Output Power Analysis

Figure 12a,b show the variation of output power for linear and nonlinear EIWECs under different ω and Δl, respectively. The mapping distribution of output power in the (ω, Δl) space is illustrated in Figure 12c,d. When the wave frequency is less than 1 rad/s, the output power of the linear EIWEC is nearly 0 W because of the small relative displacement z_3. The system exhibits a single resonance peak, and the output power initially increases and then decreases with the increase in wave frequency (Figure 12a). The effective frequency bandwidth of a linear EIWEC strongly correlates with the linear spring. The frequency bandwidth of the linear EIWEC gradually increases as initial elongation Δl increases (Figure 12c). Therefore, selecting a larger Δl facilitates the linear EIWEC to harvest energy from a wider range of wave frequencies and improves the system's energy capture performance. However, it is unreasonable and difficult to select a linear spring with excessive Δl for the EIWEC due to the limited internal space of the buoy.

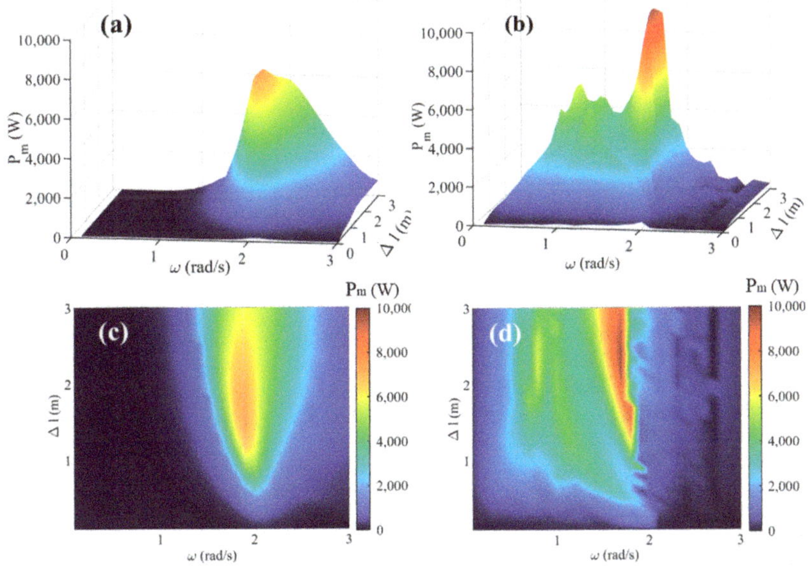

Figure 12. Distribution of generation power in the (ω, Δl) space. (**a**,**c**): Linear EIWEC; (**b**,**d**) nonlinear EIWEC. Simulation conditions: m_2 = 6670 kg, D_m= 80 cc/rev, R = 50 Ω, p_{pre} = 30 bar, and V_{g0} = 40 L.

As shown in Figure 12b, the output power of nonlinear EIWEC is noticeably improved compared to linear EIWEC. The negative stiffness of NSM reduces the intrinsic frequency of the EIWEC system, which facilitates the system to resonate more easily under low frequency wave conditions. The effective frequency band of the system shifts to a lower frequency range, and the output power and frequency bandwidth of the EIWEC is considerably enhanced. Additionally, the NSM decreases the correlation between the linear spring stiffness and the frequency bandwidth. In the frequency range from 0.4 to 1.5 rad/s, the output power of the system is insensitive to changes in the wave frequencies and linear spring stiffness. These characteristics increase the power output robustness of the EIWEC against sea state changes and reduce the design difficulty of the linear spring.

3.3. *Effect of Mass Body and Linear Spring*

The mass body and linear spring of EIWEC are the critical components for energy harvesting and conversion. This section investigates the effect of mass ratio γ of the internal mass body and initial elongation Δl of the linear spring on the performance of the nonlinear EIWEC. The mass ratio γ is defined by Equation (27). Ocean waves typically have periods between 5 and 15 s (0.42 rad/s to 1.26 rad/s). Figure 13 shows the output power of nonlinear EIWEC as a function of γ and Δl under typical wave period conditions (T = 5 s, 7.5 s, 10 s, 12.5 s, and 15 s).

$$\gamma = m_2/(m_1 + m_2) \tag{27}$$

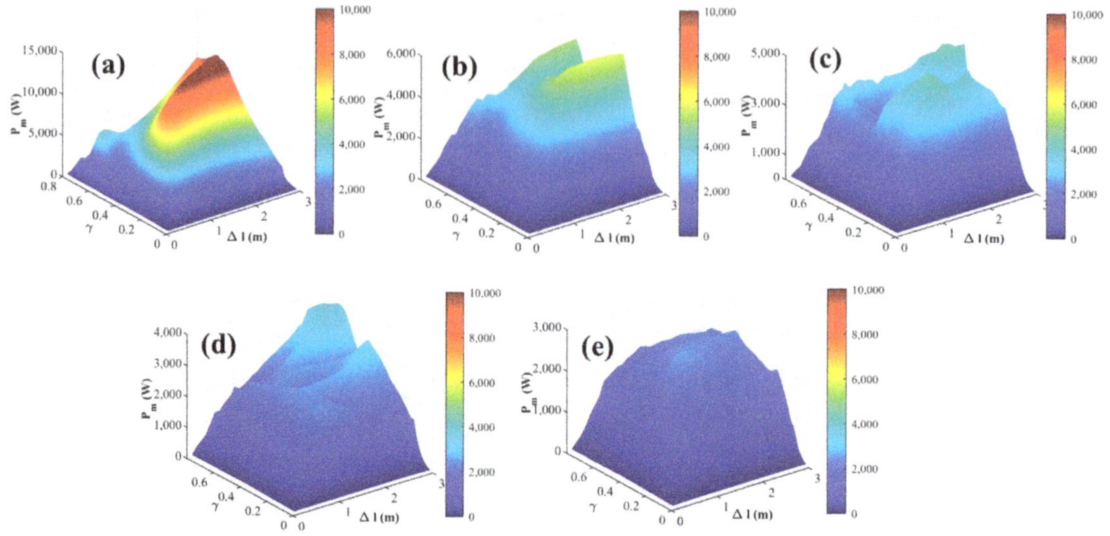

Figure 13. Output power of nonlinear EIWECs as a function of γ and Δl under typical wave period conditions. (**a**) T = 5 s, (**b**) T = 7.5 s, (**c**) T = 10 s, (**d**) T = 12.5 s, and (**e**) T = 15 s.

There are some differences in the power distribution of the system under various wave periods. As illustrated in Figure 13a, when the wave period is 5 s, nonlinear EIWEC can obtain higher output power by selecting a larger mass ratio γ and initial elongation Δl. Within a reasonable parameter range, increasing the mass of the internal mass body and decreasing the stiffness of the linear spring will have a positive effect on enhancing the energy conversion performance of the system. According to Figures 9 and 12, the system is close to resonance at this time, and the influence of the NSM on the motion response and output power is reduced. The motion response of the system is close to optimal and increasing the mass of the internal mass body can convert more wave energy into available kinetic energy. Therefore, in short-period wave scenarios, the nonlinear EIWEC is more sensitive to parameter variations in the internal mass body and linear spring. There are differences in wave conditions in different deployment areas. It is necessary to first determine the dominant wave frequency of the target deployment area. After that, the structural parameters (γ and Δl) will be determined according to this frequency to ensure maximum power output.

As the wave period increases to 7.5 s, 10 s, and 12.5 s, multiple high-power regions (red areas) appear in Figure 13b,c, and these high-power regions extend into the parameter domain of the low mass ratio and initial elongation. These results suggest that efficient energy conversion can also be achieved by choosing low mass ratios and large spring stiffnesses in seas dominated by medium wave periods, significantly reducing the design difficulty of mass bodies and linear springs. When the wave period is 15 s, the system's average power is insensitive to changes in the mass body and the linear spring. Nonlinear EIWECs with different mass ratios and initial elongations have similar energy conversion capabilities. At this time, the NSM plays a critical role in adjusting the intrinsic frequency of the system. When different internal mass bodies and linear springs are selected, the nonlinear negative stiffness characteristics of the NSM can adaptively modify the motion response to ensure the stable output of the system. Accordingly, the choice of parameters will become more convenient under long-period wave conditions.

3.4. Influence of Hydraulic PTO Parameters

3.4.1. Effect of Pre-Charged Pressure of Accumulator

The pre-charged pressure p_{pre} and initial gas volume V_{g0} are two critical parameters of an accumulator. Choosing different p_{pre} and V_{g0} affects the buffer capacity of the hydraulic PTO system, which will further change the efficiency and stability of the energy conversion and output. Sections 3.4.1 and 3.4.2 investigated the effect of the precharged pressure and initial gas volume of the accumulator on the average output power P_m and power standard deviation σ, respectively. Among them, the power standard deviation is used to measure the stability of the output power.

Figure 14a,b present the average output power P_m and power standard deviation σ as a function of the pre-charged pressure of accumulator p_{pre}, respectively. The output power of the system remains constant and the power standard deviation is relatively low when the pre-charged pressure p_{pre} is below 30 bar. Notably, a low σ means less pressure and flow fluctuations in the hydraulic PTO system and more stable output power produced by the generator.

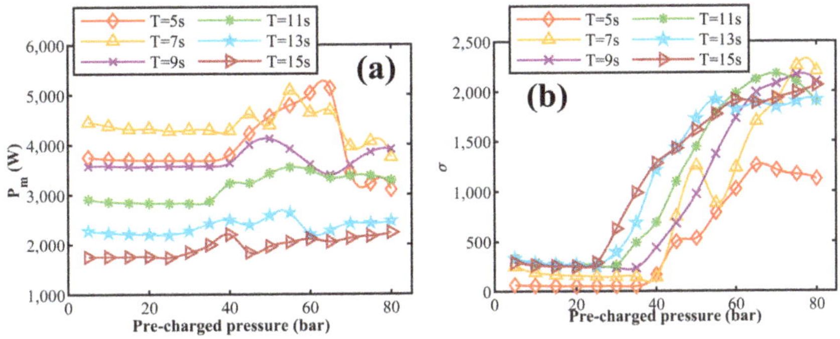

Figure 14. Average output power P_m and power standard deviation σ as a function of the pre-charged pressure of accumulator p_{pre}. (**a**) Average output power, (**b**) power standard deviation.

When p_{pre} was greater than 30 bar, the average output power changed significantly under all wave period conditions, and P_m showed an overall trend of increasing and then decreasing. As p_{pre} increases, the power standard deviation increases dramatically, and the electrical output stability of the PTO system deteriorates. These results suggest that choosing a lower pre-charged pressure of the accumulator can allow the pulsations in the hydraulic PTO system to be fully absorbed, which facilitates the improvement of the electric energy quality in nonlinear EIWEC. Although increasing the pre-charged pressure can improve the power generation of the system to some extent, it weakens the stability of the power output.

To further investigate the influence mechanism of the pre-charged pressure on the stability characteristics of the output power, Figure 15 illustrates the time-varying curves of accumulator inlet pressure q_{ac}, instantaneous power P_{Gi}, and PTO force F_{PTO} when p_{pre} = 20 bar and 60 bar, respectively. Due to the introduction of the NSM, the change in relative displacement z_3 increases, resulting in sharp pressure and flow pulsations in the hydraulic PTO system. When p_{pre} is set to 20 bar, the q_{ac} is steadily maintained at 40 bar (Figure 15a). The oil in the hydraulic line can flow continuously and smoothly into the hydraulic motor, ensuring the stability of generation power (Figure 15b). Meanwhile, the constant pressure allows the hydraulic system to provide smooth PTO force (Figure 15c), helping to improve the work reliability and service life of the hydraulic cylinder. When p_{pre} = 60 bar, the accumulator can only open and release the oil to flow into the hydraulic motor when the system pressure accumulates to 60 bar. As illustrated in Figure 15a,b, the excessive p_{pre} causes discontinuity in the oil flow and intermittent operating issues in the PTO system. Consequently, the power output of the generator exhibits unstable fluctuations resembling a sawtooth pattern. Additionally, the intermittent pulsations of pressure and flow can cause high-frequency vibration in the hydraulic components potentially compromising the reliability of the sealing elements and valve body.

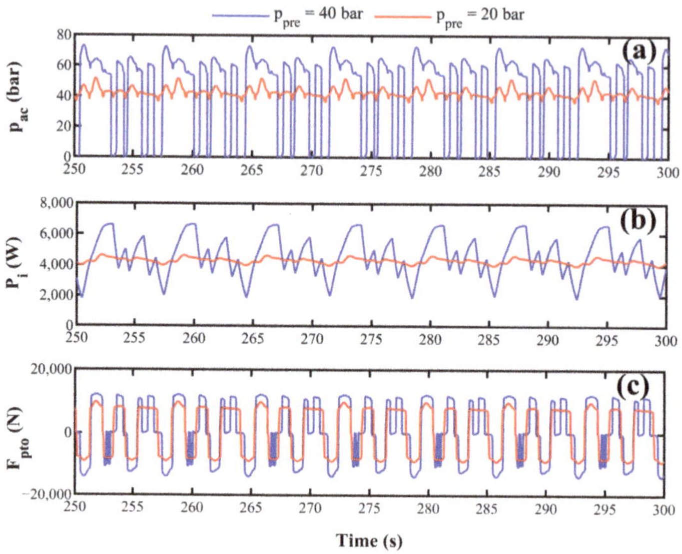

Figure 15. Time-varying curves corresponding to the pre-charged pressures of 20 bar and 60 bar when the wave height is 1 m and the period is 7 s. (**a**) accumulator inlet pressure q_{ac}, (**b**) instantaneous power P_i, and (**c**) PTO force F_{PTO}.

3.4.2. Effect of the Initial Gas Volume of the Accumulator

The initial gas volume of the accumulator V_{g0} plays the role of a buffer spring. Different initial gas volumes can affect the absorption effect of flow and pressure pulsations in the hydraulic system and change the dynamics and power output characteristics of nonlinear EIWEC. Figure 16a,b show the variation in the average output power and power standard deviation with the initial gas volume of accumulator V_{g0}, respectively.

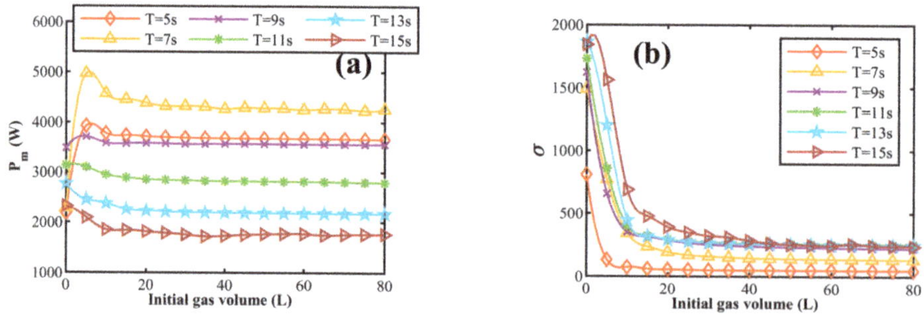

Figure 16. Variation in the average output power and power standard deviation with the initial gas volume of accumulator V_{g0}. (**a**) Average output power, (**b**) power standard deviation.

As shown in Figure 16a, under short period wave conditions (T = 5 s, 7 s, and 9 s), as V_{g0} increases, the output power of the system first increases and then decreases and finally tends to a constant. For medium and long period waves (T = 11 s, 13 s, and 15 s), the effect of initial gas volume on output power is relatively small. The increase in V_{g0} causes the average output power to decrease slightly and quickly tends towards a constant value. According to Figure 16b, when V_{g0} is less than 10 L, the power standard deviation is relatively high, which indicates that the output power generated by the PTO system has a high degree of volatility. This is mainly because the accumulators with the small V_{g0} are of limited buffer capacity and incapable of adequately absorbing pressure and flow fluctuations in a hydraulic PTO system. The pressure and flow fluctuations will cause the hydraulic motor and generator speeds to vary dramatically, thus generating an unstable power output.

With the increase in V_{g0}, the buffer capacity of the accumulator is enhanced, and the power standard deviation decreases rapidly to a constant value. This indicates that increasing the initial gas volume of the accumulator V_{g0} helps reduce electricity fluctuations and improve output power stability in the hydraulic PTO system. Accordingly, under the premise of meeting the installation space requirements, selecting an accumulator with a larger initial gas volume can enhance the quality of electrical output. For the proposed nonlinear EIWEC system, an accumulator gas volume of at least 10 L is required to ensure optimal energy conversion performance.

3.4.3. Effect of Diameter of Throttle Valve

The throttle valve is installed at the inlet of the hydraulic motor to control the flow of the hydraulic system. Figure 17 shows the average output power of the nonlinear EIWEC as a function of diameter of throttle valve D_v.

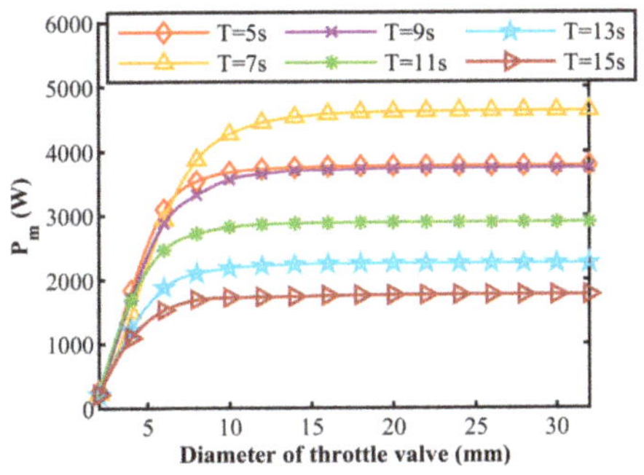

Figure 17. Average output power of the nonlinear EIWEC as a function of throttle diameter D_v. Simulation conditions: m_2 = 6670 kg, D_m = 80 cc/rev, R = 50 Ω, p_{pre} = 25 bar, and V_{g0} = 40 L.

As D_v increases, the output power gradually increases and then remains constant. When the throttle diameter is less than 10 mm, the control effect of the throttle is more pronounced, and the output power of the system is proportional to the diameter of the throttle valve. Within this parameter range, changing the value of D_v can regulate the flow rate of the hydraulic system and the speed of the hydraulic motor to control the electrical performance of the nonlinear EIWEC system. As D_v continues to increase, the through-flow area of the throttle valve increases. The local pressure drops caused by the throttle valve gradually decrease, and the flow rate and output power of the hydraulic system tend to stabilize.

3.4.4. Effect of Displacement of the Hydraulic Motor and Load Resistance

The displacement of the hydraulic motor and the load resistance directly affect the torque of the generator, which in turn changes the power output and dynamic response of the PTO system. Therefore, this subsection further investigates the effect of the displacement of hydraulic motor D_m and the load resistance R on system performance.

Figure 18 shows the variation in the output power of nonlinear EIWEC with wave frequency under different D_m and R conditions. When R = 10 Ω, increasing D_m helps to improve the energy conversion performance of the nonlinear EIWEC. As D_m increases, the output power of the system gradually rises. Meanwhile, the nonlinear EIWEC system transitions from a single-peak frequency response to a multi-peak frequency response effectively expanding the frequency bandwidth and work efficiency of the system. Consequently, the nonlinear EIWEC can exhibit stronger robustness in the face of the change in wave frequencies.

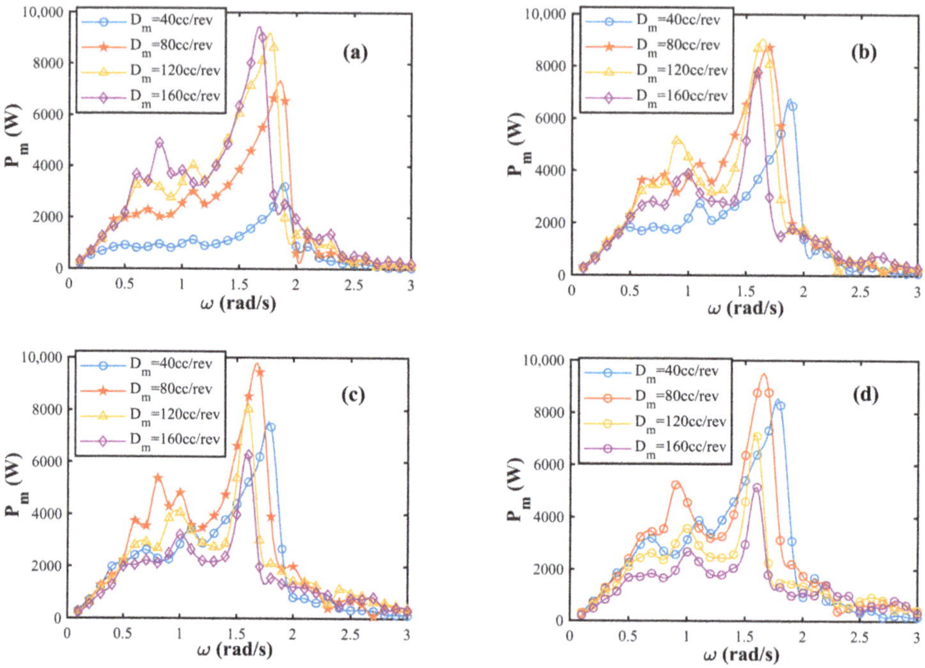

Figure 18. Variation in the output power of nonlinear EIWEC with wave frequency under different displacements of the hydraulic motor and load resistance conditions. (**a**) 10 Ω, (**b**) 30 Ω, (**c**) 50 Ω, and (**d**) 70 Ω. Simulation conditions: m_2 = 6670 kg, p_{pre} = 25 bar, and V_{g0} = 40 L.

As the load resistance increases, the output performance corresponding to small motor displacements (40 cc/rev and 80 cc/rev) gradually improves. Conversely, the energy conversion capability of large motor displacements (120 cc/rev and 160 cc/rev) gradually declines. These findings indicate the existence of optimal matching values between the displacement of the hydraulic motor and load resistance for the nonlinear EIWEC. Increasing the hydraulic motor displacement enhances the energy conversion performance when the equivalent resistance of the external load is low. Conversely, lowering D_m will be advantageous when the equivalent resistance of the external load is large.

4. Conclusions

This study proposed a nonlinear EIWEC. A mathematical model of the nonlinear EIWEC was established based on the Cummins equation and the equivalent magnetic charge method, and the joint simulations were carried out using MATLAB/Simulink 2020 and AMESim 2020 software. Performance evaluation and motion response analysis were conducted for linear and nonlinear EIWECs. The influence of different structural and hydraulic PTO parameters on the power performance of the system was analyzed to determine reasonable design parameters. The key conclusions are presented as follows:

(1) Compared to linear EIWEC, the introduction of NSM increased the motion response of nonlinear EIWEC. The large amplitude relative motion provided favorable conditions for energy conversion.

(2) The nonlinear negative stiffness property of the NSM reduced the intrinsic frequency and broadened the frequency bandwidth of the EIWEC. The effective frequency band shifted to a lower frequency range, and the output power of the nonlinear EIWEC was considerably enhanced.
(3) Within the typical wave frequency range, the output power of the nonlinear EIWEC is insensitive to changes in the wave frequencies and linear spring stiffness. This not only increases the robustness of the system but also reduces the design difficulty of the linear spring.
(4) Increasing the mass of the internal mass body and decreasing the stiffness of the linear spring will have a positive effect on enhancing the power performance of the system.
(5) Choosing an accumulator with lower pre-charged pressure and larger gas initial volume can fully absorb the pressure and flow pulsations in the hydraulic PTO system, which is favorable to the smooth power output.

This study demonstrates the advantages of nonlinear stiffness mechanisms for application in enclosed inertial wave energy converters, and it focuses on parametric and performance analysis of the nonlinear EIWEC system. In the future, we will conduct wave-tank tests to confirm the reliability of the simulation results.

Author Contributions: Conceptualization, Methodology, Writing, and Editing, J.Q.; Investigation and Validation, Z.Z.; Investigation and Validation, X.S.; Reviewing, S.H.; Reviewing and Supervision, Y.L.; Reviewing and Editing, G.X. All authors have read and agreed to the published version of the manuscript.

Funding: This research was supported by the National Natural Science Foundation of China (52171265, U22A20242, 52001186) and the Natural Science Foundation of Shandong Province (ZR2020QE292, ZR2021ZD23).

Institutional Review Board Statement: Not applicable.

Informed Consent Statement: Not applicable.

Data Availability Statement: Data are contained within the article.

Conflicts of Interest: The authors declare no conflicts of interest.

References

1. Coyle, E.D.; Simmons, R.A. *Understanding the Global Energy Crisis*; Purdue University Press: West Lafayette, IN, USA, 2014; ISBN 1557536619.
2. Bodansky, D. The United Nations framework convention on climate change: A commentary. *Yale J. Int'l L.* **1993**, *18*, 451.
3. Schleussner, C.; Rogelj, J.; Schaeffer, M.; Lissner, T.; Licker, R.; Fischer, E.M.; Knutti, R.; Levermann, A.; Frieler, K.; Hare, W. Science and policy characteristics of the Paris Agreement temperature goal. *Nat. Clim. Chang.* **2016**, *6*, 827–835. [CrossRef]
4. Davis, M.; Moronkeji, A.; Ahiduzzaman, M.; Kumar, A. Assessment of renewable energy transition pathways for a fossil fuel-dependent electricity-producing jurisdiction. *Energy Sustain. Dev.* **2020**, *59*, 243–261. [CrossRef]
5. Vidal-Amaro, J.J.; Østergaard, P.A.; Sheinbaum-Pardo, C. Optimal energy mix for transitioning from fossil fuels to renewable energy sources—The case of the Mexican electricity system. *Appl. Energy* **2015**, *150*, 80–96. [CrossRef]
6. Halkos, G.E.; Gkampoura, E. Reviewing usage, potentials, and limitations of renewable energy sources. *Energies* **2020**, *13*, 2906. [CrossRef]
7. Guo, B.; Wang, T.; Jin, S.; Duan, S.; Yang, K.; Zhao, Y. A review of point absorber wave energy converters. *J. Mar. Sci. Eng.* **2022**, *10*, 1534. [CrossRef]
8. Yue, W.; Wang, Z.; Ding, W.; Sheng, S.; Zhang, Y.; Huang, Z.; Wang, W. Feasibility of Co-locating wave energy converters with offshore aquaculture: The Pioneering case study of China's Penghu platform. *Ocean Eng.* **2023**, *288*, 116039. [CrossRef]
9. Clemente, D.; Rosa-Santos, P.; Ferradosa, T.; Taveira-Pinto, F. Wave energy conversion energizing offshore aquaculture: Prospects along the Portuguese coastline. *Renew. Energy* **2023**, *204*, 347–358. [CrossRef]
10. Clemente, D.; Rosa-Santos, P.; Taveira-Pinto, F. On the potential synergies and applications of wave energy converters: A review. *Renew. Sustain. Energy Rev.* **2021**, *135*, 110162. [CrossRef]
11. Gallutia, D.; Fard, M.T.; Soto, M.G.; He, J. Recent advances in wave energy conversion systems: From wave theory to devices and control strategies. *Ocean Eng.* **2022**, *252*, 111105. [CrossRef]
12. Tiron, R.; Mallon, F.; Dias, F.; Reynaud, E.G. The challenging life of wave energy devices at sea: A few points to consider. *Renew. Sustain. Energy Rev.* **2015**, *43*, 1263–1272. [CrossRef]

13. Xue, G.; Qin, J.; Zhang, Z.; Huang, S.; Liu, Y. Experimental Investigation of Mooring Performance and Energy-Harvesting Performance of Eccentric Rotor Wave Energy Converter. *J. Mar. Sci. Eng.* **2022**, *10*, 1774. [CrossRef]
14. Cordonnier, J.; Gorintin, F.; De Cagny, A.; Clément, A.H.; Babarit, A. SEAREV: Case study of the development of a wave energy converter. *Renew. Energy* **2015**, *80*, 40–52. [CrossRef]
15. Pozzi, N.; Bracco, G.; Passione, B.; Sirigu, S.A.; Mattiazzo, G. PeWEC: Experimental validation of wave to PTO numerical model. *Ocean Eng.* **2018**, *167*, 114–129. [CrossRef]
16. Khedkar, K.; Nangia, N.; Thirumalaisamy, R.; Bhalla, A.P.S. The inertial sea wave energy converter (ISWEC) technology: Device-physics, multiphase modeling and simulations. *Ocean Eng.* **2021**, *229*, 108879. [CrossRef]
17. The Penguin Wave Energy Converter. Available online: https://wello.eu/the-penguin-2/ (accessed on 9 January 2024).
18. Clemente, D.; Rosa-Santos, P.; Taveira-Pinto, F.; Martins, P. Influence of platform design and power take-off characteristics on the performance of the E-Motions wave energy converter. *Energy Conv. Manag.* **2021**, *244*, 114481. [CrossRef]
19. Scarcity of Resources, Climate Change. Available online: https://seaturns.com/ (accessed on 9 January 2024).
20. Crowley, S.; Porter, R.; Taunton, D.J.; Wilson, P.A. Modelling of the WITT wave energy converter. *Renew. Energy* **2018**, *115*, 159–174. [CrossRef]
21. Maheen, M.H.; Yang, Y. Wave energy converters with rigid hull encapsulation: A review. *Sustain. Energy Technol. Assess.* **2023**, *57*, 103273.
22. Clemente, D.; Rosa-Santos, P.; Taveira-Pinto, F.; Martins, P. Experimental performance assessment of geometric hull designs for the E-Motions wave energy converter. *Ocean Eng.* **2022**, *260*, 111962. [CrossRef]
23. Ding, W.; Wang, K.; Mao, Z.; Cao, H. Layout optimization of an inertial energy harvester for miniature underwater mooring platforms. *Mar. Struct.* **2020**, *69*, 102681. [CrossRef]
24. Chen, Z.; Zhang, L.; Yeung, R.W. Analysis and optimization of a Dual Mass-Spring-Damper (DMSD) wave-energy convertor with variable resonance capability. *Renew. Energy* **2019**, *131*, 1060–1072. [CrossRef]
25. Falnes, J.; Kurniawan, A. *Ocean Waves and Oscillating Systems: Linear Interactions Including Wave-Energy Extraction*; Cambridge University Press: Cambridge, UK, 2020; ISBN 1108481663.
26. Khasawneh, M.A.; Daqaq, M.F. Response behavior of bi-stable point wave energy absorbers under harmonic wave excitations. *Nonlinear Dyn.* **2022**, *109*, 371–391. [CrossRef]
27. Khasawneh, M.A.; Daqaq, M.F. Experimental assessment of the performance of a bi-stable point wave energy absorber under harmonic incident waves. *Ocean Eng.* **2023**, *280*, 114494. [CrossRef]
28. Qin, J.; Zhang, Z.; Zhang, Y.; Huang, S.; Liu, Y.; Xue, G. Design and performance evaluation of novel magnetic tristable wave energy converter. *Ocean Eng.* **2023**, *285*, 115424. [CrossRef]
29. Kurniawan, A.; Greaves, D.; Chaplin, J. Wave energy devices with compressible volumes. *Proc. R. Soc. A Math. Phys. Eng. Sci.* **2014**, *470*, 20140559. [CrossRef] [PubMed]
30. Pecher, A.; Kofoed, J.P. *Handbook of Ocean Wave Energy*; Springer Nature: Berlin, Germany, 2017.
31. Bracco, G.; Canale, M.; Cerone, V. Optimizing energy production of an inertial sea wave energy converter via model predictive control. *Control Eng. Pract.* **2020**, *96*, 104299. [CrossRef]
32. Vissio, G.; Valério, D.; Bracco, G.; Beirão, P.; Pozzi, N.; Mattiazzo, G. ISWEC linear quadratic regulator oscillating control. *Renew. Energy* **2017**, *103*, 372–382. [CrossRef]
33. Salcedo, F.; Ruiz-Minguela, P.; Rodriguez, R.; Ricci, P.; Santos, M. Oceantec: Sea trials of a quarter scale prototype. In Proceedings of the 8th European Wave and Tidal Energy Conference, Uppsala, Sweden, 7–10 September 2009; pp. 460–465.
34. Sirigu, S.A.; Bracco, G.; Bonfanti, M.; Dafnakis, P.; Mattiazzo, G. On-board sea state estimation method validation based on measured floater motion. *IFAC-PapersOnLine* **2018**, *51*, 68–73. [CrossRef]
35. Scapolan, M.; Tehrani, M.G.; Bonisoli, E. Energy harvesting using parametric resonant system due to time-varying damping. *Mech. Syst. Signal Proc.* **2016**, *79*, 149–165. [CrossRef]
36. Giorgi, G.; Faedo, N. Performance enhancement of a vibration energy harvester via harmonic time-varying damping: A pseudospectral-based approach. *Mech. Syst. Signal Proc.* **2022**, *165*, 108331. [CrossRef]
37. Yurchenko, D.; Alevras, P. Dynamics of the N-pendulum and its application to a wave energy converter concept. *Int. J. Dyn. Control* **2013**, *1*, 290–299. [CrossRef]
38. Giorgi, G. Embedding parametric resonance in a 2:1 wave energy converter to get a broader bandwidth. *Renew. Energy* **2023**, *222*, 119928. [CrossRef]
39. Schubert, B.W.; Sergiienko, N.Y.; Cazzolato, B.S.; Robertson, W.S.; Ghayesh, M.H. The true potential of nonlinear stiffness for point absorbing wave energy converters. *Ocean Eng.* **2022**, *245*, 110342. [CrossRef]
40. Guo, B.; Ringwood, J.V. Non-linear modeling of a vibro-impact wave energy converter. *IEEE Trans. Sustain. Energy* **2020**, *12*, 492–500. [CrossRef]
41. Zhang, H.; Xi, R.; Xu, D.; Wang, K.; Shi, Q.; Zhao, H.; Wu, B. Efficiency enhancement of a point wave energy converter with a magnetic bistable mechanism. *Energy* **2019**, *181*, 1152–1165. [CrossRef]
42. Zhang, X.; Tian, X.; Xiao, L.; Li, X.; Chen, L. Application of an adaptive bistable power capture mechanism to a point absorber wave energy converter. *Appl. Energy* **2018**, *228*, 450–467. [CrossRef]
43. Liu, B.; Yi, H.; Levi, C.; Estefen, S.F.; Wu, Z.; Duan, M. Improved bistable mechanism for wave energy harvesting. *Ocean Eng.* **2021**, *232*, 109139. [CrossRef]

4. Wang, L.; Tang, H.; Wu, Y. On a submerged wave energy converter with snap-through power take-off. *Appl. Ocean Res.* **2018**, *80*, 24–36. [CrossRef]
5. Todalshaug, J.H.; ásgeirsson, G.S.; Hjálmarsson, E.; Maillet, J.; Möller, P.; Pires, P.; Guérinel, M.; Lopes, M. Tank testing of an inherently phase-controlled wave energy converter. *Int. J. Mar. Energy* **2016**, *15*, 68–84. [CrossRef]
6. Corpower. Available online: https://www.corpowerocean.com/ (accessed on 9 January 2024).
7. Li, M.; Jing, X. A bistable X-structured electromagnetic wave energy converter with a novel mechanical-motion-rectifier: Design, analysis, and experimental tests. *Energy Conv. Manag.* **2021**, *244*, 114466. [CrossRef]
8. Wang, Y.; Huang, S.; Xue, G.; Liu, Y. Influence of Hydraulic PTO Parameters on Power Capture and Motion Response of a Floating Wind-Wave Hybrid System. *J. Mar. Sci. Eng.* **2022**, *10*, 1660. [CrossRef]
9. Choi, K.; Yang, D.; Park, S.; Cho, B. Design and performance test of hydraulic PTO for wave energy converter. *Int. J. Precis. Eng. Manuf.* **2012**, *13*, 795–801. [CrossRef]
0. Gao, H.; Xiao, J. Effects of power take-off parameters and harvester shape on wave energy extraction and output of a hydraulic conversion system. *Appl. Energy* **2021**, *299*, 117278. [CrossRef]
1. Liu, C.; Hu, M.; Zhao, Z.; Zeng, Y.; Gao, W.; Chen, J.; Yan, H.; Zhang, J.; Yang, Q.; Bao, G. Latching control of a raft-type wave energy converter with a hydraulic power take-off system. *Ocean Eng.* **2021**, *236*, 109512. [CrossRef]
2. Xue, G.; Zhang, Z.; Qin, J.; Huang, S.; Liu, Y. Control Parameters Optimization of Accumulator in Hydraulic Power Take-Off System for Eccentric Rotating Wave Energy Converter. *J. Mar. Sci. Eng.* **2023**, *11*, 792. [CrossRef]
3. Zhang, N.; Zhang, X.; Xiao, L.; Wei, H.; Chen, W. Evaluation of long-term power capture performance of a bistable point absorber wave energy converter in South China Sea. *Ocean Eng.* **2021**, *237*, 109338. [CrossRef]
4. Newman, J.N. The exciting forces on fixed bodies in waves. *J. Ship Res.* **1962**, *6*, 10–17. [CrossRef]
5. Cummins, W.E. *The Impulse Response Function and Ship Motions*; David Taylor Model Basin: Washington, DC, USA, 1962.
6. Ogden, D.; Ruehl, K.; Yu, Y.; Keester, A.; Forbush, D.; Leon, J.; Tom, N. Review of WEC-Sim development and applications. *Int. Mar. Energy J.* **2022**, *5*, 5–9. [CrossRef]
7. Tampier, G.; Grueter, L. Hydrodynamic analysis of a heaving wave energy converter. *Int. J. Mar. Energy* **2017**, *19*, 304–318. [CrossRef]

Journal of Marine Science and Engineering

Article

Study on Mooring Design of 15 MW Floating Wind Turbines in South China Sea

Mingsheng Chen [1,2], Jiale Jiang [1,2], Wei Zhang [3], Chun Bao Li [1,2,*], Hao Zhou [4], Yichen Jiang [5] and Xinghan Sun [3]

1 Key Laboratory of High Performance Ship Technology (Wuhan University of Technology), Ministry of Education, Wuhan 430063, China; mschen@whut.edu.cn (M.C.); le962359857@163.com (J.J.)
2 School of Naval Architecture, Ocean and Energy Power Engineering, Wuhan University of Technology, Wuhan 430063, China
3 Powerchina (Xi'an) Port Navigation Shipbuliding Technology Co., Ltd., Xian 710089, China; zhangwei@nwh.cn (W.Z.); sunxingh@nwh.cn (X.S.)
4 China Ship Scientific Research Center, Wuxi 214000, China; zhouhao@cssrc.com.cn
5 School of Naval Architecture, Dallian University of Technology, Dalian 116024, China; ycjiang@dlut.edu.cn
* Correspondence: leecubao@whut.edu.cn; Tel.: +86-186-7291-6807

Abstract: Wind turbines and floating platform upsizing are major trends in the current offshore wind development. However, harsh environmental conditions increase the risk of anchor dragging and mooring failure when deploying large offshore floating wind turbines. Therefore, it is necessary to design a mooring system for the specific deployment site. This study aims to perform the mooring system design of a floating offshore wind turbine (FOWT) operated in the South China Sea, which is a combination of the IEA 15 MW wind turbine and UMaine VolturnUS-S floating platform. Hydrodynamic coefficients were calculated based on the potential flow theory, considering the environmental loads in the South China Sea. Additionally, the hydrodynamic coefficients were imported into AQWA to calculate the time-domain mooring tension. The mooring design parameters, such as mooring line length, nominal sizes, and anchor point, were determined using the criterion of anchor uplift, maximum breaking strength, and fatigue life, respectively. The design criterion required that the anchor uplift is not more than the allowable value, the long-term breaking limit of mooring with a 100-year return period should be less than the maximum breaking limit, and the fatigue damage accumulation in 50 years should be safe. The mooring design procedure provides a reference for mooring system design and safe operation of large floating wind turbines in the South China Sea.

Keywords: 15 MW offshore wind turbine; mooring design; floating foundation; fatigue damage; maximum breaking limit

Citation: Chen, M.; Jiang, J.; Zhang, W.; Li, C.B.; Zhou, H.; Jiang, Y.; Sun, X. Study on Mooring Design of 15 MW Floating Wind Turbines in South China Sea. *J. Mar. Sci. Eng.* **2024**, *12*, 33. https://doi.org/10.3390/jmse12010033

Academic Editor: José António Correia

Received: 13 November 2023
Revised: 17 December 2023
Accepted: 19 December 2023
Published: 22 December 2023

1. Introduction

The development and utilization of offshore wind energy around the world in order to promote the further development of global renewable energy and satisfy the requirements of carbon emission limitation. According to the Global Wind Report 2023, China will continue to lead the global wind energy development [1]. Without a doubt, FOWT will flourish in China. Meanwhile, the subsidy of electricity prices for offshore wind power generation is reduced. Wind turbines and floating platform upsizing are major trends in the current offshore wind development to reduce costs and improve power generation efficiency [2]. According to the statistical data of the International Energy Agency (IEA) on wind turbines in the last 30 years, the parameters such as tower height and blade length of wind turbines are constantly increasing with the increase in rated power. Firstly, the tower height is increased to capture wind resources with less turbulence and higher speed. Secondly, the larger blades can generate more electricity to reduce the power generation costs. However, the larger wind turbines and more complex environmental conditions have higher requirements for the mooring system. For the above reasons, many scholars have carried out research on floating wind turbine mooring systems. Chen et al. present a simulation technology that controls the mooring force in real time based on the displacement of mooring nodes and calculates the mooring stress response using the finite element method. It is beneficial to further investigate the impacts of the fluid loads, the mooring geometry, and material nonlinearity on the real-time structural responses of mooring lines during their service life [3]. Barrera et al. have presented a groundbreaking technique for evaluating the fatigue of mooring systems for floating offshore wind turbines that takes into account the entire lifespan of the structure [4]. Li and Choung used an artificial neural network to predict fatigue damage in the mooring lines of a floating offshore wind turbine, in which fatigue damage can be accurately predicted in the time domain [5]. Wang et al. proposed a new method for predicting mooring tension based on a neural network [6]. Campanile and colleagues study mooring design and selection for NREL 5 MW floating offshore wind turbines, considering ultimate, accidental, and fatigue limit states and providing mooring configuration selection recommendations [7]. Hall and Goupee proposed a lumped mass mooring line model, demonstrating its effectiveness in predicting the load of DeepCwind semi-submersible mooring lines [8]. Pillai and colleagues investigated the impact of single and shared anchor loads on various mooring arrangements based on the IEA 15 MW wind turbine and the VolturnUS-S reference platform subjected to environmental conditions representative of the Celtic Sea. The importance of considering operational and parked design loads is highlighted, but the fatigue analysis of mooring lines is not considered [9,10]. Xu Sheng et al. employed the T-N curve to calculate the short-term fatigue damage of point absorption wave energy devices [11]. Ahn et al. investigated the loads on a 15 MW floating wind turbine tower based on wave conditions, which in turn analyzed the relationship between the pitching moments at the top and base of the tower with the wave conditions without regard to the effect of wave conditions on the mooring system [12]. Zhao et al. analyzed the reliability of a floating wind turbine mooring system based on the environmental isoline method and studied the relationship between the extreme tension of the mooring line and environmental parameters [13]. However, harsh environmental conditions increase the risk of anchor dragging and mooring failure when deploying large offshore floating wind turbines. Therefore, it is necessary to design a mooring system for the specific deployment site [14].

The response of the floating platform determines the response of the mooring lines [15]. Chueh et al. proposed a new power control strategy. The 15 MW floating wind turbine is validated using offshore wind and wave conditions, and it was demonstrated that the control strategy can decrease the effect of platform motion on power generation [16]. Mazarakos studied the response of the 15 MW floating wind turbine under regular or irregular waves based on Mediterranean sea wave conditions [17]. Benassai et al. compared the response of semi-submersible wind turbines under catenary and tension mooring based on 5-year and 50-year return period wave data in the southern Mediterranean Sea [18]. The depth of water in the South China Sea is typically shallower compared to the North Sea in Europe, leading to distinct dynamic response features for mooring and platform [19]. For the design and safety evaluation of mooring lines, hydrodynamic dynamics analysis of the floating platform is necessary [20]. In this paper, the motion of the floating platform supporting the wind turbine is analyzed using the environmental conditions in the South China Sea. The safety of the mooring system is evaluated by the fully coupled analysis method, including current force, wind load, wave force, and mooring forces. It is important to design a reasonable mooring line length, which can effectively ensure the quasi-static response of the mooring floating structure within an acceptable level and reduce the cost of the mooring line in the overall project [21]. The mooring design parameters, such as mooring line length, nominal sizes, and anchor point, were determined by the criterion of anchor uplift, maximum breaking strength, and fatigue life, respectively. The design criterion required that the anchor uplift is zero, the long-term breaking limit of mooring with a 100-year return period should be less than the maximum breaking limit, and the fatigue damage accumulation in the full lifecycle should be safe. The mooring lines maximum tension range in the return period of 100 years is predicted using the Weibull distribution and cumulative fatigue damage based on the design life of 50 years calculated using linear cumulative damage of Miner's rule. The mooring design procedure provides a reference for mooring system design and safe operation of large floating wind turbines in the South China Sea.

2. Methodology and Design Procedure

2.1. Methodology

2.1.1. Time Domain Motion Equation

The time domain motion response of the floating offshore wind turbine depends on the wind load, hydrodynamic load, and mooring force. The time domain motion equation is as follows [22]:

$$\sum_{i=1}^{6}\left(M_{ij} + M_{ij}(\infty)\cdot x_j''(t) + \int_0^t R(t-\tau)\cdot x_j'(t)\mathrm{d}\tau + K_{ij}\cdot x_j(t)\right) = F_A^j + F_H^j + F_M^j \quad (1)$$

where M_{ij} is the structural mass matrix, $M_{ij}(\infty)$ is the fluid-added mass matrix at infinite frequency, x_{ij}'', x_{ij}' and x_{ij} are, respectively the acceleration, velocity, and displacement of the floating platform. R is the velocity impulse function matrix, and K_{ij} is the total stiffness matrix, F_A^j, F_H^j, and F_M^j are, respectively, the wind load, hydrodynamic force, and mooring force on the floating platform.

2.1.2. Weibull Distribution

The Weibull distribution is a versatile continuous probability distribution capable of fitting a wide array of distribution shapes. Similar to the normal distribution, the Weibull distribution is unimodal and describes probabilities linked to continuous data [23]. Its three parameters include shape, scale, and threshold. If the threshold is set to zero, it becomes the two-parameter Weibull distribution.

The double-parameter Weibull distribution is as follows:

$$f(R) = \frac{\beta}{q}\left(\frac{R}{q}\right)^2 \quad (2)$$

where R is the tension range, β is the shape parameter, q is the scale parameter.

2.1.3. Rain Flow Counting Method and Goodman Correction

The mooring chain tension can be obtained from the following formula:

$$A_c = \left(\pi d_c^2\right)/2 \quad (3)$$

$$\sigma_t = \frac{F_t}{A_c} \quad (4)$$

where σ_t is the stress of the mooring chain, F_t is the tension of the mooring chain at the fairlead point, d_c is the mooring chain diameter A_c is the nominal cross-sectional area of the mooring chain, which is twice the cross-sectional area of the mooring chain [4].

The rain flow counting method is commonly considered the most dependable approach for fatigue data analysis in engineering. The rain flow counting method simplifies stress-time data into complete cycles, using the two parameters of stress amplitude and mean stress to analyze fatigue damage. This is achieved using the double-parameter counting method [24].

Figure 1a shows the cycle of the rain flow counting method. The mean stress, stress amplitude, and corresponding cycle times are derived from the mooring tension curve calculated in the time domain by the rain flow counting method.

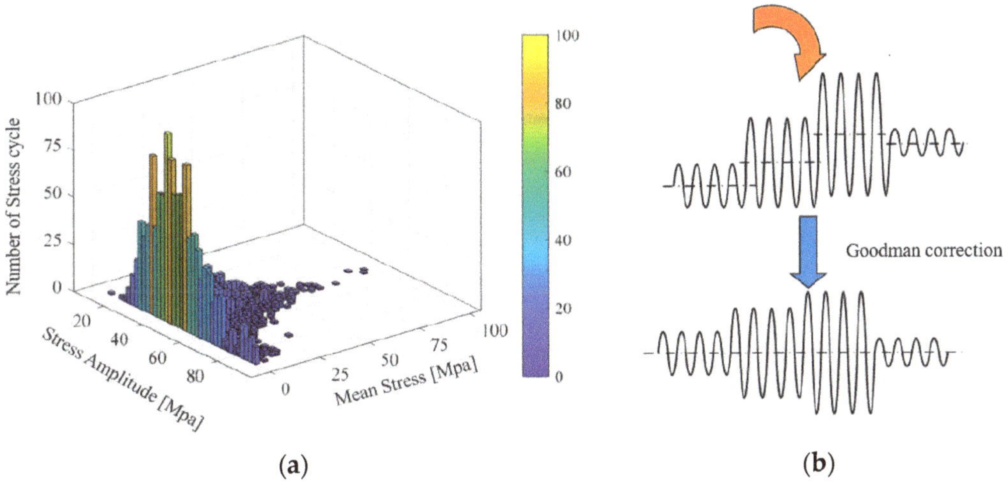

Figure 1. The rain flow counting method (**a**) and Goodman correction (**b**).

The floating offshore wind turbine is exposed to alternating loads during its operation. Mean stress and stress amplitude greatly affect the fatigue damage to the structure. To remove the effect of non-zero mean stress, the correction known as the Goodman method is applied [25]. A visual representation of the correction process is shown in Figure 1b.

$$\sigma^{RF} = \sigma_i^R \cdot \left(\frac{\sigma^{ult} - |\sigma^{MF}|}{\sigma^{ult} - |\sigma_i^M|} \right) \tag{5}$$

where σ^{RF} is the stress amplitude after Goodman correction, σ^{ult} is the ultimate stress, σ^{MF} is the mean stress, σ_i^M is the mean stress in the i-th cycle, and σ_i^R is the stress amplitude under the mean stress of σ_i^M.

2.1.4. Linear Fatigue Cumulative Damage Rule

From a macroscopic perspective, it is widely accepted that fatigue failure results from the accumulation of damage caused by fluctuating loads, specifically by each hysteresis cycle in the local stress–strain diagram. These cycles typically exhibit varying mean and amplitude, necessitating the accumulation of their respective fatigue damage. At present, the primary theories for the accumulation of fatigue damage comprise linear cumulative damage theory, bilinear cumulative damage theory, nonlinear cumulative damage theory, and additional semi-empirical formulas [26].

In this study, the fatigue damage of the mooring line of the floating offshore wind turbine was evaluated using the time-domain fatigue analysis method. The advantage of this approach is that it is not based on any assumptions. In addition, it is considered to be the most accurate approach for evaluating fatigue damage and serves as a benchmark for other methods [27]. The linear fatigue cumulative damage rule ignores the influence of loading sequence on cumulative damage, and the total damage to the structure can be approximately equal to the linear accumulation of each part of structure damage [28]. First, the mooring line tension of the floating offshore wind turbine under each sea state is obtained by coupling dynamic analysis, and then the short-term fatigue damage rate under each sea state is calculated, and then the long-term fatigue damage of the mooring lines is obtained by probability accumulation.

$$N_i = \left(\frac{\sigma^{ult} - |\sigma_i{}^{MF}|}{0.5\sigma_i{}^{RF}} \right)^m \tag{6}$$

$$D = \sum_{i=1}^{n} \frac{n_i}{N_i} / T_{\mathrm{ST}} \quad (7)$$

$$D_{\mathrm{LT}} = D \cdot T_{\mathrm{LT}} \cdot p \quad (8)$$

where N_i is the i-th stress failure cycle number, m is the Wöhler index, which is taken as 4 because of the non-corrosive single slope S-N curve. D is the short-term cumulative fatigue damage rate, n_i is the i-th stress cycle number, and n is the iteration number of the stress cycle. D_{LT} is the long-term cumulative fatigue damage, T_{ST} and T_{LT} are, respectively, the time of short-term simulation and full lifecycle, and p is the probability of environmental conditions.

2.2. Design Procedure

The design process of mooring lines is shown in Figure 2. First, after determining the basic specifications of the platform, Diffraction analysis is carried out along with the verification of the model hydrodynamic coefficient. Additionally, the initial design of mooring lines considers the type of mooring lines, initial tension, and safety of mooring lines. Based on the stability analysis of the platform with mooring lines, the tension histories of mooring lines are calculated using the motion analysis in the time domain and verification is carried out by calculating and comparing each data regarding structural strength and fatigue strength. Finally, the mooring lines are optimized by changing the anchor point position, the mooring lines radius, and length.

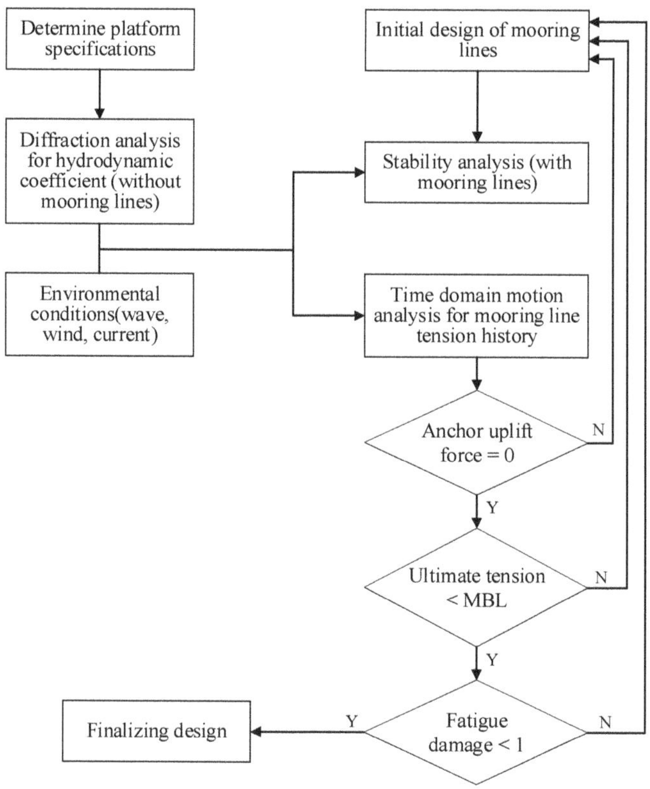

Figure 2. Design procedure of mooring lines.

3. Numerical Model and Hydrodynamic Coefficient Verification

3.1. Turbine and Platform Characteristics

In this paper, 15 MW wind turbine was selected in the wind technology cooperation project Task 37 of the International Energy Agency. Similar to the NREL 5 MW turbine, this IEA 15 MW turbine, based in part on the DTU 10 MW reference turbine, is completely defined, including the control system and using new blade construction technologies. Table 1 shows the main features of this turbine [29].

Table 1. Parameters of the IEA 15 MW Turbine.

Parameter	Value
Power	15 MW
Rotor diameter	240 m
Hub height	150 m
Hub diameter	6 m
Blade mass	65.7 t
Rotor nacelle assembly mass	1446 t
Tower mass	1211 t
Tower diameter at base	10 m
Cut-in, Rated, Cut-out Speed	3 m/s, 10.59 m/s, 25 m/s
Cut-in, Cut-out speed	4.6 rpm, 7.6 rpm

The floating platform of the present study on a four-column steel semi-submersible platform, which was developed by the University of Maine and NREL, is applied. Table 2 shows the basic characteristics of structure [30]. The hull arrangement consists of three buoyant columns with a diameter of 12.5 m, spaced radially and with centers located 51.75 m away from the vertical axis of the tower. At the center of the platform in the surge-sway plane, there is a fourth buoyant column where the platform-tower interface is located. This central column connects to the outer columns via three rectangular bottom pontoons measuring 12.5 m in width and 7.0 m in height, as well as three radial struts measuring 0.9 m in diameter, attached to the bottom and top of the buoyant columns, respectively. The floating platform model considers the quasi-static drag force on the wind turbine blade and does not consider the aerodynamic load because AQWA mainly calculates the motion of the floating platform. The mass and inertial moment of the platform are designated. The left diagram of Figure 3 shows the VolturnUS-S platform and IEA 15 MW turbine, and the right diagram shows the model of a floating platform with a mesh size of 0.8 m.

Table 2. The main characteristics of the VolturnUS-S Platform.

Parameter	Value
Hull displacement	20.206 m^3
Hull steel mass	3.914 t
Tower interface mass	100 t
Draft	20 m
Freeboard	14 m
Vertical Center of Gravity form SWL	−14.94 m
Vertical Center of Buoyancy form SWL	−13.63 m
Roll Inertia about Center of Gravity	1.251×10^{10} kg-m^2
Pitch Inertia about Center of Gravity	1.251×10^{10} kg-m^2
Yaw Inertia about Center of Gravity	2.367×10^{10} kg-m^2

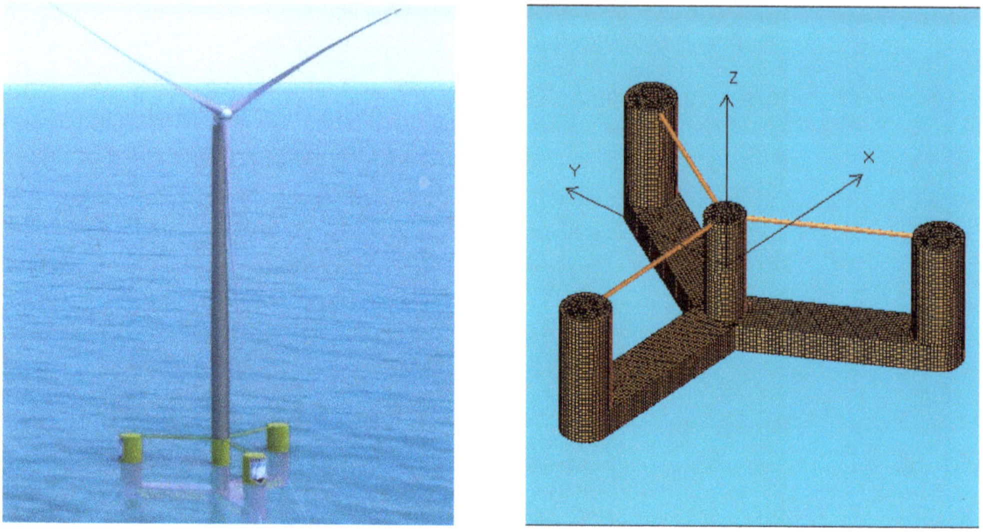

Figure 3. VolturnUS-S platform and IEA 15 MW turbine.

Figure 4 is the mesh quality validation; a comparison of near-field and far-field is carried out in this paper. In the frequency domain range of 0.05–2.55 rad/s, 25 wave frequency domains were selected with 0.1 rad/s as the discrete step, and the results of the near-field and far-field methods were carried out, respectively.

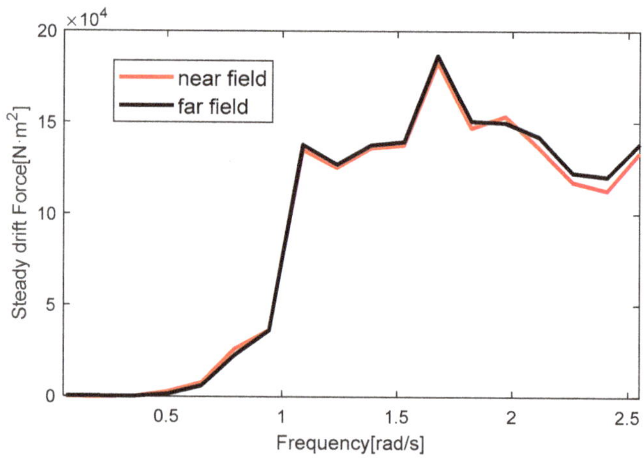

Figure 4. Mesh quality validation.

From the above figure, it can be seen that the steady drift force of the platform in the surge direction has an excellent tendency under both the near-field method and far-field method, which indicates that the platform with a mesh size of 0.8 m has better mesh quality.

3.2. Initial Mooring Design

In this paper, the depth of the water is set to 70 m, and the chain's nominal diameter is considered to be 185 mm. The mooring system selected is a studless chain, and detailed specifications are given in Table 3. According to DNV rules, the offshore structure drag coefficient of studless chain for the longitudinal and transverse 2 and 1.15 is encouraged to apply [31]. The 120-degree interval mooring line configuration is shown in Figure 5.

Table 3. Parameters of the Mooring System.

Parameter	Value
Mooring System Type	Chain Catenary
Line Type	R3 Studless Mooring Chain
Number of Lines	3
Line Breaking Strength	22,286 kN
Nominal Chain Diameter	185 mm
Dry Line Linear Density	685 kg/m
Extensional Stiffness	3270 MN
Fairlead Pretension	564 kN
Anchor Type	Gravity anchor
Anchor Weight	20 t

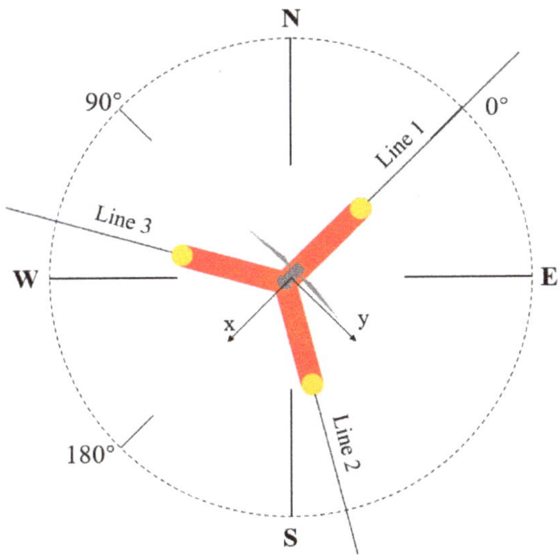

Figure 5. The configuration of mooring lines.

To determine the initial length, the equation for the catenary is applied, and the proof load of chain grade R3 is selected to determine the maximum tension according to DNV rules [20]. The initial mooring lines configuration shown in Table 4 is derived from Hazelton's guidance and iterative design using the methodology outlined in Connolly and Hall [32,33]. Additionally, the target declination angle is maintained by extending the anchor radial distance and adjusting the lengths of the lines.

Table 4. The initial Mooring configuration.

Line	Anchor Coordinates			Line Length [m]
	x [m]	y [m]	z [m]	
1	−330	0	−70	300.68
2	165	285.79	−70	300.68
3	165	−285.79	−70	300.68

The distance between the initial contact point between the mooring line and the seabed and its anchor point is known as the laid length of the mooring line. According to DNV regulations, the laid length should be designed long enough to safeguard against anchor uplift force at the anchoring point. Hence, the laid length is adjusted to prevent the anchor uplift force from being greater than the anchor gravity when environmental loads are applied. The anchor uplift force is generally stipulated not to be greater than one-fiftieth of the total gravity of the anchor and the laid mooring lines [34].

4. Mooring Design and Optimization

4.1. Environmental Condition

According to the long-term wind direction data statistics, the South China Sea is northeast wind from October to March, southwest wind from May to August, and April and September are the transition months of monsoon; the wind direction in the South China Sea is shown in Figure 6. In order to improve the efficient power of floating offshore wind turbines, it is expected to be highly utilized of the wind condition when installing wind turbines. It is most reasonable to design an angle of zero between the wind turbines and the prevailing wind direction and take the northeast as the direction of 0°. Therefore the possibility of incident direction is set as follows: 50% of the northeast wind, 33.33 of the southwest wind, and other wind directions have the same probability, which is 8.33%. The wind direction distribution and its occurrence probability are shown in Table 5.

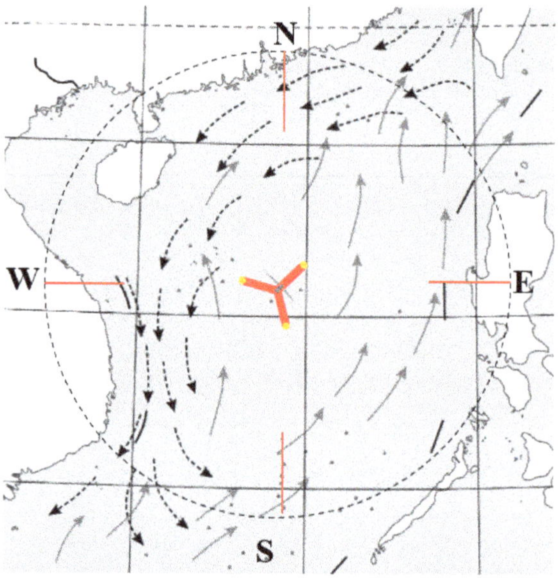

Figure 6. The wind direction in South China Sea.

Table 5. The wind direction and probability.

Direction	0°	30°	60°	90°	120°	150°	180°
Probability	50%	8.33%	8.33%	8.33%	8.33%	8.33%	33.33%

For an offshore platform operating in a specific sea area, the long-term distribution of sea conditions can be determined from the wave scatter diagram of the sea area [35]. Table 6 shows the wave scatter diagram in the South China Sea. Based on the wave scatter diagram obtained from long-term observation, significant wave height and zero-crossing period are divided into 33 girds. Each significant wave height corresponds to multiple zero-crossing periods, and the number of occurrences during the observation period is entered into the wave scatter diagram.

Table 6. Wave scatter diagram of South China Sea.

Significant Wave Height $H_S[m]$	Spectral Peak Period $T_Z[s]$				
	4.2	**5.6**	**7**	**8.4**	**9.8**
0.5	4	25	10	4	0.6
1	2	13	9.5	2	1.2
1.5		6	7	2	0.2
2		0.6	5.5	1	0.1
2.5			2	2	0.1
3			0.5	1	0.05
3.5				0.45	0.05
4				0.02	0.05
4.5				0.02	0.03
5				0.01	0.01
5.5					0.01

Waves and currents are usually caused by wind loads. To simplify the simulation calculation, it is assumed that the direction of the wind, wave, and current is the same, and the wind speed is taken as the rated wind speed of 10.59 m/s when the 15 MW wind turbine is in normal operation. Additionally, the current is a constant flow with a velocity of 1 m/s.

4.2. Optimization Based on Mooring Maximum Breaking Limit and Fatigue Damage

Due to the intricacy of the multi-body system, it is necessary to come up with comprehensive time domain simulation and statistical analysis for obtaining dependable outcomes [36]. After a high volume of time-domain simulations of the irregular waves, the extreme distribution model is employed for assessing the extreme responses. Time domain analysis was carried out using AQWA. In the wave scatter diagram of the South China Sea, a total of 33 short-term sea states exist. As a result, 264 time analyses were conducted with eight kinds of wind directions. Simulation of external forces using the floating platform motion is carried out for 3600 s, and the initial 400 s were eliminated to avoid the startup effects. Finally, the time history of mooring tension is drawn by applying the time domain analysis.

The peak of ultimate tension leads to structural failure or greatly shortens the service life [37,38]. In order to investigate ultimate strength, the effective range of mooring line tension is calculated from the tension history using the rain flow cycle counting method. Tension ranges for long-term wave conditions are extended from the probability of the wave scatter diagram. The Weibull distribution and linear regression model are used to calculate the maximum tension for a 100-year return period, and the maximum tension of each mooring line is summarized in Table 7.

Table 7. Maximum tension for a return period of 100 years.

	Tension [kN]
Line 1	1037.99
Line 2	803.54
Line 3	908.85

The cumulative fatigue damage can be assessed using tension range data, linear damage accumulation rules, and S-N curves. Assuming a 50-year design life, each mooring line's accumulation damage is calculated for 50 years and summarized in Table 8. According to the prediction results, the maximum mooring tension for a 100-year return period is less than the mooring breaking force, and the fatigue damage is far less than 1, so the structure is safe and can be optimized further.

Table 8. Accumulative damage for 50 years.

	Tension [kN]
Line 1	4.36×10^{-7}
Line 2	3.10×10^{-7}
Line 3	3.33×10^{-7}

4.3. Mooring Optimization—Anchor Point

For the initial mooring design, the maximum tension range for a return of 100 years and the fatigue damage for a lifetime of 50 years are in the safe range. Keeping a fixed angle for the configuration of mooring lines, the reanalysis is performed by changing the anchor point location and mooring line laid length. As shown in Figure 7, mooring lines 2 and 3 are taken as the research objects, and the environmental conditions condition with the maximum mooring tension are selected for optimization. The optimization method is to shorten the length of the mooring lines laid length towards the floating platform, which is 5 m of the length of the mooring lines, and the iterative calculation is carried out with the anchor uplift force as the optimization parameter. The length of each mooring line is optimized using an iterative process until the anchor uplift force does not satisfy the condition the total length of the end moorings is established.

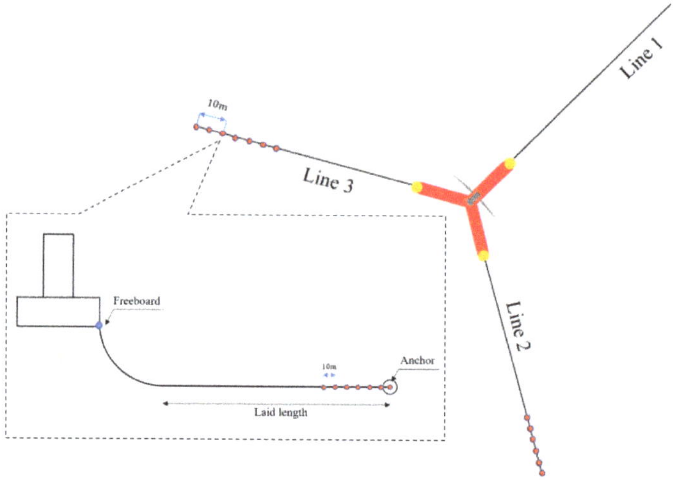

Figure 7. Iterative diagram of mooring.

As shown in Figure 8, the optimization result of the anchor point is obtained through many iterations, and the allowable force is calculated using the total gravity of the anchor and the laid mooring line. Considering the cost and safety, the intersection point is taken as the anchor point to optimize the optimal solution. The final mooring lines configuration is shown in Table 9.

Table 9. The final Mooring configuration.

Line	Anchor Coordinates			Line Length [m]
	x [m]	y [m]	z [m]	
1	−330	0	−70	300.68
2	135	233.83	−70	238.68
3	140	−242.49	−70	248.68

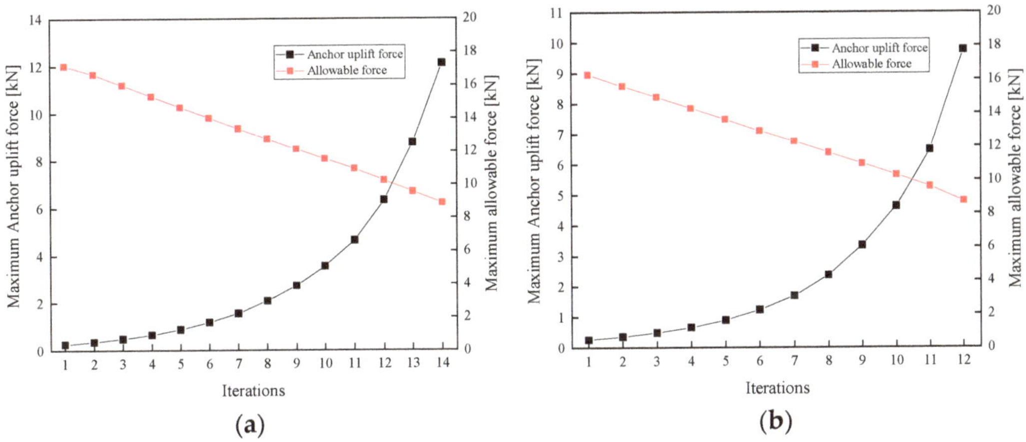

Figure 8. The optimization iterative process of mooring lines (**a**) Mooring line 2 (**b**) Mooring line 3.

Figure 9 shows the change of the mooring lines laid length with the optimization of the anchor point position. With the change in the total length of mooring line 2 and mooring line 3, the laid length of mooring line 1 was reduced by 10 m due to the greater horizontal tension, while the laid lengths of other mooring lines had little change.

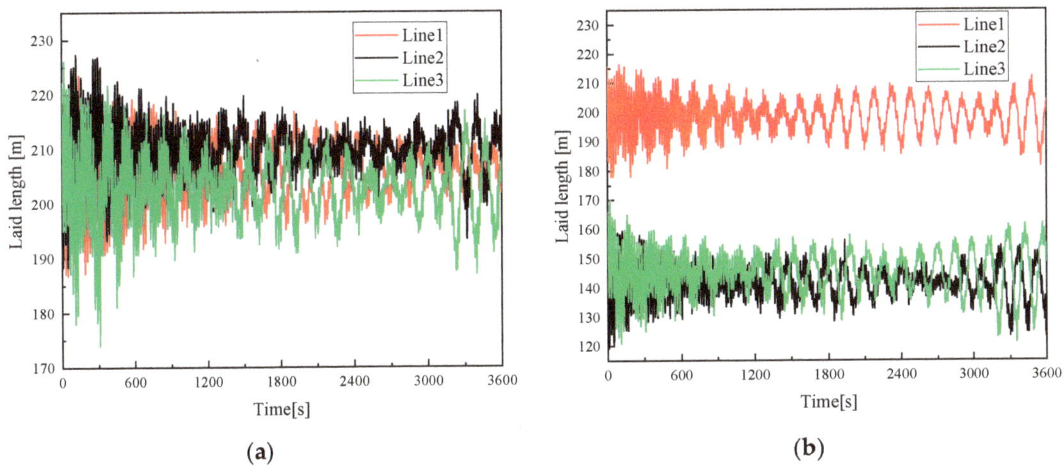

Figure 9. The laid length of mooring lines (**a**) Initial laid length (**b**) Final laid length.

As shown in Figure 10, the mooring line tension increases with the optimization of the mooring anchor point position and the decrease in mooring line length. The previous research shows that when the chain characteristics and pretension of mooring lines remain the same, the shorter mooring lines lead to stronger nonlinearity of mooring resilience and stiffness. The results of this paper are similar [39,40]. In the normal operation of floating wind turbines, the platform motion is required to be less than 5% of the water depth, and the increase in mooring line tension helps to reduce the floating platform motion. In addition, the mooring line tension and the anchor uplift force in the final design of the mooring satisfied the design requirements. The mooring line tension is less than the minimum breaking force of mooring lines, and the anchor uplift force is within a safe range.

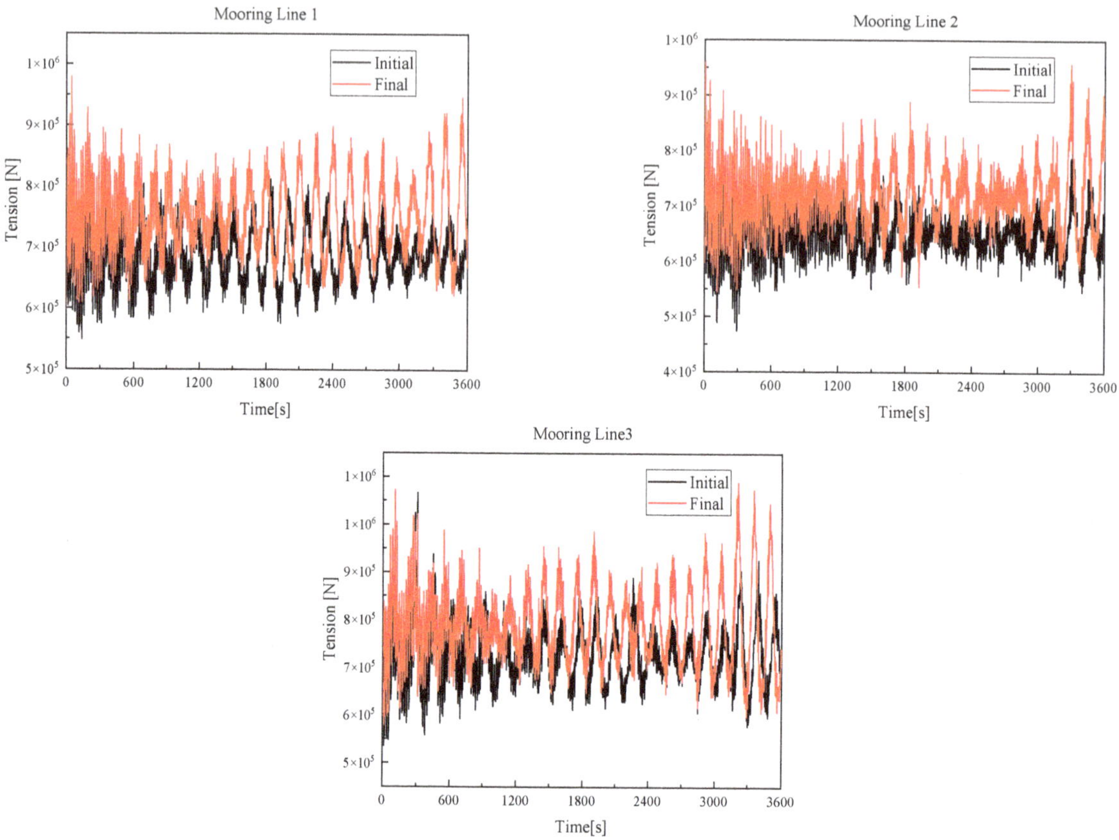

Figure 10. The tension of Moring lines.

4.4. Mooring Optimization—Nominal Diameter

In mooring optimization, three kinds of nominal chain diameter, 152 mm, 185 mm, and 210 mm, are considered to determine the influence of mooring diameter on the ultimate strength and fatigue damage. For different nominal diameters of mooring chain lines the effective elastic modulus and breaking load can be found in the DNV OS E302 and ISO 20438 [41,42]. Three kinds of mooring lines with studless chain detailed specifications are given in Table 10.

Table 10. Parameters of the Mooring chain lines.

Parameter	Value		
Line Type	**R3 Studless Mooring Chain**		
Line Breaking Strength	16,405 kN	22,286 kN	26,749 kN
Nominal Chain Diameter	152 mm	185 mm	210 mm
Dry Line Linear Density	462 kg/m	685 kg/m	882 kg/m
Extensional Stiffness	1973.08 MN	3270 MN	3766.14 MN

Regarding the nominal diameters of three types of mooring lines, a comparison and examination of total weight and maximum tension at a 100-year return period were conducted. The results are shown in Figures 11 and 12; the total mass and the maximum tension increase with the increase in the chain's nominal diameter while keeping the chain length unchanged.

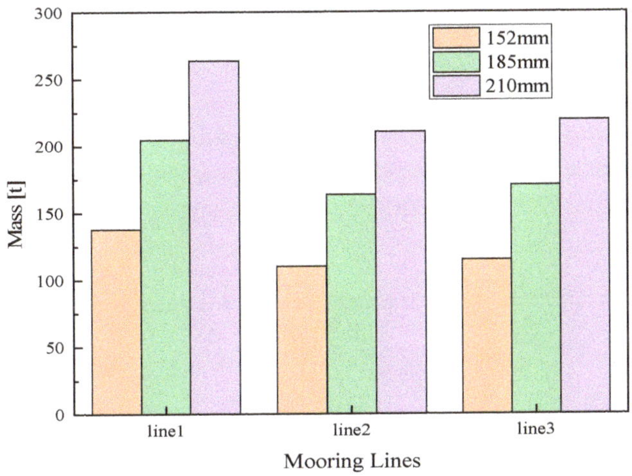

Figure 11. The total mass of Mooring lines.

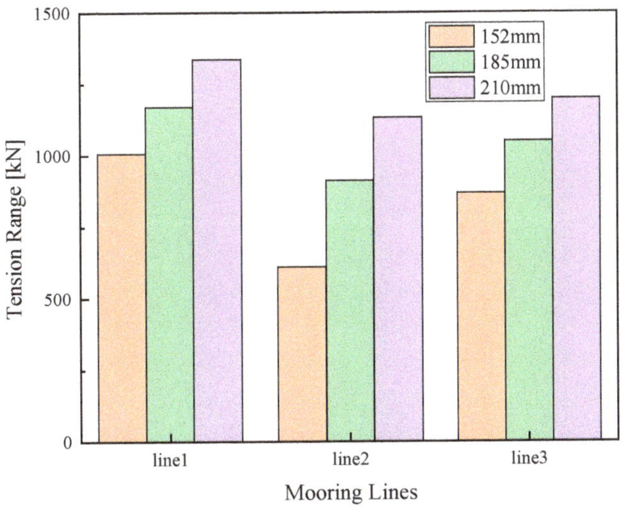

Figure 12. The 100-year return period maximum tension range.

Offshore mooring chains are available in various grades, including R3, R3S, R4, R4S and R5. Out of these grades, R5 offers the highest strength. Figure 13 illustrates the minimum breaking loads (MBL) against chain diameter for studless chains [43]. Keeping the same safety factor, the minimum breaking load of mooring chains of different grades is compared. For ultimate limit state considerations, the R5 grade chain is recommended. Because it is the most economical model, and the nominal diameter of the chain is only 147 mm. In addition, when using the R4S chain, the chain with a nominal diameter greater than 152 mm should be used.

Figure 14 shows that the more the nominal diameter increases, the more cumulative fatigue damage decreases. From the cumulative fatigue damage during the 50 years of life, mooring lines of three kinds of nominal diameter are satisfied with design criteria. Therefore, it is recommended to determine the nominal diameter and grade of the chain from the point of view of fatigue damage.

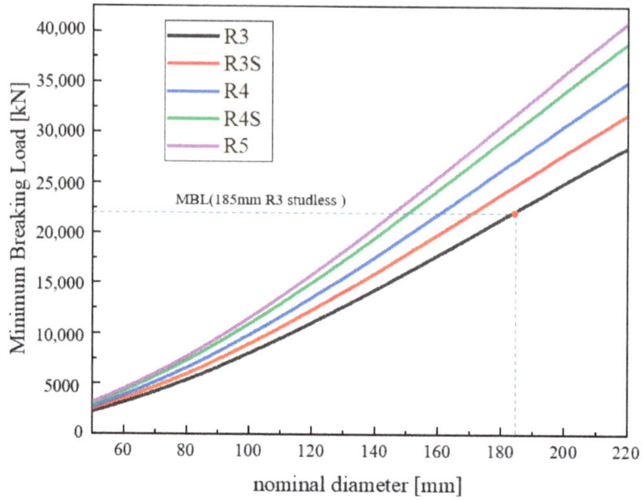

Figure 13. The minimum breaking load of different nominal diameters.

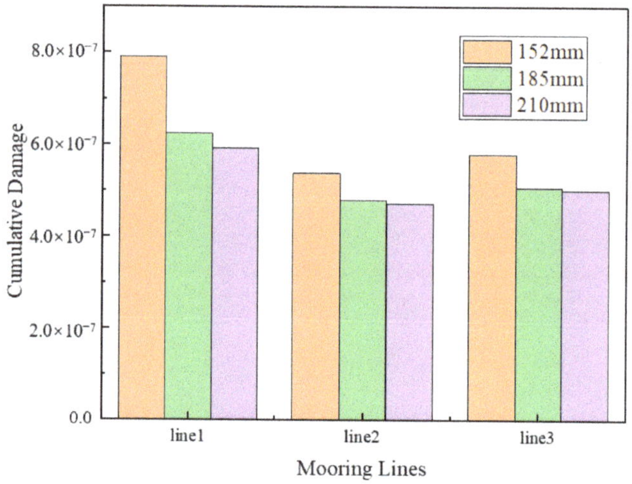

Figure 14. The cumulative damage for 50 years.

5. Conclusions

In this paper, the wind, wave, and current data in the South China Sea are collected as environmental conditions because the South China Sea has abundant wind energy IEA 15 MW wind turbine and UMaine VolturnUS-S floating platform are selected as research models, and the accuracy of the model is verified by comparing the hydrodynamic coefficient between AQWA and the technical report. The R3 class studless chain is selected as the design mooring type, and the nominal diameter, anchor uplift force, and total length of the mooring lines are used as the design parameters. Thus, the catenary equation is applied to determine the initial length of mooring lines. Using time domain hydrodynamic analysis, the anchor uplift force at the anchor point under the initial mooring line design will be checked, and the mooring line tension time history will be output on the premise that the anchor uplift force satisfies the design requirements. In the ultimate strength analysis, Weibull distribution is used to predict the maximum tension range in the return period of 100 years, and it is compared with the minimum breaking limit of mooring lines In the fatigue damage analysis, 50 years of cumulative fatigue is determined using mooring line tension amplitude data, linear cumulative fatigue damage rules, and the S-N curve. Thus, the design and optimization program of floating wind turbine mooring lines suitable for the South China Sea is established.

This paper describes the design, optimization, and safety assessment process of a floating wind turbine mooring system. Based on this research, the influence of aerodynamic load on platform motion and mooring system will be studied using the joint simulation of OPENFAST and AQWA in future research. The combination of floating wind turbines and wave energy is also one of the ways to improve power generation in the future. Many scholars have performed relevant research, but there is little research on mooring [44,45]. In addition, more cost-effective new mooring lines are also within the scope of future research, including fatigue problems of different grades of mooring lines and new mooring materials [46–48].

Author Contributions: Conceptualization, C.B.L.; methodology, C.B.L. and Y.J.; software, J.J. and H.Z.; validation, J.J. formal analysis, J.J.; investigation, W.Z. and X.S.; resources, C.B.L. and M.C.; data curation, W.Z. and X.S.; writing—original draft preparation, J.J.; writing—review and editing, M.C. and C.B.L.; supervision, H.Z.; project administration, M.C.; funding acquisition, M.C. All authors have read and agreed to the published version of the manuscript.

Funding: This research was funded by National Natural Science Foundation of China: 52171275.

Institutional Review Board Statement: Not applicable.

Informed Consent Statement: Not applicable.

Data Availability Statement: Data are contained within the article.

Conflicts of Interest: Authors Xinghan Sun and Wei Zhang were employed by the company Powerchina (Xi'an) Port Navigation Shipbuliding Technology Co., Ltd. The remaining authors declare that the research was conducted in the absence of any commercial or financial relationships that could be construed as a potential conflict of interest.

References

1. Global Wind Report 2023. Available online: https://gwec.net/globalwindreport2023/ (accessed on 15 July 2023).
2. Lantz, E.; Wiser, R.; Hand, M. *IEA Wind Task 26 the Past and Future Cost of Wind Energy*; Technical Report; National Renewable Energy Lab.(NREL): Golden, CO, USA, 2012.
3. Chen, M.; Li, C.B.; Lee, J. A Simulation Technique for Monitoring the Real-time Stress Responses of Various Mooring Configurations for Offshore Floating Wind Turbines. *Ocean. Eng.* **2023**, *278*, 114366. [CrossRef]
4. Barrera, C.; Battistella, T.; Guanche, R.; Losada, I.J. Mooring system fatigue analysis of a floating offshore wind turbine. *Ocean. Eng.* **2020**, *195*, 106670. [CrossRef]
5. Li, C.B.; Choung, J. Fatigue damage analysis for a floating offshore wind turbine mooring line using the artificial neural network approach. *Ships Offshore Struct.* **2017**, *12*, 288–295. [CrossRef]
6. Wang, Z.; Qiao, D.; Yan, J.; Tang, G.; Li, B.; Ning, D. A new approach to predict dynamic mooring tension using LSTM neural network based on responses of floating structure. *Ocean. Eng.* **2022**, *249*, 110905. [CrossRef]
7. Campanile, A.; Piscopo, V.; Scamardella, A. Mooring design and selection for floating offshore wind turbines on intermediate and deep water depths. *Ocean. Eng.* **2018**, *148*, 349–360. [CrossRef]
8. Hall, M.; Goupee, A. Validation of a lumped-mass mooring line model with DeepCwind semisubmersible model test data. *Ocean. Eng.* **2015**, *104*, 590–603. [CrossRef]
9. Pillai, A.C.; Gordelier, T.J.; Thies, P.R.; Dormenval, C.; Wray, B.; Parkinson, R.; Johanning, L. Anchor loads for shallow water mooring of a 15 MW floating wind turbine—Part I: Chain catenary moorings for single and shared anchor scenarios. *Ocean. Eng.* **2022**, *266*, 111816. [CrossRef]
10. Pillai, A.C.; Gordelier, T.J.; Thies, P.R.; Cuthill, D.; Johanning, L. Anchor loads for shallow water mooring of a 15 MW floating wind turbine—Part II: Synthetic and novel mooring systems. *Ocean. Eng.* **2022**, *266*, 112619. [CrossRef]
11. Xu, S.; Soares, C.G. Guedes Soares. Experimental investigation on short-term fatigue damage of slack and hybrid mooring for wave energy converters. *Ocean Eng.* **2019**, *195*, 106618. [CrossRef]
12. Ahn, H.; Ha, Y.-J.; Kim, K.-H. Load Evaluation for Tower Design of Large Floating Offshore Wind Turbine System According to Wave Conditions. *Energies* **2023**, *16*, 1862. [CrossRef]
13. Zhao, G.; Zhao, Y.; Dong, S. System reliability analysis of mooring system for floating offshore wind turbine based on environmental contour approach. *Ocean. Eng.* **2023**, *285*, 115157. [CrossRef]

14. Ding, W.-W.; Jiang, J.-Q.; Yue, W.-Z.; Li, Y.-Z.; Wang, W.-S.; Sheng, S.-W.; Chen, M. Numerical Study on Hydrodynamic Performance of a New Semi-Submersible Aquaculture Platform. *Appl. Sci.* **2023**, *13*, 12652. [CrossRef]
15. Trubat, P.; Molins, C.; Gironella, X. Wave hydrodynamic forces over mooring lines on floating offshore wind turbines. *Ocean. Eng.* **2020**, *195*, 106730. [CrossRef]
16. Chueh, C.-J.; Chien, C.-H.; Lin, C.; Lin, T.-Y.; Chiang, M.-H. Dynamic Co-Simulation Analysis and Control of an IEA 15MW Offshore Floating Semi-Submersible Wind Turbine under Offshore Wind-Farm Conditions of Wind and Wave. *J. Mar. Sci. Eng.* **2023**, *11*, 173. [CrossRef]
17. Mazarakos, T.P. Wind Energy Calculations of a 15 MW Floating Wind Turbine System in the Mediterranean Sea. *Environ. Sci. Proc.* **2023**, *26*, 191. [CrossRef]
18. Benassai, G.; Campanile, A.; Piscopo, V.; Scamardella, A. Mooring control of semi-submersible structures for wind turbines. *Procedia Eng.* **2014**, *70*, 132–141. [CrossRef]
19. Chen, M.; Zhou, H.; Li, C.B. Fully Coupled Dynamic Analysis of the OO-STAR Floating Wind Turbine in Different Water Depths. In Proceedings of the 2022 ISOPE International Ocean and Polar Engineering Conference, Shanghai, China, 6–10 June 2022.
20. Kim, H.; Jeon, G.-Y.; Choung, J. A Study on Mooring System Design of Floating Offshore Wind Turbine in Jeju Offshore Area. *Int. J. Ocean. Syst. Eng.* **2013**, 3, 209–217. [CrossRef]
21. Li, C.B.; Chen, M.; Choung, J. The Quasi-Static Response of Moored Floating Structures Based on Minimization of Mechanical Energy. *J. Mar. Sci. Eng.* **2021**, *9*, 960. [CrossRef]
22. ANSYS Inc. *ANSYS AQWA Theory Manual*; ANSYS Inc.: Canonsburg, PA, USA, 2023.
23. Zhao, Y.; Liao, Z.; Dong, S. Estimation of characteristic extreme response for mooring system in a complex ocean environment. *Ocean. Eng.* **2021**, *225*, 108809. [CrossRef]
24. Kebir, T.; Correia, J.; Benguediab, M.; Jesus, A.M.P.D. Numerical study of fatigue damage under random loading using rainflow cycle counting. *Int. J. Struct. Integr.* **2021**, *12*, 149–162. [CrossRef]
25. Yang, Y.; Bashir, M.; Wang, J.; Michailides, C.; Loughney, S.; Armin, M.; Hernández, S.; Urbano, J.; Li, C. Wind-wave coupling effects on the fatigue damage of tendons for a 10 MW multi-body floating wind turbine. *Ocean. Eng.* **2020**, *217*, 107909. [CrossRef]
26. Milne, I.; Ritchie, R.O.; Karihaloo, B.L. (Eds.) Comprehensive Structural Integrity: Cyclic Loading and Fatigue. Elsevier: Amsterdam, The Netherlands, 2003; Volume 4.
27. Low, Y.M. Extending a time/frequency domain hybrid method for riser fatigue analysis. *Appl. Ocean. Res.* **2011**, *33*, 79–87. [CrossRef]
28. Du, J.; Wang, H.; Wang, S.; Song, X.; Wang, J.; Chang, A. Fatigue damage assessment of mooring lines under the effect of wave climate change and marine corrosion. *Ocean. Eng.* **2020**, *206*, 107303. [CrossRef]
29. Gaertner, E.; Rinker, J.; Sethuraman, L.; Zahle, F.; Anderson, B.; Barter, G.; Abbas, N.; Meng, F.; Bortolotti, P.; Skrzypinski, W.; et al. Definition of the IEA Wind 15-Megsestt Offshore Reference Wind Turbine. 2020, Technical Report. Available online: https://www.nrel.gov/docs/fy20osti/75698.pdf (accessed on 15 July 2023).
30. Allen, C.; Viscelli, A.; Dagher, H.; Goupee, A.; Gaertner, E.; Abbas, N.; Hall, M.; Barter, G. *Definition of the UMaine VolturnUS S Reference Platform Developed for the IEA Wind 15Megawatt Offshore Reference Wind Turbine*; 2020 Technical Report; National Renewable Energy Lab.(NREL): Golden, CO, USA, 2020.
31. DNV. *Offshore Standard DNV_OS_E301, Position Mooring*; DNV: Bærum, Norway, 2010.
32. Hazelton, M. Blyth Offshore Demonstration Project Phase 2—Supporting Environmental Information Blyth Offshore Demonstrator Phase 2 Works 1 Document Control, 2020, Technical Report, EDF Renewables. Available online: https://www.edf-re.uk/ (accessed on 18 December 2023).
33. Connolly, P.; Hall, M. Comparison of pilot-scale floating offshore wind farms with shared moorings. *Ocean. Eng.* **2019**, *171*, 172–180. [CrossRef]
34. Guo, Y.; Wang, H.; Lian, J. Review of integrated installation technologies for offshore wind turbines: Current progress and future development trends. *Energy Convers. Manag.* **2022**, *255*, 115319. [CrossRef]
35. Hsu, W.-T.; Thiagarajan, K.P.; Manuel, L. Extreme mooring tensions due to snap loads on a floating offshore wind turbine system. *Mar. Struct.* **2017**, *55*, 182–199. [CrossRef]
36. Chen, M.; Ouyang, M.; Li, T.; Zou, M.; Ye, J.; Tian, X. Numerical modelling of a catamaran float-over deck installation for a spar platform with complex hydrodynamic interactions and mechanical couplings. *Ocean. Eng.* **2023**, *287*, 115905. [CrossRef]
37. Wang, Z.; Zhou, L.; Dong, S.; Wu, L.; Li, Z.; Mou, L.; Wang, A. Wind wave characteristics and engineering environment of the South China Sea. *J. Ocean. Univ. China* **2014**, *13*, 893–900. [CrossRef]
38. Chen, M.; Zou, M.; Zhu, L.; Ouyang, M.; Liang, Q.; Zhao, W. A Fully Coupled Time Domain Model Capturing Nonlinear Dynamics of Float-over Deck Installation. Available at SSRN 4583209. Available online: https://papers.ssrn.com/sol3/papers.cfm?abstract_id=4583209 (accessed on 18 December 2023).
39. Yan, X.; Chen, C.; Yin, G.; Ong, M.C.; Ma, Y.; Fan, T. Numerical investigations on nonlinear effects of catenary mooring systems for a 10-MW FOWT in shallow water. *Ocean. Eng.* **2023**, *276*, 114207. [CrossRef]
40. Ghafari, H.; Dardel, M. Parametric study of catenary mooring system on the dynamic response of the semi-submersible platform. *Ocean. Eng.* **2018**, *153*, 319–332. [CrossRef]
41. *DNV-OS-E302*; Offshore Mooring Chain. DNV: Bærum, Norway, 2022.
42. *ISO20438*; Ships and Marine Technology—Offshore Mooring Chains. ISO International Standards: Geneva, Switzerland, 2018.

3. Ma, K.-T.; Luo, Y.; Kwan, T.; Wu, Y. *Mooring System Engineering for Offshore Structures*; Gulf Professional Publishing: Houston, TX, USA, 2019.
4. Chen, M.; Ren, W.; Xiao, P.; Zhu, L.; Li, F.; Sun, L. Numerical analysis of a floating semi-submersible wind turbine integrated with a point absorber wave energy convertor. In Proceedings of the Thirtieth International Ocean and Polar, Virtual, 11 October 2020.
5. Chen, M.; Xiao, P.; Zhou, H.; Li, C.B.; Zhang, X. Fully Coupled Analysis of an Integrated Floating Wind-Wave Power Generation Platform in Operational Sea-States. *Front. Energy Res.* **2022**, *10*, 931057. [CrossRef]
6. Liu, H.; Chen, M.; Han, Z.; Zhou, H.; Li, L. Feasibility Study of a Novel Open Ocean Aquaculture Ship Integrating with a Wind Turbine and an Internal Turret Mooring System. *J. Mar. Sci. Eng.* **2022**, *10*, 1729. [CrossRef]
7. Harrold, M.J.; Thies, P.R.; Newsam, D.; Ferreira, C.B.; Johanning, L. Large-scale testing of a hydraulic non-linear mooring system for floating offshore wind turbines. *Ocean. Eng.* **2020**, *206*, 107386. [CrossRef]
8. Xu, S.; Wang, S.; Soares, C.G. Experimental study of the influence of the rope material on mooring fatigue damage and point absorber response. *Ocean. Eng.* **2021**, *232*, 108667. [CrossRef]

Journal of *Marine Science and Engineering*

MDPI

Article

Annular Electromagnetic Generator for Harvesting Ocean Wave Energy

Chunjie Wang [1,2], Linghao Guo [1,2], Peng Chen [1,2], Qiang Fu [1,2] and Lin Cui [3,*]

[1] Tianjin Key Laboratory of New Energy Power Conversion, Transmission and Intelligent Control, Tianjin 300384, China; chunjie_wang@tjut.edu.cn (C.W.); glhtjut@stud.tjut.edu.cn (L.G.); chenpeng@email.tjut.edu.cn (P.C.); fuqiang@email.tjut.edu.cn (Q.F.)
[2] School of Electrical Engineering and Automation, Tianjin University of Technology, Tianjin 300384, China
[3] National Ocean Technology Center, Tianjin 300112, China
* Correspondence: cuilin_oceanenergy@126.com

Abstract: To sustainably power ocean sensors by harvesting ocean wave energy, an annular electromagnetic generator (A-EMG) based on the principle of Faraday electromagnetic induction is proposed in this paper. The specific structure and working principle of the generator are introduced. The distribution of the magnetic field in the coil, the variation in the induced voltage and the influence of the coil parameters on the output were simulated by the COMSOL Multiphysics software version 6.0. At the same time, an experimental platform was built to test the output characteristics of the generator. Through a comparative study of the capacitor's charging characteristics, the optimal connection mode between the multiple groups of coils of the generator was preliminarily verified. Finally, the six-degree-of-freedom (6-DOF) platform was used to simulate various wave motion parameters, and the feasibility of the generator for supplying power to ocean sensors was verified.

Keywords: A-EMG; ocean sensors; COMSOL Multiphysics; ocean wave energy harvesting

Citation: Wang, C.; Guo, L.; Chen, P.; Fu, Q.; Cui, L. Annular Electromagnetic Generator for Harvesting Ocean Wave Energy. *J. Mar. Sci. Eng.* **2023**, *11*, 2266. https://doi.org/10.3390/jmse11122266

Academic Editor: Mariano Buccino

Received: 30 October 2023
Revised: 26 November 2023
Accepted: 28 November 2023
Published: 29 November 2023

1. Introduction

Ocean environmental monitoring is very important for the exploitation and utilization of ocean resources. Various types of ocean sensors are used to collect ocean environmental data, but there is always a lack of a reliable power supply in this field. At present, batteries, wind energy, solar energy, etc., are the most common power supply methods applied to ocean sensors. However, battery replacement is difficult and is often polluted, and solar and wind energy are both seriously affected by weather factors and their reliability is poor [1]. The ocean itself is rich in energy. In order to solve the bottleneck problem of supplying power, the most ideal way of supplying power is undoubtedly to be able to use the energy in the ocean [2–4]. Many wave-energy-harvesting devices have already been proposed and designed, and the most widely used method of wave energy harvesting is based on rotating electromagnetic generators (EMG). In other words, in these generators, wave energy is converted into mechanical energy, pneumatic energy or potential energy by the power take-off (PTO) device. The PTO is then connected to a rotating generator, and the energy is converted into electrical energy [5]. However, in recent years, with the rapid development of microelectronics, micro-electro-mechanical systems (MEMS) and mobile internet technologies, more and more sensors have widely been used in ocean equipment such as ocean environmental monitoring and position tracking. The traditional wave-energy-harvesting devices have the disadvantages of having a complex mechanical structure, they are easily corroded and they have high costs. In addition, these devices always have a large volume and a low energy harvesting efficiency, which is not suitable for powering ocean sensors [6]. Providing sustainable, low-cost and easy-to-maintain power energy for ocean sensors has become a major challenge. The development of energy-harvesting devices with a small size and high output power is of great significance for the further development and application of various ocean devices [7].

As early as 2007, the American scholar Arnold, D.P. discussed the practical significance and feasibility of small electromagnetic generators, and he believes that small power supply devices are of great value for powering smart intelligent sensors, robots, wireless communication networks and other distributed microelectronic devices [8]. Harvesting energy from the environment is now increasingly becoming a promising method for powering small devices. Several studies have demonstrated the feasibility and superiority of small EMGs for harvesting energy from the surrounding environment [9]. Bowers et al. invented the rolling spherical magnet generator, which was used to harvest energy generated by human motion [10]. Samad et al. proposed a curved wearable electromagnetic energy harvesting system that can be used to harvest energy from the human body while walking or running from two directions [11]. Tu et al. invented a bistable vibration energy harvester based on spherical moving magnets for harvesting the vibration energy of a moving vehicle, which is applied to power safety detection equipment related to rail and road transportation [12]. Similarly, the research on the application of EMGs in wave energy harvesting is becoming more intensive, and more EMGs with different structures have been developed for wave energy harvesting. Wave energy harvesting is usually performed in an inertial pendulum structure [13,14], a gyroscopic structure or an eccentric structure [15,16]. Guo et al. proposed an eccentric pendulum structure consisting of a transmission gear and an electromagnetic power module, in which the pendulum converts low-frequency wave energy into mechanical rotational energy, and the transmission gear amplifies the low rotational speed of the pendulum. The electromagnetic power module is used to convert rotating mechanical energy into electrical energy. The maximum output power is 122 mW, and the output power density is 0.24 mW/cm^3 [17].

However, the horizontal single-axis pendulum structure can only harvest wave energy from a specific direction, and the wave direction of wave motion is always random, so it is more advantageous to harvest wave energy from multiple directions [18]. Lou et al. proposed a novel two-axis pendulum vibration energy harvester. Harvesting vibration energy based on oscillation is a promising energy-harvesting technology for buoys and small ocean vehicles. However, energy harvesters with a single-axis pendulum require the vibration excitation by the waves to be generated in a specific direction. A new vibration energy harvester for a two-axis pendulum has been developed. The hemispherical pendulum can rotate around two axes simultaneously to match to the direction of the vibration excitation. In addition, the magnets and coils are distributed on the surface of the pendulum and the housing, respectively, to achieve a compact structure, and its output can reach a peak voltage of 14.25 V and a power of 2.03 W under certain conditions [19]. Wang et al. proposed an EMG that simulates the motion of a dual hula-hoop for wave energy harvesting and presented a two-degree-of-freedom wave energy converter (WEC), which is a biaxial toroidal and eccentric generator composed of a circular Halbach array disk and an iron core coil. This WEC can convert the pitching, rolling and rotating motions of an unmoored buoy into electrical energy, and the device can output 0.56 W of power when the buoy's motion frequency is between 0.7 Hz and 1.0 Hz [20]. Carandell et al. proposed a kinetic energy harvester (KEH) applied to an ocean drifter, which combines a pendulum system with a gyroscope system. When waves act on the device, the relative motion is induced by an internal articulated arm, and it then drives a flywheel to drive a small generator via a gear system. When the external excitation reaches 1.5 Hz, its output power is 2 mW [21]. Pan et al. proposed a tumbler electromagnetic generator (T-EMG) with a Halbach array, in which coils, Halbach magnet units and magnetostrictive structures are installed in two hollow semi-elliptical enclosures. Under the conditions of adjusting the load to reach the optimal power matching point, a T-EMG has a maximum peak to peak voltage of 3 V and a maximum average output power of 65 mW under an external excitation of 2.6 Hz. In real wave environments of less than 1 Hz, its maximum peak power reaches 120 mW, and it can successfully drive wireless temperature and humidity sensors and LEDs [22].

In order to achieve efficient and reliable energy harvesting for powering small marine equipment or sensors, an annular electromagnetic generator (A-EMG) is proposed in this paper. Compared with the structure proposed herein, in other existing works, such as in [21], the structures are more complex and are formed by a combination of a plurality of mechanical structures or by the connection of mechanical devices. The structures are also still relatively complex, and the more complex the structure is, the higher the probability of it malfunctioning, which may reduce the reliability of the device and reduce the energy conversion efficiency of.

The A-EMG proposed in this study only consists of a magnetic ball, an annular shell and three coils without any other mechanical structures and devices. The energy conversion process is simple; the wave motion that drives the movement of the magnetic ball can be converted into electricity, which improves the reliability of the device and reduces the energy loss. References [12,23], which are similar to the principle of the A-EMG, involve linear or curved tubes with a rolling magnetic ball that generate an induced current in separate winding coils. They place the magnetic ball in curved and straight tubes, respectively. When the magnetic ball moves to one end, the speed of the moment is reduced to 0, which undoubtedly increases the speed loss and energy loss, whereas the annular shell of the A-EMG allows for the continuous movement of the spherical magnet without any obstruction, which avoids the speed loss and energy loss to the greatest extent. The structure and working principle of the A-EMG are explained in this paper. The principle was simulated using the COMSOL Multiphysics software version 6.0, and the influence of the coil parameters on the output was analyzed. Finally, an experimental prototype was built, and the experiment was verified by using the six-degree-of-freedom (6-DOF) platform.

2. Structure Principle

As shown in Figure 1, the A-EMG consists of an annular shell, a magnetic ball that can move inside the annular shell and three groups of coils wrapped around the outer part of the annular shell at equal intervals. When the device is placed in an ocean environment, due to the fluctuating fluctuations of the ocean waves, the device will shake with the waves, thereby driving the magnetic ball to move. And when the magnetic ball passes through the coil, the magnetic flux changes, generating an induced electromotive force in the coil.

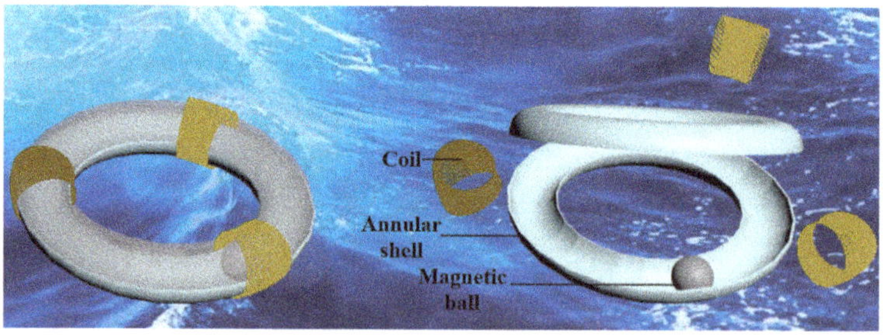

Figure 1. Three-dimensional structure of the A-EMG.

Figure 2 shows the working principle of the A-EMG. After being excited by a wave, the magnetic ball moves inside the annular shell, and according to the law of electromagnetic induction, the expression of the induced electromotive force can be obtained by using V to denote the induced electromotive force produced by the coil and Φ to denote the magnetic flux in one turn of the coil:

$$V = -N\frac{d\Phi}{dt} \tag{1}$$

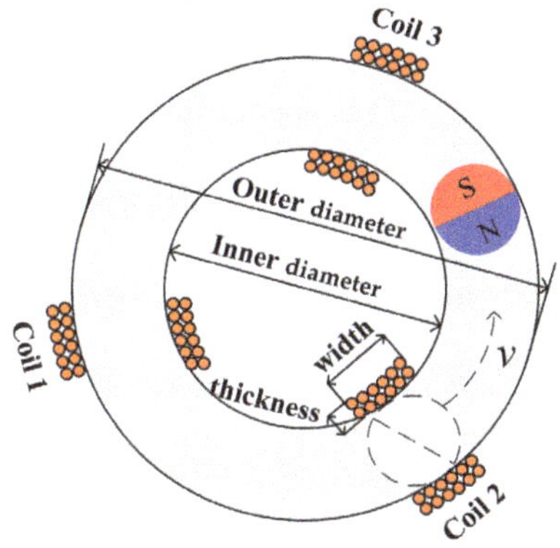

Figure 2. Working principle of the A-EMG.

Assuming that the time for the spherical magnet to move inside the annular shell at the speed v is t and that the displacement of the movement is x, and denoting the magnetic flux density through the coil and the area of the coil by $B(x)$ and S, respectively, the induced electromotive force can be further expressed as:

$$V = -N\frac{d\Phi}{dt} = -NS\frac{dB(x)}{dx}\frac{dx}{dt} = -NS\frac{dB(x)}{dx}v \tag{2}$$

The angular velocity ω of the motion of the magnet ball can be expressed as:

$$\omega = \frac{v}{r} \tag{3}$$

The frequency f of the magnet ball moving around a ring can be expressed as:

$$f = \frac{1}{T} = \frac{1}{2\pi/\omega} = \frac{\omega}{2\pi} = \frac{v}{2\pi r} \tag{4}$$

Based on the above analysis, the relationship between the frequency and the induced electromotive force can be deduced:

$$V = -NS\frac{dB(x)}{dt}2\pi rf \tag{5}$$

According to Equation (2), it is clear that the magnitude of the induced voltage is related to the speed of motion of the magnetic ball through the coil; the faster the speed, the greater the induced voltage produced in the coil. From Equation (5), when the wave motion resonates with the ring generator, the magnetic ball will make a complete circular motion, and the induced voltage in the coil is related to the frequency of the circular motion, where the greater the frequency, the greater the induced electromotive force that is produced [24].

3. Principle Analysis and Parameter Effect Analysis Based on COMSOL

The working principle of the A-EMG was analyzed above. In order to further analyze and intuitively understand the working characteristics of the A-EMG, the COMSOL Multiphysics software was used to simulate and analyze the A-EMG. The parameter settings in the simulation are shown in Table 1. The A-EMG was modeled in a two-dimensional plane, and in the simulation, we set the outer diameter of the annular shell to 70 mm and the inner diameter to 35 mm. Considering that the larger the magnetic ball, the more favorable it is for the output power, and to ensure its movement in the annular shell, it was necessary to leave a certain air gap between the annular shell and the magnet ball. The diameter of the magnetic ball was set to 30 mm, the magnetic field module was selected as the simulation background and the number of coil turns was set to 1000 turns. The variation in the flux density, magnetic field and induced voltage in the coil of the EMG section was investigated.

Table 1. Parameter settings in simulation.

Item	Value
Outer diameter × inner diameter of annular shell	140 mm × 70 mm
Magnetic ball diameter	30 mm
Turns of coil	1000

In the above principal explanation, we only considered the movement of the magnetic ball around the annular shell as "sliding" (the magnetic ball only moves around the annular shell, and the magnetic ball itself does not rotate). However, in reality, the magnetic ball may move in a "rolling" (the magnetic ball itself rotates) manner or in a combination of "sliding" and "rolling" during the process of passing through the coils. Similar to the references [23,25], we simulated and analyzed the magnetic field changes caused by the magnetic ball rotating 360° at a fixed position and the flux density transformation at point 1 on coil 1, as shown in Figure 3a. The protrusion on the annular shell in the figure indicates the coils set in the simulation. The simulation range was set to 0–360°, and the simulation step size was 5°.

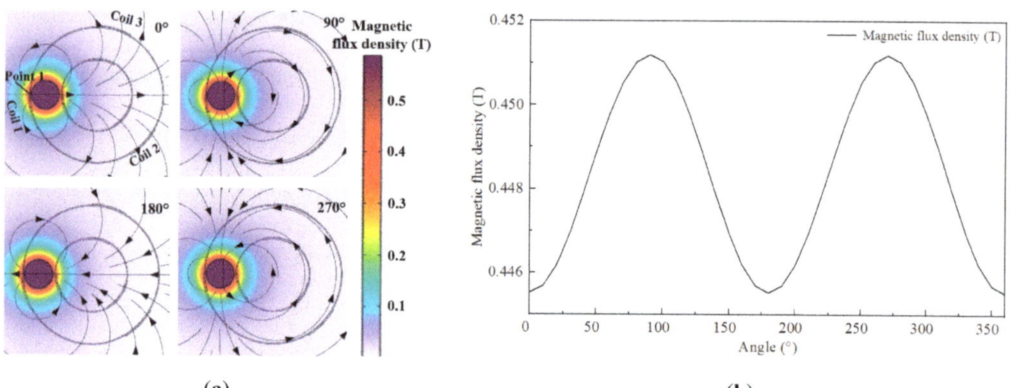

Figure 3. Magnetic field and flux density variations when magnetic ball is rotated in a fixed position (**a**) The distribution of the magnetic field when the magnetic ball is in a different angle; (**b**) law of change in magnetic flux density at point 1 with the angle of the magnetic ball.

When the magnet ball rotates in the annular shell, it causes a change in the surrounding magnetic field and a change in the magnetic flux density at point 1 (same point as in Figure 3a). The simulation range was set to 0–360°, and the simulation step size was 5°. As shown in Figure 4, it can be seen that the motion of the magnet ball has an effect on the magnetic field and the flux produced.

Comparing Figure 3b with Figure 4b, it can be seen that the change in the magnetic flux density due to the rotation of the ball itself was not dominant compared to the movement of the ball around the annular shell, and considering that even if the ball passed through the coil in a "rolling" manner, the angle of rotation of the ball itself was difficult to determine So, in the principle explanation and simulation, we simplified the motion form of the magnetic ball, ignored the rotation of the magnetic ball itself [26] and only considered the "sliding" motion of the magnetic ball.

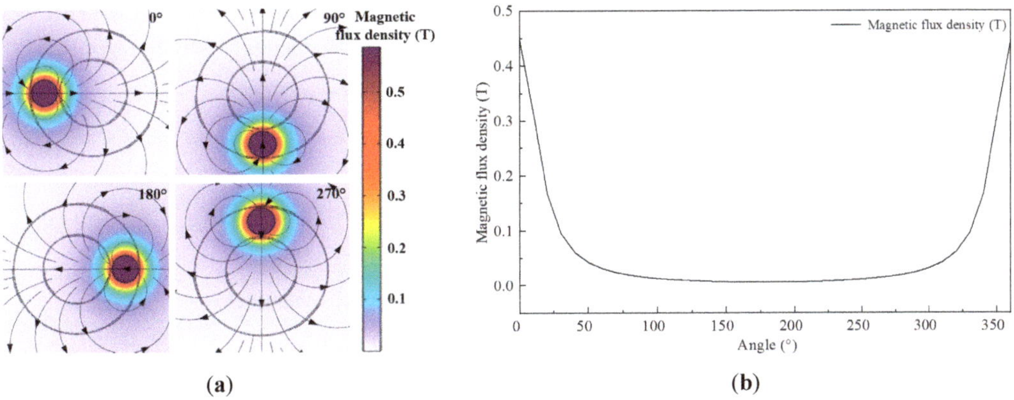

Figure 4. Magnetic field and flux density variations when magnetic ball is "sliding" around annular shell. (**a**) The distribution of the magnetic field when the magnetic ball is in different positions; (**b**) law of change in magnetic flux density at a point 1 with the position of the magnetic ball.

The induced electric potential generated in the coil when the magnetic ball moves in the annular shell at an angular velocity of 2π rad/s was studied and analyzed by using COMSOL. The simulation time was set to 5 s with a step of 0.1 s so as to verify the rule of change and the trend of the induced voltage in the coil during the passage of the magnetic ball through the coil. Figure 5 shows the induced electromotive force of the magnetic ball moving through the annular shell for five cycles, and in conjunction with Figure 4, it is clear that during a cycle, the magnetic flux in the coil increased as the ball rolled toward the coil and decreased as the ball moved away from the coil, generating a concomitant change in the electromotive force [23].

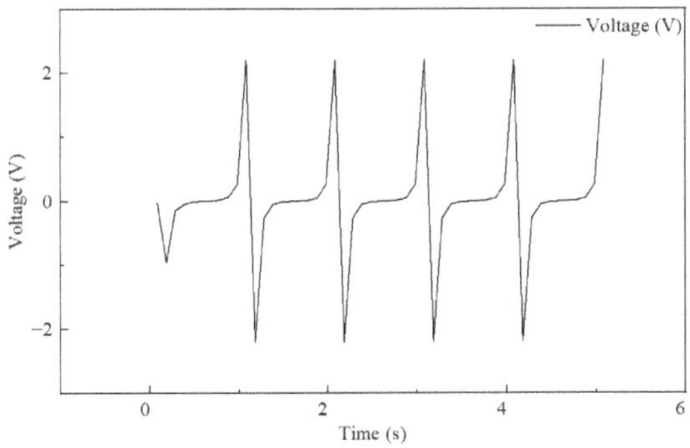

Figure 5. The law of change in the induced voltage in the coil.

The coil parameters have an important effect on the output of the coil. In addition to the number of turns of the coil, the difference in the width and thickness of the coil will also change the interaction of the coil with the magnetic field of the magnetic ball, so the magnetic flux change in the coil caused by the movement of the magnetic ball will be different, which will affect the power output [27,28]. The definitions of the coil width and thickness are shown in Figure 2. Using the control variable method, the effect of varying the thickness and width of the coil on the output was analyzed.

As shown in Figure 6, when the number of turns and the width of the coil were kept constant, the thickness of the coil was increased from 1 mm to 6 mm, and the peak value of the induced voltage was decreased from 2.57 V to 2.17 V. The results indicate that the thicker the coil, the more unfavorable it is for the output [11,29]. If the number of turns and the thickness of the coil were kept constant and the width of the coil was increased from 0.75 cm to 4.5 cm, the peak value of the induced voltage decreased from 2.18 V to 1.48 V. The wider the coil, the more unfavorable it is to the output [20].

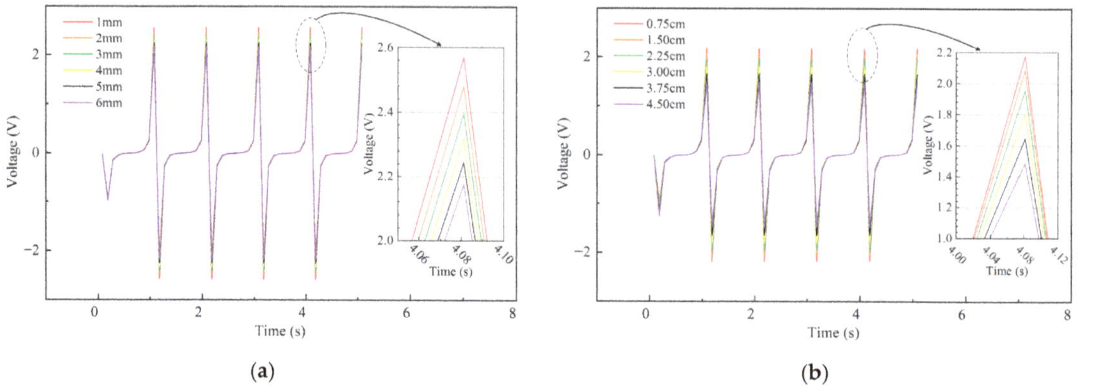

Figure 6. Law of the effect of coil thickness and width on output. (**a**) Output voltage at different coil thicknesses; (**b**) output voltage at different coil widths.

4. Experiment

4.1. Motion Analysis of A-EMG under Wave Action

The motion of the A-EMG under wave action was analyzed in order to develop an experimental program. In fact, the motion of waves in the ocean is particularly complex and variable, and it is not a simple periodic movement. The generation of waves in the ocean is essentially a vibration process in the ocean, and its motion can be regarded as a combination of countless disordered regular fluctuations. Due to the complexity of the causes of wave motion and the difficulty of using accurate mathematical models to make a precise description, in the analysis and design of machinery related to ocean engineering, researchers have always simplified the motion of waves to simple harmonic motion or to the superposition of simple harmonic motion in their studies. And, they often use the Airy linear wave theory to characterize the wave motion law in a general steady state situation [30,31]. Here, we also used the Airy linear wave theory to explain the wave principle. The profile of a wave is described by a simple harmonic curve [32], as show in Figure 7.

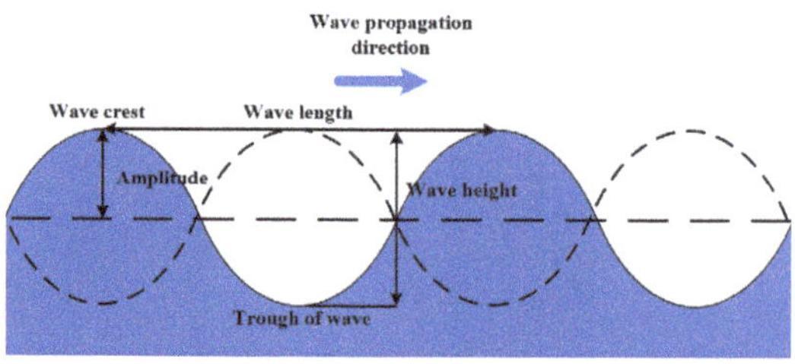

Figure 7. Airy linear wave theory model.

According to the Airy linear wave theory, the motion of a wave can be described by Equation (6).

$$\begin{cases} \eta(s,t) = \frac{H}{2}\cos(ks - \omega t) \\ k = \frac{2\pi}{L} \\ \omega = \frac{2\pi}{T} \end{cases} \quad (6)$$

where s is the displacement of the wave in the direction of horizontal motion, η is the elevation of the wave from the still water surface, H is the wave height, k is the wave number, t is the time, ω is the frequency of the wave and T and L are the period and wavelength, respectively.

When a floating body is on the water surface, it will move with the waves, and the floating body will produce different forms of motion with the change in the wave parameters. Taking the XOZ plane as an example, Figure 8 shows a schematic diagram of waves with different parameters acting on the A-EMG in the XOZ plane, where φ is the angle between the wave's direction of action and the X-direction, so $tan\varphi$ is the derivative of the line between the zero point and the position of the A-EMG. The wave will cause the A-EMG to produce different forms of motion, which are mainly manifested in the form of reciprocal displacement motion in the direction of wave action, so φ also denotes the direction of motion of the A-EMG, and the tilt motion around the Y-axis, producing a constantly changing tilt angle θ with respect to the X-axis, and $tan\theta$ is the derivative of the wave at this point [17,33].

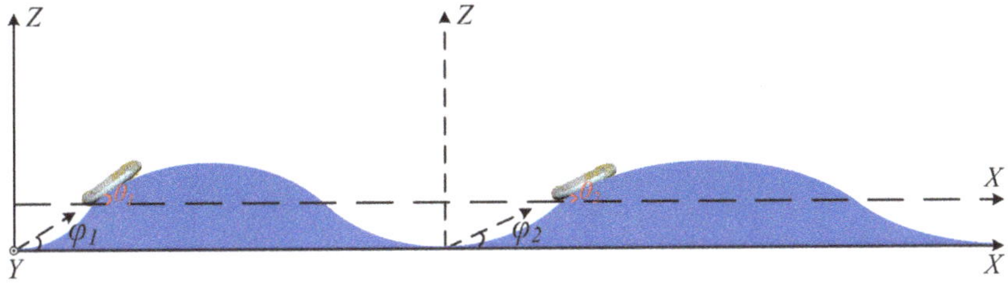

Figure 8. Wave action on the A-EMG.

Therefore, in this paper, a 6-DOF platform was used to simulate the motion of a floating body under the action of waves with different parameters. The control variable method was used to investigate the effects of the motion direction φ, amplitude, frequency and tilt angle θ of the A-EMG on the output and to explore the regular pattern.

4.2. Experimental Platform Construction

Based on the laws derived from the theoretical analysis and simulation results, an annular electromagnetic generator was fabricated. The specific parameters are shown in Table 2.

Table 2. Parameters in fabrication.

Item	Value
Outer diameter × inner diameter of annular shell	140 mm × 70 mm
Annular shell thickness	1 mm
Magnetic ball (N35) diameter	30 mm
Turns of coil	1000
Thickness × width of coil	3 mm × 15 mm
Volume of A-EMG	0.539×10^{-3} m^3

As shown in Figure 9, two identical semi-annular shells with an inner diameter of 70 mm and an outer diameter of 140 mm were fabricated by 3D printing. The annular shell needed to have sufficient hardness and to also leave room for the movement of the magnetic ball. The thickness of the shell was set to 1 mm in the 3D printing, so its internal space was 33 mm. After placing a 30 mm diameter N35 magnetic ball inside, three 0.1 mm enameled wire coils were wound along the radial direction of the annular shell, and the coils were equally spaced on the annular shell. With the coil width and number of turns set to 1.5 cm and 1000 turns, respectively, the thickness of the coil was approximately 3 mm. The volume of the A-EMG was calculated to be approximately 0.539×10^{-3} m^3, which includes the inner diameter of the annular shell and is the volume of the actual space occupied by the A-EMG. A TH2690 electrostatic meter was used as the data acquisition device, and the 6-DOF platform was used as the external excitation device to simulate the wave motion. Figure 9 shows a schematic diagram of the overall experimental platform. The TH2690 electrometer is an integrated instrument that functions as a voltmeter, ammeter, and electrometer, among other things. When measuring an open-circuit voltage and a short-circuit current, the probe of the instrument should simply be connected to the output terminal of the measured target, and the corresponding testing function should be selected.

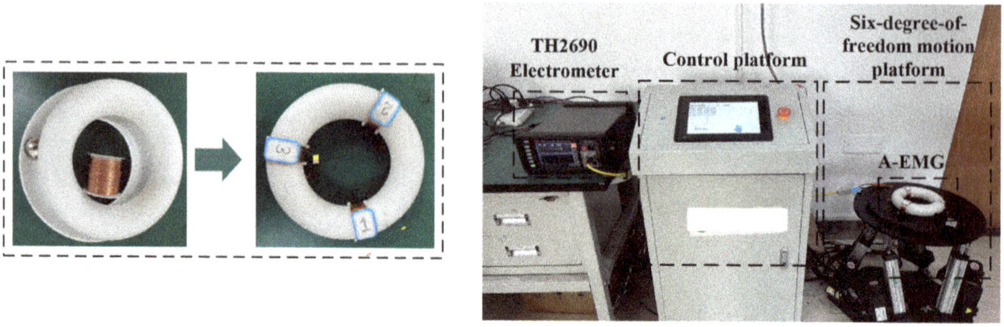

(a) (b)

Figure 9. A-EMG production and experimental platform construction. (**a**) A-EMG physical object (**b**) experimental platform.

The 6-DOF platform consisted of two parts: a motion platform and a software control platform. The software control platform controlled the displacement and frequency of six linear motors in the motion platform, which could achieve linear motion along the *X*, *Y* and *Z* axes, respectively, in three-dimensional coordinates as well as rotational motion around the *X*, *Y* and *Z* axes, are denoted by α, β, γ respectively. It could also couple individual motion into composite motion. In the output testing section of this article, the motion effect of the waves in different directions, φ, acting on the A-EMG in the *XOZ* plane was achieved by coupling the motion of the platform along the *X* and *Z* axes, and the tilt angle of θ was achieved by setting the rotation around the *Y*-axis on the control platform. The method in this paper is more accurate and convenient than the method in reference [25,34], which used a single linear motor and manually changed the excitation direction.

4.3. Output Performance of the A-EMG

It should be noted that since the parameters of each group of coils were the same their outputs ere similar, and the output of coil 1 will be used as an example to illustrate this. Firstly, the influence of the tilt angle was neglected, and the tilt angle was set to 0 using the 6-DOF platform. Through the coupling of the *X*-direction and the *Z*-direction, the motion form of the waves acting on the A-EMG in different directions in the *XOZ* plane was simulated, and the reciprocal displacement motions in the different directions were generated.

We collected the wave parameters from a certain sea area in China, where most of the wave heights are higher than 5 cm and the wave frequency near 5 cm is 1.2 Hz. By verifying that the A-EMG is able to work with small vibrations, we can show that it is able to work under most sea conditions. So, the frequency was set to 1.2 Hz, and the amplitude was set to 5 cm. The variation in the A-EMG with an open-circuit voltage and a short-circuit current with the motion direction φ is shown in Figure 10. As φ increased, the open-circuit voltage and short-circuit current of the A-EMG gradually decreased. When the value of φ reached 60°, the A-EMG could not generate electrical energy because the test only considered the influence of the reciprocating displacement motion, but the larger the wave action, φ, the easier it is to generate a larger tilt angle, θ [34]. Therefore, we also conducted an experimental test on the influence of the tilt angle on the output.

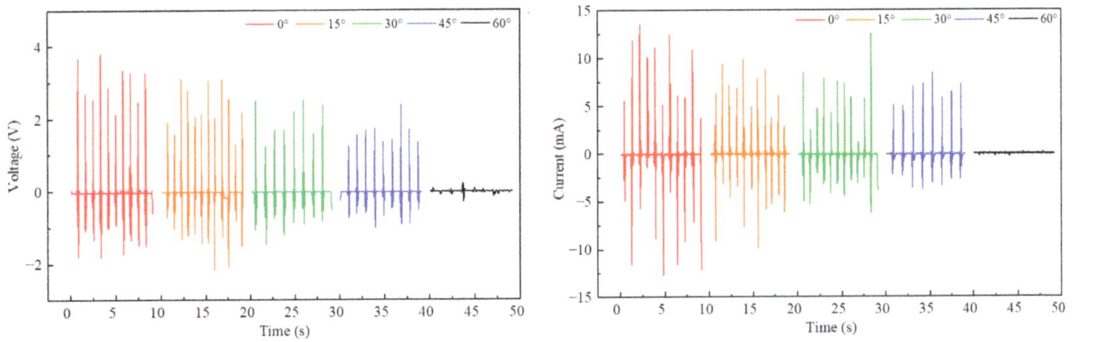

(**a**) (**b**)

Figure 10. Open-circuit voltage and short-circuit current under the influence of motion direction, φ. (**a**) Open-circuit voltage; (**b**) short-circuit current.

If we want to make sure that the A-EMG can adapt to most wave conditions, we should first make sure that the A-EMG can work properly in a small angle range under smaller fluctuations. As shown in Figure 11, at a frequency of 1.2 Hz, when the tilt angle increased from 3° to 12°, there was no obvious effect on the output because the A-EMG is a centrosymmetric structure and the magnetic ball could easily resonate with the external excitation, making the magnetic ball perform a circular motion with the same frequency in the annular shell, which is in accordance with the theoretical derivation of Equation (5). However, it can be obviously observed in the experimental process that the larger the tilt angle, the more the ball in the A-EMG was stimulated to move, which compensated for the situation that the output of the reciprocating displacement motion was weakened when the wave action, φ, was increased; so, the device can collect wave energy from every direction.

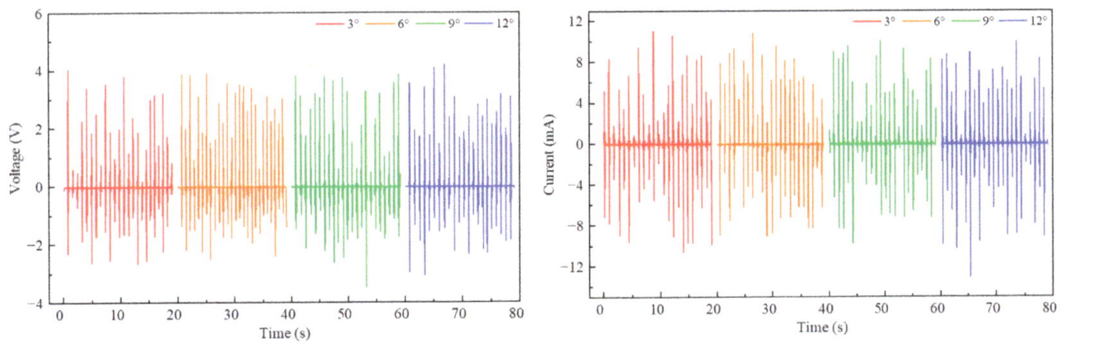

(**a**) (**b**)

Figure 11. Open-circuit voltage and short-circuit current under the influence of tilt angle, θ. (**a**) Open-circuit voltage; (**b**) short-circuit current.

The frequency and amplitude are the most fundamental elements of waves. We tested the influence of the frequency and amplitude on the output of the A-EMG in the *X* direction (φ and θ were zero). As shown in Figure 12, when the amplitude was fixed, the maximum open-circuit voltage increased from 1.52 V to 7.2 V; as the frequency was increased from 1.0 Hz to 1.8 Hz, the maximum short-circuit current increased from 10.08 mA to 21.2 mA. As shown in Figure 13, when the frequency was fixed, the open-circuit voltage and short-circuit current were almost unchanged in the test range of 60 mm–120 mm. This was for the same reason that θ affected the output, which was also due to resonance.

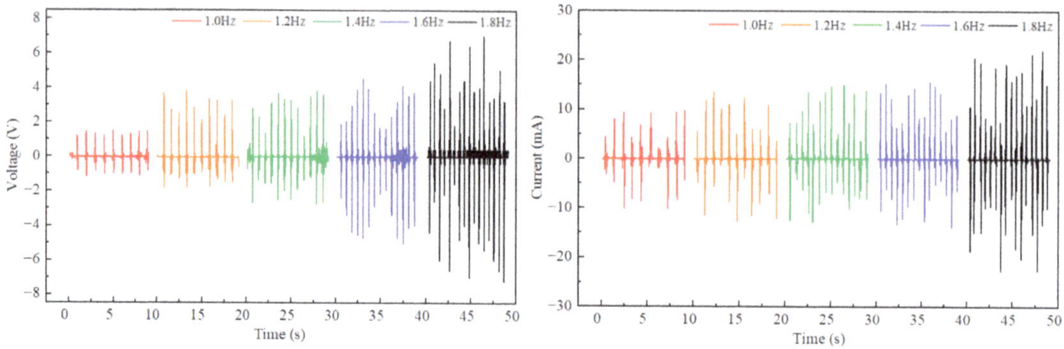

(a) (b)

Figure 12. Open-circuit voltage and short-circuit current under the influence of frequency. (**a**) Open circuit voltage; (**b**) short-circuit current.

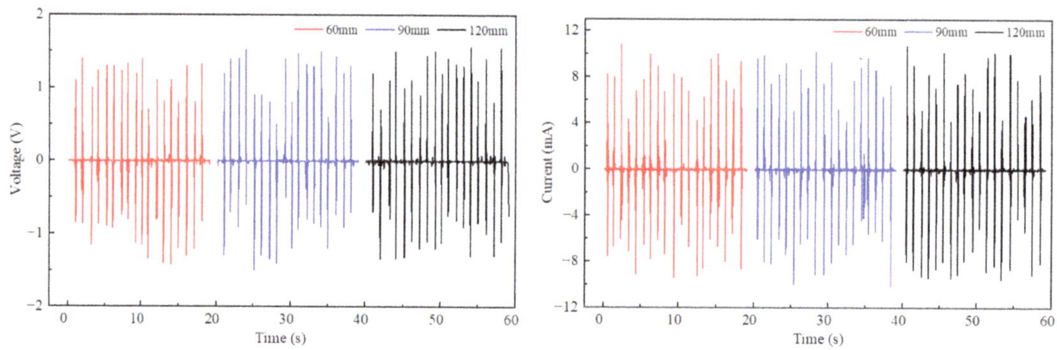

(a) (b)

Figure 13. Open-circuit voltage and short-circuit current under the influence of amplitude. (**a**) Open circuit voltage; (**b**) short-circuit current.

The optimum output power of the electromagnetic generator was found by impedance matching, where the output voltage was measured across various external resistors to study the optimum output power. Theoretically, when the load resistance is equal to the internal resistance of the electromagnetic generator, for the optimal output power, at this time, the relationship between the load voltage, V_L, and the open-circuit voltage, V_{OC}, is given by Equation (7).

$$V_L = \frac{1}{2}V_{OC} \tag{7}$$

In this paper, the voltage at different loads was tested to calculate the output power of the electromagnetic generator, as per Equation (8) [22,35].

$$P_{RMS} = \frac{U_{RMS}}{R_L} \tag{8}$$

where U_{RMS} is the root mean square (RMS) voltage on different load resistors, R_L is the load resistor and P_{RMS} is the optimal output power.

Tested at frequencies of 1 Hz and 1.8 Hz, respectively, as shown in Figure 14, the value of U_{RSM} of the external load increased with the increase in the load resistance, and the value of P_{RSM} first increased and then decreased with the increase in the load. Under both conditions, the maximum values of P_{RSM} obtained at an external load resistance of 150 Ω were 3.42 mW and 80.87 mW, respectively.

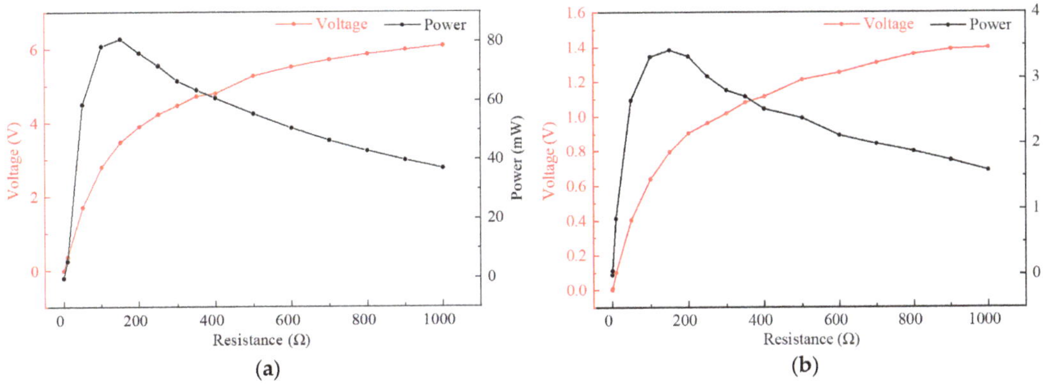

Figure 14. U_{RMS} and P_{RMS} at different external load resistances. (**a**) At 1.8 Hz excitation; (**b**) at 1 Hz excitation.

The maximum power density of the A-EMG can reach 150.174 W/m^3. Table 3 compares the performance of the A-EMG with existing wave energy harvesters [22]. Through comparison, it was found that the power density of the A-EMG has significant advantages.

Table 3. Comparison of different works.

Ref.	Type	Frequency (Hz)	Output Power (mW)	Power Density (W/m^3)
[32]	EMG + TENG	2.5	1.25	9
[35]	EMG + PZT	1.8	21.95	3.914
[22]	EMG	2.6	65	43.5
This work	EMG	1.8	80.87	150.174

Since the device generates AC power, which cannot directly power electronic devices, it must be rectified to DC power and stored in capacitors or batteries in order to provide stable power. The ability of energy harvesters to charge capacitors is an important index for evaluating their performance. At the same time, in order to verify the best connection mode between the coils, the output of each coil is rectified by a rectifier bridge and then connected in series or parallel to compare and study the effect of the two different connections on the charging characteristics of the capacitor. As shown in Figure 15, under the experimental conditions of the X direction and with 1.8 Hz excitation, the characteristics of charging the 22 μF capacitor were compared under three charging modes; one coil, three coils in series and three coils in parallel were compared. The circuit diagrams of the parallel and series charging modes of the three coils are shown in Figure 15a,b, respectively. Figure 15c. compares the three charging modes. It can be seen that multiple coils can charge the capacitor to a higher voltage in a relatively shorter time than a single coil, where three coils in parallel can charge the capacitor in the best way, being capable of charging the 22 μF capacitor to 8.6 V in 3.8 s.

To verify the application performance of the A-EMG, as shown in Figure 16, three coils were rectified and then connected in parallel to 25 parallel-connected commercial LED arrays, and single-degree-of-freedom wave motion modes, including the experimental conditions in this paper, and multiple-degree-of-freedom coupled wave motion modes, i.e., those shown in Table 4, were simulated by using a 6-DOF platform, and the LED arrays could be instantaneously lit under all these excitations.

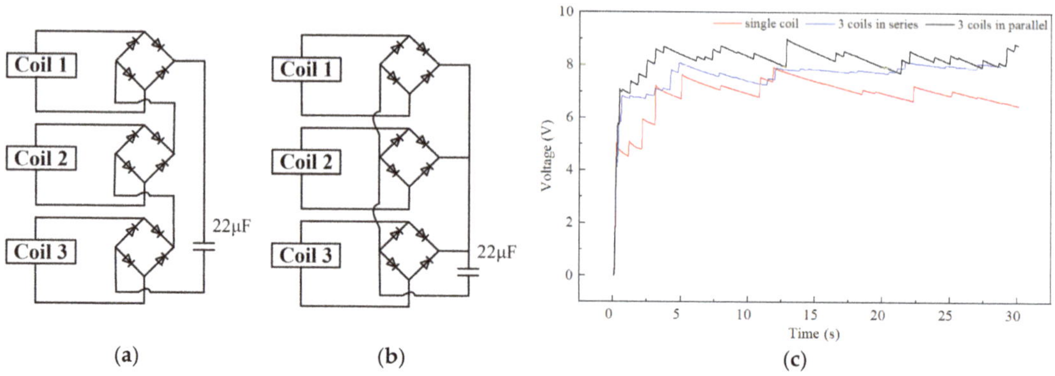

Figure 15. Capacitor charging with different connections between coils. (**a**) Series connection between coils; (**b**) parallel connection between coils; (**c**) comparison of different connections for capacitor charging.

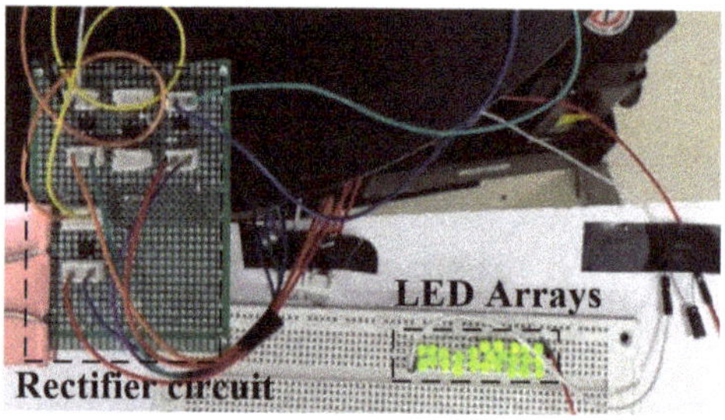

Figure 16. Application performance verification: lighting LED arrays.

Table 4. Parameters of simulating wave.

Direction	Frequency (Hz)	Amplitude (mm)	Tilt (°)
X, β coupling	1	50	3
Z, β coupling	1	50	3
X, Y coupling	1	50	-

5. Conclusions

In order to effectively harvest wave energy, this paper presents an annular electromagnetic generator for harvesting wave energy. The output performance of the three electromagnetic coil generating units was investigated. Under an external excitation of 1.8 Hz, the maximum open-circuit voltage of a single group of coils could reach 7.2 V, and the short-circuit current could reach 21.2 mA. An optimum output power of 80.87 mW was obtained when the external load was 150 Ω. The power density could reach 150.174 W/m^3. The output performance is superior to most wave energy collectors in the same field. Comparing the characteristics of the three coils after rectifying them to charge the capacitor in series or parallel, it was found that the parallel connection mode was the best, and a capacitor of 22 μF could be charged to 8.6 V in 3.8 s. Finally, the 6-DOF platform was used to simulate the typical motion characteristics of a wave, such as single-degree-of-freedom and multi-degree-of-freedom coupling, and to simulate lighting up an LED array. In summary, the device is simple in structure, low in terms of fabrication cost, responds well to the external excitation of wave energy, can continuously charge the capacitor and supply power to LED loads and has good application prospects in supplying power to ocean sensors.

Author Contributions: Conceptualization, C.W. and L.G.; methodology, C.W. and Q.F.; software, L.G.; validation, C.W., L.G. and L.C.; formal analysis, Q.F.; investigation, L.G.; resources, C.W.; data curation, L.G.; writing—original draft preparation, L.G.; writing—review and editing, C.W. and Q.F.; visualization, L.G. and P.C.; supervision, C.W.; project administration, L.C.; funding acquisition, C.W. All authors have read and agreed to the published version of the manuscript.

Funding: This research was funded by the National Key Research and Development Program of China: Efficient and Safe Electrical System Design and Study on weather resistance of offshore floating photovoltaics, grant number 2022YFB4200703, and the APC was funded by the National Key Research and Development Program of China: Efficient and Safe Electrical System Design and Study on weather resistance of offshore floating photovoltaics.

Institutional Review Board Statement: Not applicable.

Informed Consent Statement: Not applicable.

Data Availability Statement: Data are contained within the article.

Conflicts of Interest: The authors declare no conflict of interest.

Abbreviations

The following abbreviations are used in this manuscript:

A-EMG	Annular electromagnetic generator
6-DOF	Six-degree-of-freedom

References

1. Khan, F.A.; Pal, N.; Saeed, S.H. Review of solar photovoltaic and wind hybrid energy systems for sizing strategies optimization techniques and cost analysis methodologies. *Renew. Sustain. Energy Rev.* **2018**, *92*, 937–947. [CrossRef]
2. Nagura, M. Annual Rossby waves below the pycnocline in the Indian Ocean. *J. Geophys. Res. Ocean* **2018**, *123*, 9405–9415. [CrossRef]
3. Gemme, D.A.; Bastien, S.P.; Sepe, R.B.; Montgomery, J.; Grilli, S.T.; Grilli, A. Experimental testing and model validation for ocean wave energy harvesting buoys. In Proceedings of the 2013 IEEE Energy Conversion Congress and Exposition, Denver, CO, USA, 28 October 2013; pp. 337–343.
4. Pirisi, A.; Grimaccia, F.; Mussetta, M.; Zich, R.E.; Johnstone, R.; Palaniswami, M.; Rajasegarar, S. Optimization of an energy harvesting buoy for coral reef monitoring. In Proceedings of the 2013 IEEE Congress on Evolutionary Computation, Cancun, Mexico, 20–23 June 2013; pp. 629–634.
5. Zhang, Y.; Zhao, Y.; Sun, W.; Li, J. Ocean wave energy converters: Technical principle, device realization, and performance evaluation. *Renew. Sustain. Energy Rev.* **2021**, *141*, 110764. [CrossRef]
6. Ahamed, R.; McKee, K.; Howard, I. Advancements of wave energy converters based on power take off (PTO) systems: A review. *Ocean Eng.* **2020**, *204*, 107248. [CrossRef]
7. Hu, Y.; Yang, J.; Niu, S.; Wu, W.; Wang, Z.L. Hybridizing triboelectrification and electromagnetic induction effects for high-efficient mechanical energy harvesting. *ACS Nano* **2014**, *8*, 7442–7450. [CrossRef] [PubMed]
8. Arnold, D.P. Review of Microscale Magnetic Power Generation. *IEEE Trans. Magn.* **2007**, *43*, 3940–3951. [CrossRef]
9. Carneiro, P.; Soares Dos Santos, M.P.; Rodrigues, A.; Ferreira, J.A.; Simões, J.A.; Marques, A.T.; Kholkin, A.L. Electromagnetic energy harvesting using magnetic levitation architectures: A review. *Appl. Energy* **2020**, *260*, 114191. [CrossRef]
10. Bowers, B.J.; Arnold, D.P. Spherical, rolling magnet generators for passive energy harvesting from human motion. *J. Micromech. Microeng.* **2009**, *19*, 94008. [CrossRef]
11. Samad, F.A.; Karim, M.F.; Paulose, V.; Ong, L.C. A Curved Electromagnetic Energy Harvesting System for Wearable Electronics. *IEEE Sens. J.* **2016**, *16*, 1969–1974. [CrossRef]
12. Tu, D.; Zhang, Y.; Zhu, L.; Fu, H.; Qin, Y.; Liu, M.; Ding, A. A bistable vibration energy harvester with spherical moving magnets: Theoretical modeling and experimental validation. *Sens. Actuators A Phys.* **2022**, *345*, 113782. [CrossRef]
13. Graves, J.; Kuang, Y.; Zhu, M. Counterweight-pendulum energy harvester with reduced resonance frequency for unmanned surface vehicles. *Sensors Actuators A Phys.* **2021**, *321*, 112577. [CrossRef]
14. Yerrapragada, K.; Ansari, M.H.; Karami, M.A. Enhancing power generation of floating wave power generators by utilization of nonlinear roll-pitch coupling. *Smart Mater. Struct.* **2017**, *26*, 94003. [CrossRef]
15. Townsend, N.C. Self-powered autonomous underwater vehicles: Results from a gyroscopic energy scavenging prototype. *IET Renew. Power Gener.* **2016**, *10*, 1078–1086. [CrossRef]
16. Townsend, N.C.S.R. Gyrostabilizer Vehicular Technology. *Appl. Mech. Rev.* **2011**, *64*, 10801. [CrossRef]

17. Guo, Q.; Sun, M.; Liu, H.; Ma, X.; Chen, Z.; Chen, T.; Sun, L. Design and experiment of an electromagnetic ocean wave energy harvesting device. In Proceedings of the 2018 IEEE/ASME International Conference on Advanced Intelligent Mechatronics (AIM), Auckland, New Zealand, 9–12 July 2018; pp. 381–384.
18. Ding, W.; Song, B.; Mao, Z.; Wang, K. Experimental investigations on a low frequency horizontal pendulum ocean kinetic energy harvester for underwater mooring platforms. *J. Mar. Sci. Technol.* **2016**, *21*, 359–367. [CrossRef]
19. Lou, H.; Wang, T.; Zhu, S. Design, modeling and experiments of a novel biaxial-pendulum vibration energy harvester. *Energy* **2022**, *254*, 124431. [CrossRef]
20. Wang, Y.; Lee, C. Dynamics and power generation of wave energy converters mimicking biaxial hula-hoop motion for mooringless buoys. *Energy* **2019**, *183*, 547–560. [CrossRef]
21. Carandell, M.; Toma, D.M.; Gasulla, M.; Rio, J.D. Experimental Validation of a Kinetic Energy Harvester Device for Oceanic Drifter Applications. In Proceedings of the IEEE OES/MTS OCEANS 2019-Marseille, Marseille, France, 17–20 June 2019; pp. 1–7.
22. Pan, X.; Ling, P.; Bao, H.; He, W.; Li, Q.; Yan, B. Tumbler-inspired electromagnetic generator for low-frequency ocean wave energy harvesting. *Energy Convers. Manag.* **2023**, *294*, 117569. [CrossRef]
23. Maharjan, P.; Toyabur, R.M.; Park, J.Y. A human locomotion inspired hybrid nanogenerator for wrist-wearable electronic device and sensor applications. *Nano Energy* **2018**, *46*, 383–395. [CrossRef]
24. Gao, L.; Lu, S.; Xie, W.; Chen, X.; Wu, L.; Wang, T.; Wang, A.; Yue, C.; Tong, D.; Lei, W.; et al. A self-powered and self-functional tracking system based on triboelectric-electromagnetic hybridized blue energy harvesting module. *Nano Energy* **2020**, *72*, 104684. [CrossRef]
25. He, J.; Fan, X.; Mu, J.; Wang, C.; Qian, J.; Li, X.; Hou, X.; Geng, W.; Wang, X.; Chou, X. 3D full-space triboelectric-electromagnetic hybrid nanogenerator for high-efficient mechanical energy harvesting in vibration system. *Energy* **2020**, *194*, 116871. [CrossRef]
26. Hao, C.; He, J.; Zhai, C.; Jia, W.; Song, L.; Cho, J.; Chou, X.; Xue, C. Two-dimensional triboelectric-electromagnetic hybrid nanogenerator for wave energy harvesting. *Nano Energy* **2019**, *58*, 147–157. [CrossRef]
27. Gao, S. Design and Implementation of a Piezoelectric and Magnetoelectric Vibration Energy Harvester. Master's Thesis, Harbin Institute of Technology, Harbin, China, 2017.
28. Wu, K. Research On Magnetic Spring Vertical Vibration Energy Harvesting Technology. Master's Thesis, Xidian University, Xi'an, China, 2013.
29. Vidal, J.V.; Rolo, P.; Carneiro, P.M.; Peres, I.; Kholkin, A.L.; dos Santos MP, S. Automated electromagnetic generator with self-adaptive structure by coil switching. *Appl. Energy* **2022**, *325*, 119802. [CrossRef]
30. Zhang, Y. Design and Analysis of Multi Degree of Freedom Wave Energy Power Generation Device. Master's Thesis, Shantou University, Shantou, China, 2021.
31. Tao, Y. Design and Analysis of Multi-Degree-of-Freedom Wave Power Generation System. Master's Thesis, Tianjin University, Tianjin, China, 2017.
32. Chen, X.; Gao, L.; Chen, J.; Lu, S.; Zhou, H.; Wang, T.; Wang, A.; Zhang, Z.; Guo, S.; Mu, X.; et al. A chaotic pendulum triboelectric-electromagnetic hybridized nanogenerator for wave energy scavenging and self-powered wireless sensing system. *Nano Energy* **2020**, *69*, 104440. [CrossRef]
33. Li, Y.; Ma, X.; Tang, T.; Zha, F.; Chen, Z.; Liu, H.; Sun, L. High-efficient built-in wave energy harvesting technology: From laboratory to open ocean test. *Appl. Energy* **2022**, *322*, 119498. [CrossRef]
34. Gao, X. Triboelectric Nanogenerator for Harvesting Multidirectional Water Wave Energy. Master's Thesis, Lanzhou University, Lanzhou, China, 2022.
35. Shi, G.; Zeng, W.; Xia, Y.; Xu, J.; Jia, S.; Li, Q.; Wang, X.; Xia, H.; Ye, Y. A floating piezoelectric electromagnetic hybrid wave vibration energy harvester actuated by a rotating wobble ball. *Energy* **2023**, *270*, 126808. [CrossRef]

Journal of
Marine Science
and Engineering

MDPI

Article

Modeling and Simulation of a Turbine Access System with Three-Axial Active Motion Compensation

Jingfu Wang [1], Songtao Zhang [1,*], Jiahuan Cheng [2], Yunfei Li [1,*], Yan Shen [1] and Zihao Wu [1]

1 College of Intelligent Systems Science and Engineering, Harbin Engineering University, Harbin 150001, China; ahwjf@hrbeu.edu.cn (J.W.); shenyan0715@163.com (Y.S.); xiaocuihuaer@163.com (Z.W.)
2 China Ship Development and Design Center, Shanghai 201108, China; cjh198808@163.com
* Correspondence: hrbzst@126.com (S.Z.); liyunfei@hrbeu.edu.cn (Y.L.); Tel.: +86-139-4603-2279 (S.Z.); +86-176-3486-6593 (Y.L.)

Abstract: As an essential transportation equipment for personnel to access offshore wind plants, the safety and stability of the turbine access system (TAS) have gained increasing significance. However, when sea conditions deteriorate, the end of the TAS will experience large-angle shaking and an increase in the heave height. A novel 3-DOFs TAS with active motion compensation for the Fujian coastal area is designed to solve the problem of the stability of the end of the TAS under sea conditions with a significant wave height of 2.2 m, including structural design, kinematic analysis, hydraulic drive design, and control system design. In the research, a new stacking compensation method is proposed to compensate for the roll angle, pitch angle, and heave height at the end of the TAS. The 3-DOFs TAS is modeled mathematically by the Denavit–Hartenberg parameters, and the hydraulic system model is established. In order to improve the compensation effect, a fuzzy PID controller with feedforward compensation is designed based on fuzzy PID control, and the model simulation experiment is carried out by MATLAB/Simulink. Finally, the experimental simulation shows that under the compensation of the feedforward compensation fuzzy PID control algorithm and the new compensation algorithm, the roll angle of the TAS is reduced by a maximum of 84.8%, the pitch angle is reduced by 75%, and the heave height is reduced by 73.6%. This validates the effectiveness of the proposed scheme and provides a reference for future TAS design and development.

Keywords: active motion compensation; turbine access system; kinematic analysis; fuzzy PID control; offshore wind plants

Citation: Wang, J.; Zhang, S.; Cheng, J.; Li, Y.; Shen, Y.; Wu, Z. Modeling and Simulation of a Turbine Access System with Three-Axial Active Motion Compensation. *J. Mar. Sci. Eng.* **2023**, 11, 2237. https://doi.org/10.3390/jmse11122237

Academic Editor: José-Santos López-Gutiérrez

Received: 24 October 2023
Revised: 15 November 2023
Accepted: 22 November 2023
Published: 26 November 2023

1. Introduction

In the context of the growing tension surrounding land resources, a rising awareness has emerged regarding the abundant resources of the ocean. Offshore wind plants present several advantages over land-based wind plants, such as a smaller land use area and more abundant wind resources [1]. In 2021, China added 16.9 million kilowatts of wind turbines, which is 1.8 times as many as all other wind turbines built before. Some scholars have even proposed new hybrid devices that can capture wind and wave energy and proposed new optimization solutions [2]. However, the harsh marine environment poses a threat to the safety of maritime operators, mainly due to the roll, pitch, yaw, sway, surge, and heave caused by waves [3]. Therefore, this requires the turbine access system (TAS) to transport operators and various equipment from ships to wind turbines. This not only improves the safety of offshore operators and reduces the difficulty of offshore operations, but also improves the efficiency of offshore operations [4].

In the offshore industry, two means of transport are being used to reach offshore structures: helicopters and vessels [5].

1. Helicopter access. Helicopters have higher safety and commuting efficiency and are mainly used for the operation and maintenance of large wind farms. Helicopter

transportation and transfer are not limited by wave conditions. China's offshore wind power plants are built on the continental shelf, usually 10–20 km away from the coast, while in some European countries, such as the Netherlands, offshore wind power plants can reach a distance of 30–100 km [6]. Taking an offshore wind farm 100 km offshore as an example, the average maximum speed of crew transfer vehicles (CTVs) is 25 knots, which can be reached by boat in 2.2 h, while it only takes half an hour by helicopter [7]. However, there are limitations to the use of helicopters. It is required to install a platform on the wind turbine for helicopters and personnel to use. Although this method is fast, its drawbacks are high operating costs and the need for a lifting platform for each turbine. In addition, it is not suitable for situations with high winds or low visibility, which can easily lead to accidents. Although the probability of helicopter crashes is low, the likelihood of such accidents causing a large number of deaths is high.

2. Ship-based access. In oil and gas extraction, personnel are generally transported through personnel baskets and swing ropes. Both methods have a major drawback which is that in order to ensure personnel safety, they must be transported under relatively mild sea conditions. Both methods require the ship to be equipped with a crane, which transports personnel to the target working point through its long boom and cable. However, once encountering wind and waves that cause the ship to sway, the long boom of the crane will have a strong amplification effect, causing severe shaking at the end of the boom and cable. This poses a serious threat to the safety of personnel. Due to unpredictable wind and wave conditions, offshore lifting operations are difficult. Performing lifting operations typically requires special expensive, and scarce equipment [5,8]. In addition, the ship is propelled forward through the propeller to increase friction between the bow rubber buffer and the landing point, which can eliminate the effects of sway and yaw. The DP system (dynamic positioning system) can effectively suppress the sway and surge of the ship's hull, which can reduce the impact of waves on the ship. However, an important drawback of this access method is that it is only limited to medium wave conditions [9] Considering safety and cost-effectiveness issues, the offshore wind power industry is not keen on using helicopters as the main access method for offshore wind turbines, and ships have become the most commonly used vehicles.

After determining the vehicle, it is necessary to determine the method to compensate for interference caused by waves, which requires the use of active motion compensation technology to achieve appropriate compensation. The research on motion compensation technology originated from the needs of offshore drilling, and later evolved into various compensation methods, such as velocity compensation, displacement compensation, force compensation, and comprehensive compensation [10]. Taking seabed salvage, recovery and rescue operations as an example, Southland outlined the difficulties of handling heavy objects at sea and proposed, for the first time, the use of active and passive compensation systems to address interference caused by wind and waves [11].

In current offshore operations, whether using the TAS or other offshore tools, such as offshore cranes, the heave effect is the most affected. Therefore, researchers have conducted extensive research on the heaving effect of waves. In order to reduce the impact of heave motion on the launch and recovery of remotely operated vehicles (ROVs), Yang et al. [12] designed an active heave compensation system mainly based on hydraulic winches, which reduces the impact of the heave motion of the tether management system (TMS) on remote-controlled submersibles by controlling the extension and contraction of the cable. Sebastian et al. [13] proposed an active compensation control algorithm to control hydraulic-driven winches, which considers the delay between the length of the winch rope and the payload to compensate for the vertical motion of the ship. Huang et al. [14] used the Lagrange kinematics equation and numerical simulation to analyze the influence of the rope length and lifting speed of the floating crane wave compensation system on the system response, which has great reference significance for the design of wave compensation systems. Zhou et al. [15] proposed a genetic PID (proportional-integral-differential) controller with feedforward compensation for compensating cranes to optimize the optimal indicators and PID parameters of the system. While maintaining system stability and accuracy, it improved the system's fast response ability and achieved good control results. Xie et al. [16] proposed an active heave compensation method of an electric marine winch based on the sliding-mode control algorithm to solve the six degrees of freedom of the mother ship caused by wave fluctuations and achieved excellent results.

The research on the TAS based on wave compensation has also made significant progress. Christopher put forward the earliest concept of the gangway, which includes a support frame designed to be supported by grating positioned above horizontal structural members of the offshore structure. It mainly provides access in environments seriously affected by weather conditions [17]. There are many factors that affect the compensation effect of TAS, such as model complexity, control strategies, and structural design. For model complexity, Rong et al. proposed a discrete time transfer matrix method of a multibody system for dynamic modeling and analysis of a ship's seaborne supply (SSS) systems, which is used to solve real-time dynamic analysis of SSS operations under complex sea conditions [18]. The method can easily solve the dynamic problems of the system by only using low-order transfer equations, which have lower computational complexity and a satisfactory compensation effect. For control algorithms, Tang et al. established and simulated the wave compensation control model using particle swarm optimization (PSO) to optimize the control parameters of the controller. The result showed that this design has the optimization effects of small over-shoot and a fast response time [19]. Cai et al. proposed a sliding-mode control scheme for the ship-mounted Stewart platform to increase the workable time of carrying out these operations of offshore installations [20]. The novel velocity feedforward compensator and command-filtered-based sliding-mode backstepping controller have been proposed and used on this platform, demonstrating good compensation performance. Yin et al. proposed a novel Stewart platform with a gangway, which is equipped with a robust controller and an estimator [21]. This ensures that the control force does not increase indefinitely with position and attitude errors. The simulation results verified the effectiveness and performance of the controller. Chen proposed a three-loop control strategy based on active disturbance rejection control (ADRC), in which the internal model controller (IMC) in the current loop is designed to achieve fast control. This method has excellent performance in reducing power consumption, decoupling control, and anti-interference [22]. Zhang et al. proposed a system method for designing multivariable fuzzy logic controllers, which applies genetic algorithms to optimize fuzzy scaling factors and can reject strong disturbances [23]. Bai et al. proposed a kinematic-based Lagrangian method for generating motion equations and designing adaptive control laws for multi-body systems based on the dynamic analysis and controller design of the Stewart platform. This method can tolerate performance under large parameter errors and external interference [24]. Liu et al. designed an improved adaptive control strategy based on a radial basis function neural network (RBFNN) with fading factors to compensate for external interference on the Stewart platform. This method reduced the error by 70% and the heave compensation error by 40% [25].

For structural renovation, Chen et al. proposed a new turbine access system suitable for the Taiwan Strait [26]. This design effectively reduced the roll angle, vertical displacement, and vertical acceleration. However, the design is only suitable for marine environments with a significant wave height of 1.5 m. Huang et al. added a fourth axis mechanism to the 3-DOFs TAS, proposing a four-axis TAS to compensate for the surge displacement, heave displacement, pitch angle, and roll angle of the end effector under wave conditions and load tests in the Taiwan Strait. This study used a model reference robust adaptive controller (MRRAC) for control [27]. Tang et al. proposed a four degrees of freedom rope-driven, rigid, flexible hybrid wave compensation mechanism. This wave compensation device can achieve compensation for four degrees of freedom: heave, sway, roll, and pitch, by controlling the mobile platform, reducing, or even eliminating, the relative motion of the ship [28]. In addition, some commercial companies have also launched motion compensation systems for offshore wind power plants. The company Ampelmann launched the first 6-DOFs TAS Stewart, which measures the motion state of the ship in real time through sensors [29]. Nonetheless, the device was constructed using a platform featuring six hydraulic cylinders, resulting in a structurally intricate design that necessitates a substantial amount of space. The company Houlder (London, UK) has developed a kind of TAS, which can compensate for roll, pitch, and surge, but cannot achieve full compensation [30].

This study focuses on reducing the end-angle shaking and vertical height increase of the TAS under adverse sea conditions, and proposes a novel TAS designed for the coastal area of Fujian, which addresses the challenge of maintaining the stability of the TAS in the presence of sea conditions characterized by significant wave heights of up to 2.2 m. A new stacking compensation method is proposed to compensate for the roll angle, pitch angle, and heave height at the end of the TAS. The effects of ship roll, pitch, and heave on the end of the TAS are analyzed separately and compensation values are calculated. The sum of the compensation values is used as the final compensation value for the TAS. In addition, the mathematical modeling of a 3-DOFs TAS and hydraulic system is conducted using the D-H modeling method. A fuzzy PID controller with feedforward for the TAS is developed based on fuzzy PID control theory. The model simulation experiments are then performed using MATLAB/Simulink. Finally, the simulation shows that compared to PID controllers, the control strategy proposed in this study can reduce the roll angle, pitch angle, and heave height of the TAS by 84.8%, 75%, and 73.6%, respectively. The displacement of the TAS end in the X, Y, and Z axes is reduced by at least 65%, with a maximum reduction of 82.69%.

2. Mechanical Structure Design of the TAS

The design proposed in this study solely accounts for the influence of roll, pitch, yaw, and heave on the ship, as it is assumed that the vessel will be outfitted with a DP system.

Based on the above considerations, a TAS with three degrees of freedom active motion compensation, including a rotary joint, pitch joint, and telescopic joint, was designed. The mechanical structure and dimensions of the TAS are shown in Figure 1. The rotary joint contains a symmetrical hydraulic cylinder, the pitch joint contains two asymmetrical hydraulic cylinders, and the telescopic joint contains one asymmetrical hydraulic cylinder. As roll, pitch, and yaw represent rotational movements, the end effector of the TAS may produce varied movements in distinct dimensions, necessitating coordinated movements among different joints for compensation. Specifically, to effectively compensate for roll motion, the pitch and telescopic joints must cooperate, while other joints may need to collaborate to address pitch and yaw movements, and so on. In addition, there is a ladder at the back of the TAS for people to pass, and there is an aluminum alloy guardrail at the top to protect the workers' safety.

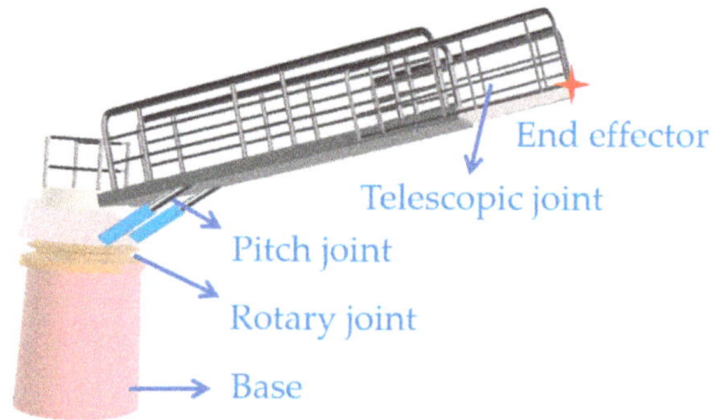

Figure 1. Mechanical structure of the TAS.

3. Ship Motion Model

In order to obtain motion of the TAS, it is necessary to obtain the motion of the ship when it is disturbed by sea waves. Therefore, modeling of random sea waves and ship motions must be carried out.

3.1. Wave Model

It is difficult to calculate the complex and irregular wave motion. However, it is generally believed that when studying the effect of sea waves on ships, the sea waves are regarded as a stationary random process [31]. The power spectral density of waves is a crucial aspect of ocean wave models, as it illustrates the distribution of wave energy across different frequencies. Now, Pierson–Moskowitz and JONSWAP spectra are extensively employed in oceanographic research because they can describe wave motion in most sea areas.

However, the wave motion in different sea areas is different. This study focuses on the coastal region of Fujian Province and adopts the Putian spectrum derived by Yang et al. [32] from the statistical analysis of the nearby sea area of Putian City, Fujian Province. It can be described as:

$$\frac{S(\omega/\overline{\omega})}{M_0} = 1.8865\left(\frac{\omega}{\overline{\omega}}\right)^{-4.5} \exp\left[-0.9632\left(\frac{\omega}{\overline{\omega}}\right)^{-2.8}\right] \times \left[1 + 0.4334\left(\frac{\omega}{\overline{\omega}}\right)^{-2.8}\right] \tag{1}$$

$$H_{1/3} = 3.75\sqrt{M_0} \tag{2}$$

where ω is the wave frequency, $\overline{\omega}$ is the average wave frequency, M_0 is the zero-order moment of the power spectral density of the wave, and $H_{1/3}$ is the significant wave height.

The sea wave can be seen as a superposition of countless harmonic components with different frequencies and phases [33]. These tiny cosines have random initial phases, which is noteworthy. Assuming the existence of the ground coordinate system, $E-\xi\eta\varsigma$, the ship coordinate system, $O-xyz$, the $E-\xi\eta$ plane parallel to the Ox axis, and the angle between the Ox axis and the $E\xi$ axis is μ, the wave height can be described as:

$$\zeta(t) = \sum_{i=1}^{\infty} \zeta_{ai}\cos(k_i\xi\cos\mu + k_i\eta\sin\mu - \omega_i t + \varepsilon_i) \tag{3}$$

where ζ_{ai} is the amplitude of the i-th harmonic component, k_i is the wave number of the i-th harmonic component, ω_i is the angular frequency of the i-th harmonic component, and ε_i is the initial phase of the i-th harmonic component.

According to the statistical results of Jian Shi et al. [34] on the sea wave data along the coast of China from 1979 to 2017, the mean significant wave height and mean period in the four seasons near Putian City, Fujian Province, from 1979 to 2017 are shown in Table 1.

Table 1. The mean significant wave height and mean period of the coast of Putian City during the four seasons from 1979 to 2017.

Season	Mean Significant Wave Height (m)	Mean Period (s)
Spring	1.5	5.3
Summer	1.2	5.7
Autumn	2.2	6.8
Winter	2.5	6.5

According to Table 1, it can be calculated that the mean significant wave height within a year is 1.85 m, and the mean wave period within a year is 6.075 s. Therefore, the mean period of waves in this paper was set to 6 s. Due to the significant differences in the height of meaningful waves in different seasons along the coast of Putian, in order to ensure that the TAS can be used for most of the year, the maximum significant wave height of the usage environment was set to 2.2 m.

From Figure 2a, the frequency of the energy spectrum concentration of waves ranges from 0.5 rad/s to 3.5 rad/s, and the peak frequency of the spectrum is around 0.8 rad/s which is not consistent with the mean frequency. This is normal because the peak frequency is indeed related to the average frequency of waves, but it is not always the same. From Figure 2b, the amplitude of the wave is between 0 m and 3.5 m, and the first one-third of the amplitude height is concentrated between 1.8 m and 3.5 m, which is consistent with the significant wave height of 2.2 m.

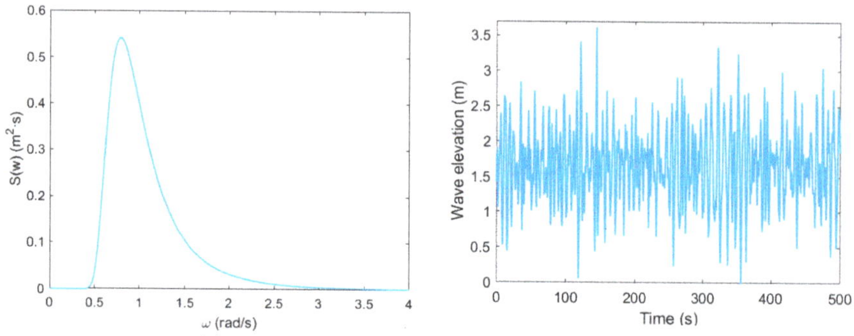

Figure 2. Wave simulation: (**a**) power spectral density of the wave and (**b**) height of the wave.

For the calculation of peak frequency, the peak frequency is the frequency at spectral density, indicating that the most wave energy is concentrated at this frequency. The average frequency is the average of all frequencies, weighted by their respective spectral density values. The average frequency of the waves in Figure 2a can be obtained by the following equation:

$$\begin{cases} S(\omega_i)_{total} = \sum\limits_{i=1}^{n} S(\omega_i) \\ \omega_0 = \sum\limits_{i=1}^{n} \frac{\omega_i S(\omega_i)}{S(\omega_i)_{total}} \end{cases} \tag{4}$$

where ω_0 is the peak frequency, $S(\omega_i)$ is the power spectrum of ω_i, and $S(\omega_i)_{total}$ is the total power spectrum.

According to Equation (4), the average wave period of the simulation model ω_0 = 1.0767 rad/s. According to the set period T_0 = 6 s, the set frequency of ocean waves can be calculated as:

$$\omega = \frac{2\pi}{T} = 1.05\ \text{rad}/s \tag{5}$$

This is very similar to the calculation results, ω_0, verifying the correctness of the wave model

3.2. Model of Ship Motion

The forces and moments acting on the ship are affected by many factors, and can usually be expressed as:

$$F = F_s + F_H + F_P + F_c + F_D \tag{6}$$

$$G = G_s + G_H + G_P + G_c + G_D \tag{7}$$

where F_s and G_s are the restoring force and restoring moment received of the ship, respectively, F_H and G_H are the hydrodynamic force and hydrodynamic moment of the ship, respectively, F_P and G_P are the propeller thrust and thrust moment of the ship, respectively, F_c and G_c are the control force and restraint moment provided by the ship control system, respectively, while F_D and G_D are the disturbing forces and moments generated by the marine environment, respectively.

It can be seen from Equation (3) that the long peak wave is composed of numerous sub-waves, so the interference force and moment of roll, pitch, yaw, and heave on the ship can be described as:

$$R = \sum_{i=1}^{n} R_i \tag{8}$$

$$P = \sum_{i=1}^{n} P_i \tag{9}$$

$$Y = \sum_{i=1}^{n} Y_i \tag{10}$$

$$H = \sum_{i=1}^{n} H_i \tag{11}$$

where R_i is the roll interference force and interference torque of the i-th harmonic component to the ship, P_i is the pitch interference force and interference torque of the i-th harmonic component to the ship, Y_i is the yaw interference force and interference torque of the i-th harmonic component to the ship, and H_i is the heave interference force and interference torque of the i-th harmonic component to the ship.

Based on the above theory, the six degrees of freedom simulation of a supply ship was carried out using the MATLAB MSS (Marine Systems Simulator) toolbox [35]. The parameters of the supply vessel are shown in Table 2.

Table 2. Parameters of the supply vessel.

Parameter	Value
Draught (m)	6.0
Breadth (m)	19.2
Length between perpendiculars (m)	82.8
Mass (kg)	6.3622×10^6
Radius of gyration in roll (m)	6.72
Radius of gyration in pitch (m)	2.70
Radius of gyration in yaw (m)	20.70
Volume displacement (m^3)	6.207×10^3
Transverse metacentric height (m)	2.1140
Lateral metacentric height (m)	103.6280

The simulation conditions were set as follows: the significant wave height was 2.2 m, the mean period was 6 s, the peak frequency was 0.8 rad/s, and the encounter angles were 45 and 90 degrees. The 6-DOFs simulation results of the ship are shown in Figure 3.

From Figure 3a,b, the maximum amplitude of roll is about 2.2 degrees, and the amplitude is constantly changing. The maximum angles of pitch and yaw are 2 and 0.7 degrees, respectively. The maximum displacement of sway and surge is 0.4 m, and the maximum displacement of heave is 0.7 m.

However, the encounter angle can affect the 6-DOFs motion of the ship. From Figure 3c,d, when the encounter angle is 90 degrees, the roll angle significantly increases, with the maximum amplitude approaching 3 degrees. This is because the angle of action of the waves has changed from 45 to 90 degrees, which increases the force on the hull, resulting in an increase in the amplitude of the roll angle. On the contrary, the amplitude of pitch decreases due to a decrease in longitudinal force on the hull. For linear motion, the increase in roll causes an increase in the shaking amplitude of the ship, resulting in an increase in vertical motion displacement, while the rest of the motion changes little.

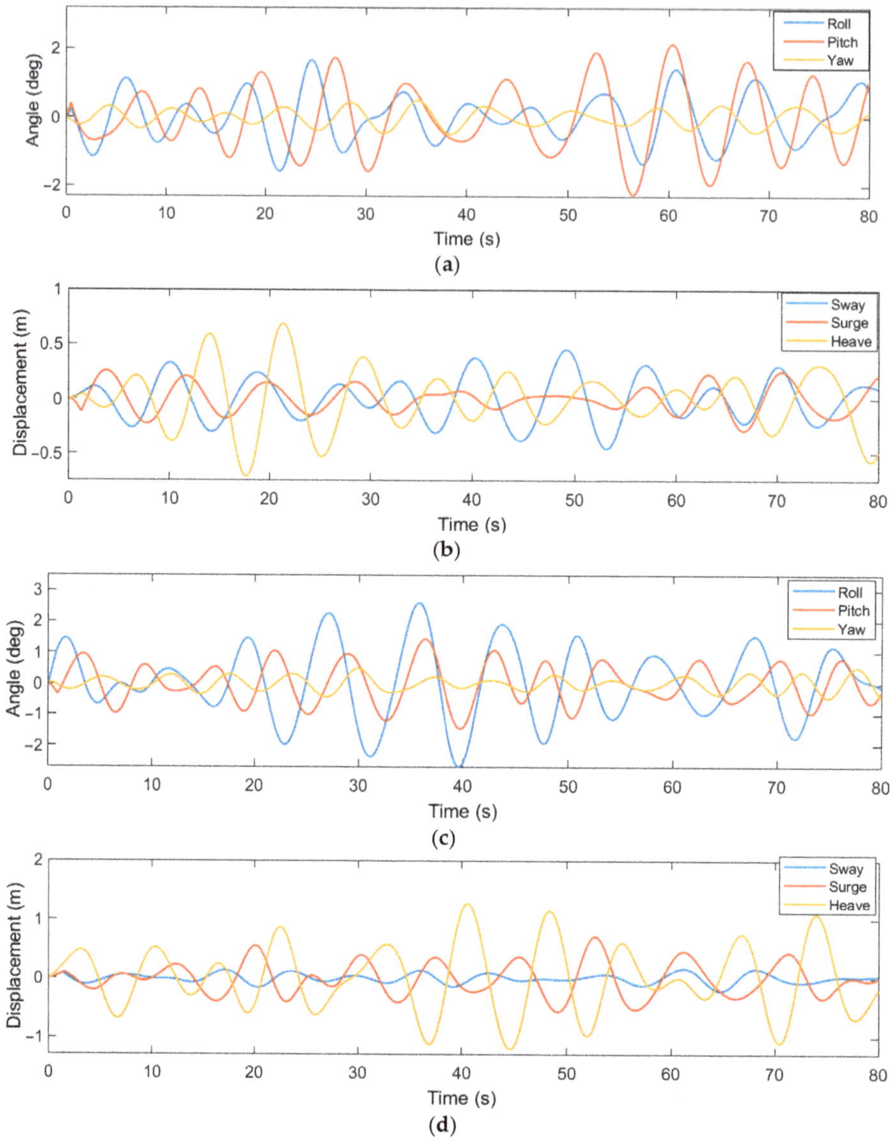

Figure 3. Six DOFs motion of the vessel. (**a**) The angle of roll, pitch, and yaw of the vessel when the encounter angle is 45 degrees. (**b**) The displacement of sway, surge, and heave of the vessel when the encounter angle is 45 degrees. (**c**) The angle of roll, pitch, and yaw of the vessel when the encounter angle is 90 degrees. (**d**) The displacement of sway, surge, and heave of the vessel when the encounter angle is 90 degrees.

4. Kinematic Model of the TAS

4.1. Forward Kinematics of the TAS

The setting of the coordinate system is shown in Figure 4. Point G is the center of the geodetic coordinate system, point V is the center of the ship's coordinate system, and point B is the geometric center of the base bottom of the TAS. In this study, G and V were set to coincide. In addition, V is the intersection of the ship's roll axis, pitch axis, and yaw axis. The installation position of the TAS is on the starboard side of the ship, parallel to and directly above the pitch axis, y_V. The distance between B and V on the y_V axis is B_y. The distance between B and V on the z_V axis is B_z.

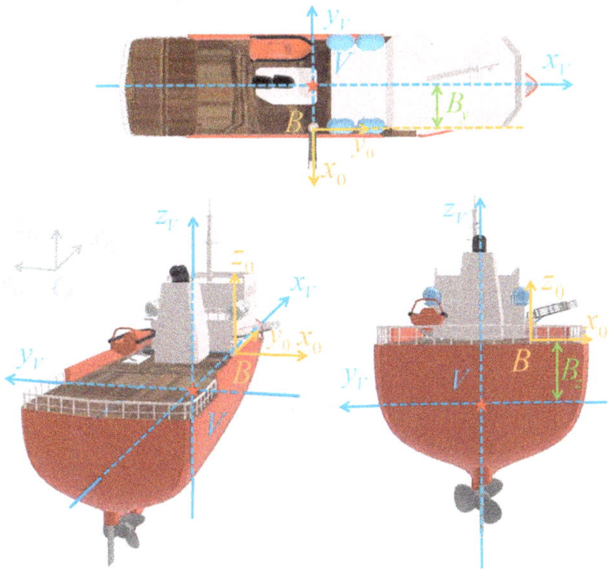

Figure 4. Coordinate system settings.

Therefore, the position of point B in the geodetic coordinate system {G} is:

$$ {}_{B}^{G}P = \begin{bmatrix} B_x \\ B_y \\ B_z \end{bmatrix} \tag{12} $$

The motion modeling of the TAS focuses on the motion state of the end of the TAS device to obtain the motion state of the end effector under different sea conditions. This study employed Denavit–Hartenberg (D-H) parameters to determine the motion state based on the connecting rod parameters.

The coordinate system for the TAS is established in Figure 5. Coordinate system {0} is positioned at the geometric midpoint of the base bottom, while coordinate system {1} is located at the intersection of the rotary and pitch joints. Additionally, coordinate system {2} is established on the pitch arm. While the Y-axis shifts in tandem with the pitch arm movement, the coordinate origin remains identical to {1}. The coordinate system {3} is established at the intersection of the pitch and telescopic arms, and the z-axis points in the telescopic direction. The coordinate system {4} is established at the end of the telescopic arm, namely the end effector of the TAS. According to the D-H parameters, the four parameters between the two joints can be determined: connecting rod length, a_{i-1}, torsion angle, α_{i-1}, connecting rod offset, d_i, and joint angle, θ_i, as shown in Table 3.

In this study, the parameters, such as the rotary joint angle, α, pitch joint angle, β, and variable telescopic length, T, are limited according to the design requirements. L_1 is the distance in the Z-axis from the origin of the coordinate system {0} to the origin of coordinate system {1}. L_2 is the distance between the origin of the coordinate system {0} and the origin of the coordinate system {1} in the X-axis. L_3 is the length of the TAS when the telescopic joint is not used. L_5 is the width of the passageway of the TAS. L_6 is the diameter of the base.

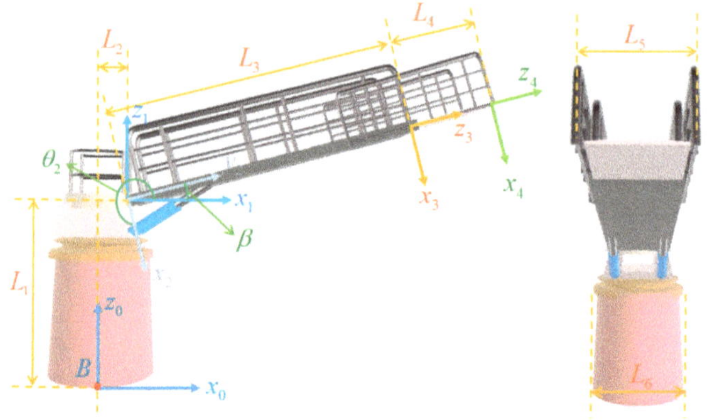

Figure 5. Coordinate system and D-H parameters of the TAS.

Table 3. D-H parameters.

i	a_{i-1}	α_{i-1}	d_i	θ_i
1	0	L_2	L_1	θ_1
2	$\pi/2$	0	0	θ_2
3	$-\pi/2$	0	L_3	0
4	0	0	L_4	0

The following equation can be obtained, as:

$$\begin{cases} \theta_1 = \alpha \\ \theta_2 = \frac{3}{2}\pi + \beta \\ L_4 = T \end{cases} \tag{13}$$

The coordinate transformation matrix of the latter coordinate system $\{i\}$ relative to the previous coordinate system $\{i-1\}$ can be obtained. It can be expressed as:

$$ {}^{i-1}_{i}T = \begin{bmatrix} {}^{i-1}_{i}R & {}^{i-1}P_{iORG} \\ 0 & 1 \end{bmatrix} = \begin{bmatrix} c\theta_i & -s\theta_i & 0 & a_{i-1} \\ s\theta_i c\alpha_{i-1} & c\theta_i c\alpha_{i-1} & -s\alpha_{i-1} & -s\alpha_{i-1}d_i \\ s\theta_i s\alpha_{i-1} & c\theta_i s\alpha_{i-1} & c\alpha_{i-1} & c\alpha_{i-1}d_i \\ 0 & 0 & 0 & 1 \end{bmatrix} \tag{14}$$

where ${}^{i-1}_{i}R$ is the rotation matrix of the coordinate system $\{i\}$ relative to the coordinate system $\{i-1\}$, and ${}^{i-1}P_{iORG}$ is the position vector of the origin of the coordinate system $\{i\}$ relative to the origin of the coordinate system $\{i-1\}$, $c\theta_i$ is the abbreviation of $\cos\theta_i$, $s\theta_i$ is the abbreviation of $\sin\theta_i$, and so on.

According to Table 3, the transformation matrix of the latter coordinate system relative to the previous coordinate system can be calculated as:

$$ {}^{0}_{1}T = \begin{bmatrix} \cos\theta_1 & -\sin\theta_1 & 0 & L_2 \\ \sin\theta_1 & \cos\theta_1 & 0 & 0 \\ 0 & 0 & 1 & L_1 \\ 0 & 0 & 0 & 1 \end{bmatrix} \tag{15}$$

$$ {}^{1}_{2}T = \begin{bmatrix} \cos\theta_2 & -\sin\theta_2 & 0 & 0 \\ 0 & 0 & -1 & 0 \\ \sin\theta_2 & \cos\theta_2 & 0 & 0 \\ 0 & 0 & 0 & 1 \end{bmatrix} \tag{16}$$

$$ {}^{2}_{3}T = \begin{bmatrix} 1 & 0 & 0 & 0 \\ 0 & 0 & 1 & L_3 \\ 0 & -1 & 0 & 0 \\ 0 & 0 & 0 & 1 \end{bmatrix} \tag{17} $$

$$ {}^{3}_{4}T = \begin{bmatrix} 1 & 0 & 0 & 0 \\ 0 & 1 & 0 & 0 \\ 0 & 0 & 1 & L_4 \\ 0 & 0 & 0 & 1 \end{bmatrix} \tag{18} $$

The transformation matrix of the end of the TAS relative to the base can be obtained as:

$$ {}^{0}_{4}T = {}^{0}_{1}T^{1}_{2}T^{2}_{3}T^{3}_{4}T = \begin{bmatrix} c\theta_1 c\theta_2 & -s\theta_1 & -c\theta_1 s\theta_2 & L_2 - (L_3 + L_4)c\theta_1 s\theta_2 \\ c\theta_2 s\theta_1 & c\theta_1 & -s\theta_1 s\theta_2 & -(L_3 + L_4)s\theta_1 s\theta_2 \\ s\theta_2 & 0 & c\theta_2 & L_1 + L_3 c\theta_2 + L_4 c\theta_2 \\ 0 & 0 & 0 & 1 \end{bmatrix} \tag{19} $$

The displacement of the TAS end effector relative to the base can be obtained from Equation (19) as:

$$ \begin{bmatrix} Tx \\ Ty \\ Tz \end{bmatrix} = \begin{bmatrix} L_2 - (L_3 + L_4)c\theta_1 s\theta_2 \\ -(L_3 + L_4)s\theta_1 s\theta_2 \\ L_1 + (L_3 + L_4)c\theta_2 \end{bmatrix} \tag{20} $$

In addition, since the base of the TAS and the center of the hull are not coincident, the coordinate transformation matrix of the base $\{B\}$ relative to the vessel center $\{V\}$ can be defined as:

$$ {}^{V}_{B}T = \begin{bmatrix} 0 & 1 & 0 & B_x \\ -1 & 0 & 0 & B_y \\ 0 & 0 & 1 & B_z \\ 0 & 0 & 0 & 1 \end{bmatrix} \tag{21} $$

where B_x, B_y, and B_z are the offsets of the base on the x-axis, y-axis, and z-axis relative to the hull center.

As the ship is always disturbed by wind, waves, and currents when sailing on the sea, the hull center moves with six degrees of freedom relative to the ground coordinate system [36]. In the research, the geodetic coordinate system $\{G\}$ was set at the position of the hydrodynamic coordinate system $\{H\}$. Equation (22) provides a description of the coordinate transformation matrix for the hull center concerning the ground coordinate system:

$$ {}^{G}_{V}T = \begin{bmatrix} c\psi c\theta & c\psi s\theta s\varphi - s\psi c\varphi & c\psi s\theta c\varphi + s\psi c\varphi & x \\ s\psi c\theta & s\psi s\theta s\varphi + c\psi c\varphi & s\psi s\theta c\varphi - c\psi s\varphi & y \\ -s\theta & c\theta s\varphi & c\theta c\varphi & z \\ 0 & 0 & 0 & 1 \end{bmatrix} \tag{22} $$

where φ is the roll angle, θ is the pitch angle, ψ is the yaw angle, x is the sway displacement, y is the surge displacement, and z is the heave displacement.

The transformation matrix of the TAS end effector relative to the ground coordinate system can be calculated as:

$$ {}^{G}_{4}T = {}^{G}_{V}T \cdot {}^{V}_{B}T \cdot {}^{0}_{4}T \tag{23} $$

Finally, according to Equation (23) and Table 4, the reachable working space of the TAS within the specified rotation joint angle, pitch joint angle, and telescopic length limits can be obtained, as shown in Figure 6.

Table 4. Motion parameters of the TAS.

Parameter	Value
α	−20°~20°
β	−20°~20°
L_1	2.32 m
L_2	0.325 m
L_3	3.625 m
L_4	0~1.2 m
L_5	0.9 m
L_6	1.47 m

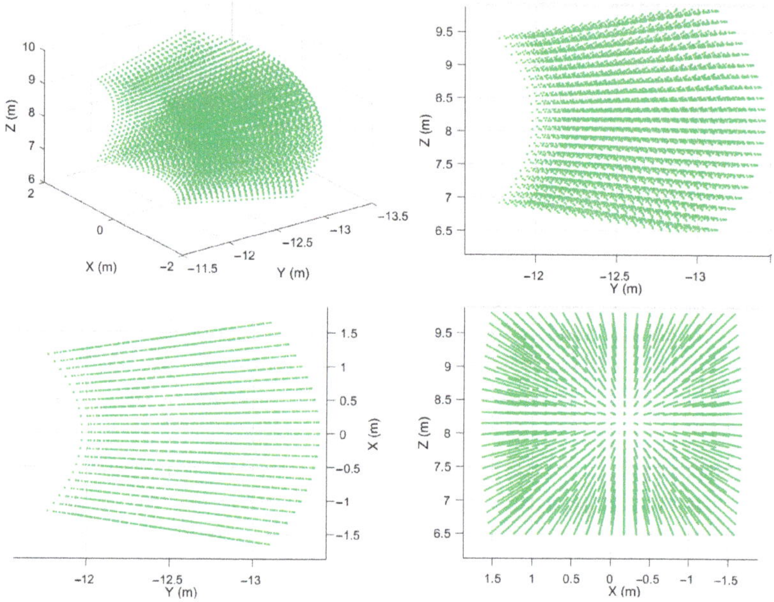

Figure 6. Reachable working space of the TAS.

It is easy to see that the TAS in this study has a wide workspace and the outermost side of the workspace is curved. The workspace ranges −1.65 m~1.65 m on the x-axis, −13.38 m~−11.77 m on the y-axis, and 6.49 m~9.79 m on the z-axis. The maximum distance of the TAS in the workspace on the x-axis is greater than its telescopic length, L_4, which is due to the curved surface formed at the edge of the workspace at the end of the TAS, and the distance between the edges of the two curved surfaces is greater than the width of the inner side, as can be seen from the first image in Figure 6.

4.2. Compensation Value

A superimposed compensation scheme is proposed to obtain appropriate compensation values. By analyzing the effects of pitch, roll, and heave on the end of the TAS separately, appropriate rotary joint angles, pitch joint angles, and telescopic lengths were determined. Finally, the sum of the three obtained rotation rotary joint angles, pitch joint angles, and telescopic lengths is the target compensation value.

In order to reduce the difficulty of solving and simplify the solving process, the simplified TAS model was used here to replace the TAS model in Figure 7.

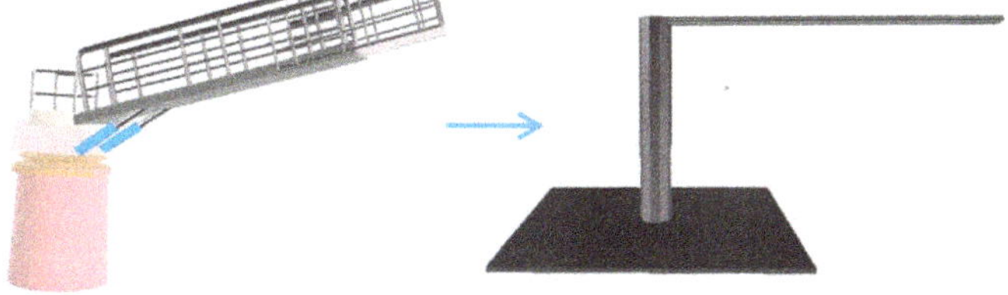

Figure 7. Model simplification.

As shown in Figure 7, the simplified model combines the base and rotary joint of the pre-simplified model into a standing cylinder. In addition, the simplified model integrates the arm of the pitch joint and the telescopic joint.

4.2.1. Pitch Compensation

In Figure 8, the position of the TAS when not affected by waves is shown in black. When the pitch angle is θ, the TAS moves to the position shown in gold in the figure. Among them, p is the position at the end of the TAS when there are no waves, and it is also the position that the end of the TAS needs to be compensated for. p_2 is the center of the {1} coordinate system of the TAS when the pitch angle is θ, p_6 is the center of the top of the column where the rotary joint is located, p_4 is the position of the end of the bridge device when the pitch angle is θ, p_3 represents the position of the TAS after rotation α_1 for compensation, which is the rotary joint angle, and p_1 represents the position of the TAS after rotation β_1 for compensation, which is the pitch joint angle.

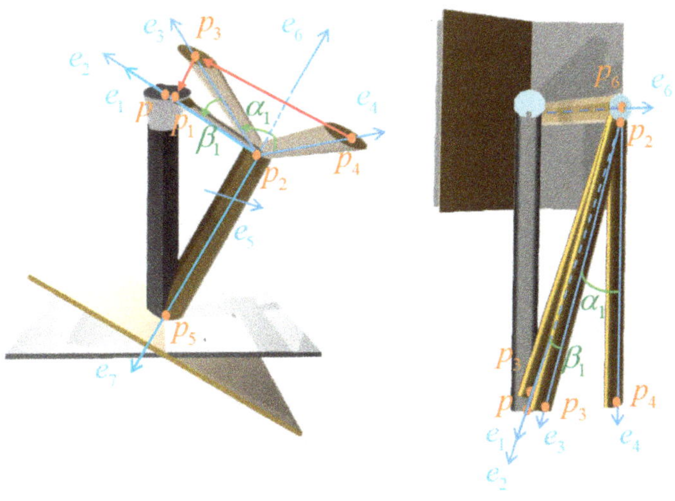

Figure 8. Pitch compensation.

The compensation process is shown by the red arrow. At the end of the bridge-crossing device, the rotation angle of the rotary joint is adjusted from position p_4 to p_3, then the rotation angle of the pitch joint is adjusted from p_3 to p_1. Finally, the expansion joint is adjusted to extend the end to p_1, and the telescopic length is T_1. Therefore, α_1, β_1, and T_1 need to be solved.

The required rotation joint angle and pitch joint angle require the position of p_3 or the direction of vector e_3. It can be seen that the straight line p_2p_3 is on plane $p_3p_4p_2$ and on plane $p_1p_2p_5$, which means that the straight line p_2p_3 is the intersection of two planes. To obtain the direction of p_2p_3, the normal vectors of these two planes are simply obtained, which are e_5 and e_6.

The coordinates of points p, p_2, p_4, and p_5 can all be calculated using Equation (23).

In order to achieve complete compensation, it is necessary to point the TAS end p_1 that has undergone rotation joint compensation and pitch joint compensation towards the target point p, where points p, p_1, and p_2 are in a straight line. The vector e_1 pointing from point p_2 to p_1 and the vector e_2 pointing from point p_2 to p are in the same direction; then vector e_2 is:

$$e_2 = \begin{bmatrix} p_x - p_{2x} \\ p_y - p_{2y} \\ p_z - p_{2z} \end{bmatrix} \tag{24}$$

where p_x is the abscissa of point p, p_{2y} is the ordinate of point p_2, and so on.

Since e_1 and e_2 are in the same direction and the initial TAS arm length is L, the coordinates of point p_1 are:

$$\begin{cases} p_1 = p_2 + \frac{e_2}{|e_2|} \cdot L \\ L = L_3 + d_0 \end{cases} \tag{25}$$

where L_3 is the length of the pitch joint, which can be obtained from Table 4, and d_0 is the length of the telescopic joint and was set to 0.5 in this study.

The vector e_7 pointing from point p_2 to p_5 is:

$$e_7 = \begin{bmatrix} p_{5x} - p_{2x} \\ p_{5y} - p_{2y} \\ p_{5z} - p_{2z} \end{bmatrix} \tag{26}$$

The vector e_5 is the cross-product of e_1 and e_7, and e_5 is also the normal vector of the surface $p_3p_2p_5$, which can be determined by the following equation:

$$e_5 = e_1 \times e_7 \tag{27}$$

The direction of the normal vector e_6 of plane $p_3p_2p_4$ is:

$$e_6 = \begin{bmatrix} p_{5x} - p_{6x} \\ 0 \\ p_{5z} - p_{6z} \end{bmatrix} \tag{28}$$

The vector e_3 is the intersection of plane $p_3p_2p_5$ and plane $p_3p_2p_4$, which can be determined by the following equation:

$$e_3 = e_5 \times e_6 \tag{29}$$

The rotation joint angle α_1 is the angle between vectors e_3 and e_4, and it can be calculated as:

$$\alpha_1 = \arccos\left(\frac{e_3 \cdot e_4}{|e_3||e_4|}\right) \tag{30}$$

The pitch joint angle β_1 is the angle between vectors e_3 and e_2, and it can be calculated as

$$\beta_1 = \arccos\left(\frac{e_3 \cdot e_2}{|e_3||e_2|}\right) \tag{31}$$

The telescopic length T_1 is the distance between points p and p_1, and it can be calculated as:

$$T_1 = \sqrt{(p_x - p_{1x})^2 + (p_y - p_{1y})^2 + (p_y - p_{1y})^2} \tag{32}$$

In addition, the α_1, β_1, and T_1 calculated above are only numerical values, and different directions need to be added for correct compensation for different movements.

When the pitch angle θ is negative (the ship tilts forward), the swing arm of the TAS should move to the right and extend forward:

$$\begin{cases} \alpha_1 = -\alpha_1 \\ \beta_1 = -\beta_1 \\ T_1 = T_1 \end{cases} \tag{33}$$

When the pitch angle is positive (the ship tilts backwards), the swing arm of the bridge-crossing device should move to the left and extend forward:

$$\begin{cases} \alpha_1 = \alpha_1 \\ \beta_1 = \beta_1 \\ T_1 = T_1 \end{cases} \tag{34}$$

4.2.2. Roll Compensation

In Figure 9, the TAS when not affected by waves is shown in black. When the roll angle is φ, the TAS moves to the position shown in gold in the figure. p_9 is the position of the TAS end when there are no waves, and it is also the target position of the TAS end. p_{10} is the center of the {1} coordinate system of TAS when the roll angle is φ. p_7 is the position of the TAS end when the roll angle is φ. p_8 represents the position of the TAS after rotating β_2 for compensation; that is, β_2 represents the pitch joint angle. It is worth mentioning that there is no need to use the rotary joint angle in roll compensation, as the TAS only performs longitudinal motion under the influence of roll, so rotary joint angle α_2 = 0.

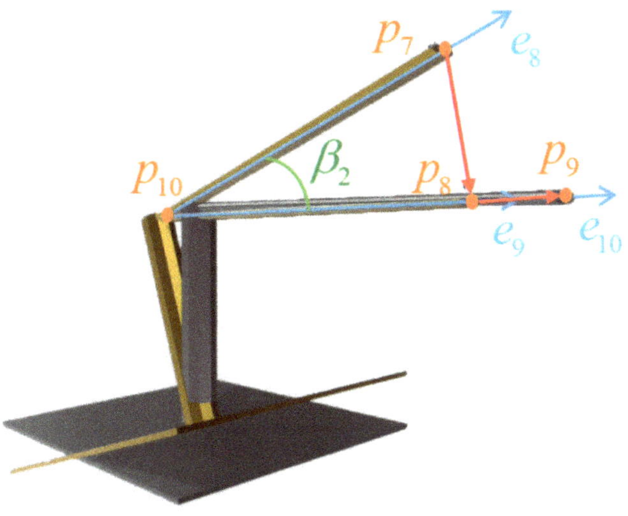

Figure 9. Roll compensation.

The compensation process is indicated by the red arrow. The pitch joint is rotated β_2 degrees to move the end from position p_7 to p_8, and then the extension joint is extended T_2 to p_9. Therefore, β_2 and p_9 need to be solved.

The coordinates of points p_7, p_9, and p_{10} can all be calculated using Equation (23).

The vector e_8 pointing from point p_{10} to p_7 is:

$$e_8 = \begin{bmatrix} p_{7x} - p_{10x} \\ p_{7y} - p_{10y} \\ p_{7z} - p_{10z} \end{bmatrix} \tag{35}$$

To achieve complete compensation, it is necessary to point the vector e_9 from point p_{10} to p_8 and to the target position p_{10}. The vector e_{10} pointing from p_{10} to p_9 is in the same direction as e_9. e_{10} can be calculated as:

$$e_{10} = \begin{bmatrix} p_{9x} - p_{10x} \\ p_{9y} - p_{10y} \\ p_{9z} - p_{10z} \end{bmatrix} \tag{36}$$

The pitch joint angle β_2 is the angle between vectors e_8 and e_{10}:

$$\beta_2 = \arccos(\frac{e_8 \cdot e_{10}}{|e_8||e_{10}|}) \tag{37}$$

Since e_9 and e_{10} are in the same direction and the initial TAS arm length is L, the coordinates of point p_8 are:

$$\begin{cases} p_8 = p_{10} + \frac{e_{10}}{|e_{10}|} \cdot L \\ L = L_3 + d_0 \end{cases} \tag{38}$$

The elongation T_2 that needs to be adjusted for the telescopic joint is the distance between points p_8 and p_9:

$$T_2 = \sqrt{(p_{8x} - p_{9x})^2 + (p_{8y} - p_{9y})^2 + (p_{8y} - p_{9y})^2} \tag{39}$$

In addition, the α_2, β_2, and T_2 calculated above are only numerical values, and different directions need to be added for correct compensation for different movements.

When the roll angle φ is negative (the ship tilts left), the swing arm of the TAS should move downward and extend forward:

$$\begin{cases} \alpha_2 = 0 \\ \beta_2 = -\beta_2 \\ T_2 = T_2 \end{cases} \tag{40}$$

When the roll angle φ is positive (the ship tilts to the right), the swing arm of the TAS should move upwards and shorten backwards:

$$\begin{cases} \alpha_2 = 0 \\ \beta_2 = \beta_2 \\ T_2 = -T_2 \end{cases} \tag{41}$$

4.2.3. Heave Compensation

In Figure 10, the TAS when not affected by waves is shown in black. When the heave height is z, the TAS moves to the position shown in gold. p_{11} is the center of the {1} coordinate system of TAS when the heave height is z. p_{12} is the position of the TAS end when the heave height is z. p_{14} is the center of the {1} coordinate system of the TAS when there are no waves. p_{15} compensates for the position of the TAS after rotating by β_3 degrees; that is, β_3 is the pitch joint angle. p_{16} is the position of the TAS end when there are no waves, and it is also the target position of the TAS. Cross-pointing p_{15} creates a perpendicular intersection line segment $p_{13}p_{14}$, with the intersection point at point p_{13}. Cross-pointing p_{15} makes a perpendicular line that intersects line segment $p_{14}p_{16}$, with the intersection point being point p_{17}. It is worth mentioning that, similar to roll compensation there is no need to retrieve the rotary joint angle α_3 in heave compensation, as the TAS only moves in the vertical plane under the influence of heave, so $\alpha_3 = 0$.

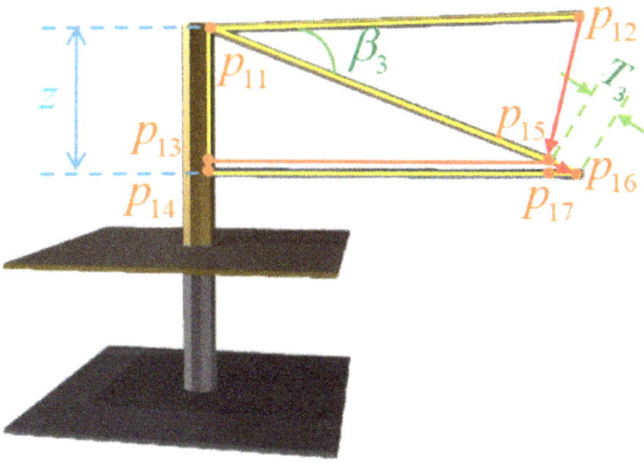

Figure 10. Heave compensation.

The compensation process is indicated by the red arrow. The pitch joint is rotated β_3 degrees to move the end from p_{12} to p_{15}, and then the telescopic joint is extended T_3 to p_{16}. Therefore, T_3 needs to be solved.

Since $p_{11}p_{12} \parallel p_{14}p_{16}$, the pitch joint angle β_3 can be calculated as:

$$\begin{cases} \angle\beta_3 = \angle p_{11}p_{16}p_{14} = \arctan(\frac{z}{L}) \\ L = L_3 + d_0 \end{cases} \tag{42}$$

As the three angles of $\triangle p_{11}p_{14}p_{16}$ and $\triangle p_{15}p_{17}p_{16}$ are completely equal, and it can be concluded that $\triangle p_{11}p_{14}p_{16} \sim \triangle p_{15}p_{17}p_{16}$, the proportion relationship of edges can be represented as:

$$\frac{\overline{p_{11}p_{13}}}{\overline{p_{11}p_{15}}} = \frac{\overline{p_{15}p_{17}}}{\overline{p_{15}p_{16}}} \tag{43}$$

Equation (43) can be represented as:

$$\frac{L\sin\beta_3}{L} = \frac{|z| - L\sin\beta_3}{T_3} \tag{44}$$

where $|z|$ is the absolute value of the heave height z.

Solving Equation (44), the telescopic length T_3 can be calculated as:

$$T_3 = \frac{|z|}{\sin\beta_3} - L \tag{45}$$

In addition, α_3, β_3, and T_3 are only numerical values, and different directions need to be added for correct compensation for different movements.

When z is positive (when the ship rises), the swing arm of the TAS should move downward and extend forward:

$$\begin{cases} \alpha_3 = 0 \\ \beta_3 = -\beta_3 \\ T_3 = T_3 \end{cases} \tag{46}$$

When z is negative (ship descending), the swing arm of the TAS should move upwards and extend forward:

$$\begin{cases} \alpha_3 = 0 \\ \beta_3 = \beta_3 \\ T_3 = T_3 \end{cases} \tag{47}$$

Finally, by adding the rotation joint angle, pitch joint angle, and telescopic joint angl from the three compensations above, the obtained sum is the final rotary joint angle α, pitc joint angle β, and telescopic joint length T, which can be represented as:

$$\begin{cases} \alpha = \alpha_1 + \alpha_2 + \alpha_3 \\ \beta = \beta_1 + \beta_2 + \beta_3 \\ T = T_1 + T_2 + T_3 \end{cases} \quad (48$$

The values of α, β, and T are limited by Table 4.

In addition, according to the results obtained in this study, ship pitch compensatio requires the participation of the rotary joint, pitch joint, and expansion joint in the comper sation. The roll compensation and heave compensation of ships require the participatio of pitch and expansion joints. That is, the rotary joint determines the pitch compensa tion ability, while the pitch joint and expansion joint determine the roll, pitch, and heav compensation results of the ship. If the rotation angle and pitch angle range of the TAS' rotation joint are larger, and the expansion joint and extension length are longer, then th TAS's workspace will be larger, and it can cope with more severe sea conditions.

5. Hydraulic Servo System Modeling of TAS

The hydraulic cylinder can be classified into two types, namely the symmetrica cylinder and the asymmetrical cylinder, based on the area being the same on both sides o the piston. The asymmetrical cylinder has an edge over the symmetrical cylinder since i does not comprise a piston rod component at the other end, thereby reducing the cylinde volume and occupying less space. In this study, among the three joints of the TAS, excep the rotary joint, which uses a four-way valve symmetrical cylinder, the pitch joint and th telescopic joint both use a four-way valve asymmetrical cylinder.

5.1. Four-Way Valve-Controlled Symmetrical Cylinder

5.1.1. Flow Equation of Slide Valve

The structure of the four-way valve-controlled symmetrical cylinder is shown ir Figure 11. The piston of a symmetrical hydraulic cylinder has the same diameter in bot directions of motion, resulting in an equal effective area on both sides. This makes th driving force and speed of the hydraulic cylinder similar in both directions. Q_1 is the tota amount of hydraulic oil flowing into the valve-controlled hydraulic cylinder, and Q_2 is th total amount of hydraulic oil flowing out of the valve-controlled hydraulic cylinder.

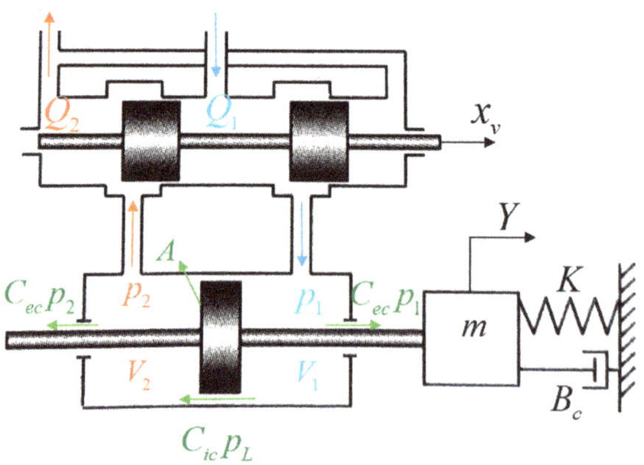

Figure 11. Structure of the four-way valve-controlled symmetrical cylinder.

For the change in oil quantity entering the valve-controlled symmetrical cylinder, ΔQ_1, and the change in oil quantity exiting the valve-controlled symmetrical cylinder, ΔQ_2, the following equations are established:

$$\Delta Q_1 = K_q \Delta x_v - 2K_c \Delta p_1 \tag{49}$$

$$\Delta Q_2 = K_q \Delta x_v + 2K_c \Delta p_2 \tag{50}$$

where K_q is the flow gain, K_c is the pressure–flow coefficient, p_1 is the left chamber pressure, p_2 is the right chamber pressure, and x_v is displacement of the spool valve.

Adding Equations (51) and (52), the change of load flow, ΔQ_L, can be expressed as:

$$\Delta Q_L = \frac{\Delta Q_1 + \Delta Q_2}{2} = K_q \Delta x_v - K_c \Delta p_L \tag{51}$$

$$p_L = p_1 - p_2 \tag{52}$$

where p_L is the load pressure drop.

Due to the piston moving slightly near the stable working point most of the time, the load flow, Q_L, can be expressed as:

$$Q_L = K_q x_v - K_c p_L \tag{53}$$

5.1.2. Flow Equation of Hydraulic Cylinder

For the flow into the hydraulic cylinder, Q_{in}, and the flow out of the hydraulic cylinder, Q_{out}, the continuous equation of compressed fluid can be expressed as:

$$Q_{\text{in}} - Q_{out} = \frac{dV}{dt} + \frac{V}{\beta} \cdot \frac{dp}{dt} \tag{54}$$

$$V_t = V_1 + V_2 = 2V_0 \tag{55}$$

$$Q_{in} = Q_1 - C_{ic}(p_1 - p_2) - C_{ec} p_1 = A\frac{dY}{dt} + \frac{V_0}{\beta_e} \cdot \frac{dp_1}{dt} \tag{56}$$

$$Q_{out} = C_{ic}(p_1 - p_2) - C_{ec} p_1 - Q_2 = -A\frac{dY}{dt} + \frac{V_0}{\beta_e} \cdot \frac{dp_2}{dt} \tag{57}$$

$$Q_L = \frac{Q_1 + Q_2}{2} \tag{58}$$

where V is the initial volume of the chamber liquid, Δp is the change in chamber pressure, β_e is the equivalent bulk modulus, V_t is the total volume of the hydraulic cylinder, V_1 and V_2 are the volumes of two hydraulic chambers, respectively, V_0 is the volume of each side when the piston is in the middle of the chamber, A is the effective area of the piston, and Y is the displacement of the hydraulic cylinder piston.

Substituting Equations (55)–(58) into Equation (54), the flow equation of the hydraulic cylinder can be obtained:

$$Q_L = A\frac{dY}{dt} + C_{tc} p_L + \frac{V_t}{4\beta_e} \cdot \frac{dp_L}{dt} \tag{59}$$

where $C_{tc} = C_{\text{ic}} + \frac{1}{2}C_{ec}$ is the total leakage coefficient, C_{ic} is the internal leakage coefficient, and C_{ec} is the external leakage coefficient.

5.1.3. Flow Equation of Slide Valve

According to Newton's second law, we can obtain:

$$F_g = Ap_L = m\frac{d^2Y}{dt^2} + B_c\frac{dY}{dt} + KY + F \tag{60}$$

where m is the total mass of piston and load, Y is the displacement of the hydraulic rod, B is the viscous damping coefficient of piston and load, K is the spring stiffness of load, F is the external force applied on the piston, and F_g is the hydraulic driving force.

According to Equations (53), (59), and (60), ignoring the spring stiffness of load K, the transfer function between the piston displacement Y and slide valve displacement x_v of a symmetric hydraulic cylinder can be calculated as:

$$\frac{Y}{x_v} = \frac{\frac{K_q}{A}}{s\left(\frac{s^2}{\omega_h^2} + \frac{2\zeta_h}{\omega_h}s + 1\right)} \tag{61}$$

$$\omega_h = \sqrt{\frac{4\beta_e A^2}{V_t m}} \tag{62}$$

$$\zeta_h = \frac{K_{ce}}{A}\sqrt{\frac{B_c m}{V_t}} \tag{63}$$

$$K_{ce} = K_c + C_{ic} + \frac{C_{ce}}{2} \tag{64}$$

where ω_h is the hydraulic natural frequency, ζ_h is the hydraulic damping ratio, and K_{ce} is the total flow–pressure coefficient.

The relevant parameters of the rotary joint are listed in Table 5.

Table 5. The relevant parameters of the rotary joint.

Parameter	Value
K_q	5.560×10^{-5} m^3/(s · V)
K_{ce}	1.8×10^{-11} m^3/(s · Pa)
A	0.001 m^3
β_e	6.9×10^8 N/m^2
V_t	8.762×10^{-4} m^3
m	1600 kg
B_c	5000 N/(m/s)

According to Table 5, the transfer function of a symmetric hydraulic cylinder can be calculated as:

$$Y_{rotray}(s) = \frac{0.056}{s\left(\frac{s^2}{44.662^2} + \frac{2\times 0.676s}{44.662} + 1\right)} \tag{65}$$

5.2. Four-Way Valve-Controlled Asymmetric Cylinder

The difference from the symmetrical cylinder of the four-way slide valve is that the effective area of the piston is different during extension and retraction. The structure of the asymmetric valve-controlled hydraulic cylinder is shown in Figure 12. This design can be used to generate motion with different speeds and forces. This is suitable for the pitch joint and the telescope joint, as the pitch joint requires greater force when lifted and the telescope joint when extended, while the opposite movement can achieve faster speed. In addition, p_s is the fuel supply pressure, and p_0 is the return pressure.

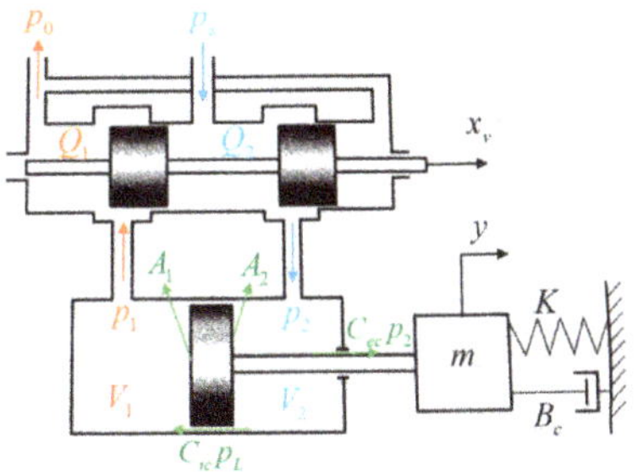

Figure 12. Structure of the asymmetric valve-controlled hydraulic cylinder.

5.2.1. Flow Equation of Slide Valve

The effective area ratio of the left and right cavities of the hydraulic cylinder can be set as $n = \frac{A_2}{A_1}$. The flow equation of the slide valve can be defined as:

$$Q_L = K_q x_v - K_c p_L \tag{66}$$

$$K_q = \begin{cases} K_{q1} = C_d W \sqrt{\frac{2}{\rho}\left(\frac{p_s - p_L}{1+n^3}\right)}, x_v > 0 \\ K_{q2} = C_d W \sqrt{\frac{2}{\rho}\left(\frac{np_s + p_L}{1+n^3}\right)}, x_v < 0 \end{cases} \tag{67}$$

$$K_c = \begin{cases} K_{c1} = C_d W x_v \sqrt{\frac{1}{2\rho(p_s - p_L)(1+n^3)}}, x_v > 0 \\ K_{c2} = -C_d W x_v \sqrt{\frac{1}{2\rho(np_s + p_L)(1+n^3)}}, x_v < 0 \end{cases} \tag{68}$$

$$p_L = p_1 - np_2 \tag{69}$$

$$Q_L = \frac{Q_1 + nQ_2}{1+n^2} \tag{70}$$

where C_d is the flow coefficient of the throttling port, generally taken as 0.60~0.65, W is the area gradient, ρ is the density of liquid, p_s is the fuel supply pressure, p_L is the load pressure drop, x_v is the spool displacement of the slide valve, K_{q1} and K_{q2} are the flow gains when $x_v > 0$ and $x_v < 0$, respectively, while K_{c1} and K_{c2} are the pressure–flow gains when $x_v > 0$ and $x_v < 0$, respectively.

5.2.2. Flow Equation of Hydraulic Cylinder

Assuming that the initial volumes of the two chambers of the piston are equal, the following equations can be obtained:

$$V_{10} = V_{20} = \frac{V_t}{2} \tag{71}$$

$$A_1 x_p \ll \frac{V_t}{2} \tag{72}$$

$$A_2 x_p \ll \frac{V_t}{2} \tag{73}$$

where V_{10} and V_{20} are the initial volumes of the two chambers of the hydraulic cylinder, and V_t is the total volume of the two chambers.

According to Equations (70), (72), and (73), the flow equation of the hydraulic cylinder can be obtained as:

$$Q_L = A_p \frac{dy}{dt} + \frac{V_t}{2(1+n^2)\beta_e}\frac{dp_L}{dt} + C_t p_L + C_{tc} p_s \tag{74}$$

where β_e is the equivalent bulk modulus, C_t is the total leakage coefficient related to the pressure load, which can be obtained from Equation (75), and C_{tc} is the leakage coefficient related to the fuel supply pressure, which can be obtained from Equation (76):

$$C_t = \frac{n^2(n^2-1)}{(1+n^2)(1+n^3)}C_{ic} \tag{75}$$

$$C_{tc} = \begin{cases} C_{tc1} = \frac{1+n}{1+n^3}C_i + \frac{C_{ec}}{1+n^2}, x_v > 0 \\ C_{tc2} = \frac{n^2-1}{(1+n^3)(1+n^2)}C_{ic}, x_v < 0 \end{cases} \tag{76}$$

where C_{ic} and C_{ec} are the internal leakage coefficient and external leakage coefficient, respectively.

5.2.3. Force Balance Equation

The balance equation of the asymmetrical cylinder force of the four-way slide valve can be described as:

$$A_1 p_L = m\frac{d^2y}{dt^2} + B_c\frac{dy}{dt} + Ky + F \tag{77}$$

where m is the total mass of the piston and load, y is the displacement of the hydraulic rod, B_c is the viscous damping coefficient of the piston and load, and F is the external force applied on the piston.

Equations (66), (74), and (77) can be transformed by Laplace transform when the spring stiffness, K, of the load can be ignored. The transfer function between the displacement of the hydraulic rod, y, and the displacement of the valve core, x_v, can be expressed as:

$$\frac{y}{x_v} = \frac{\frac{K_q}{A_1}}{s\left(\frac{s^2}{\omega_h^2} + \frac{2\zeta_h}{\omega_h}s + 1\right)} \tag{78}$$

where ω_h is the hydraulic natural frequency, which can be obtained from Equation (79), and ζ_h is the hydraulic damping ratio, which can be obtained from Equation (80):

$$\omega_h = \sqrt{\frac{2(1+n^2)\beta_e A_1^2}{mV_t}} \tag{79}$$

$$\zeta_h = \frac{K_{ce}}{2A_1}\sqrt{\frac{2(1+n^2)\beta_e m}{V_t}} \tag{80}$$

where $K_{ce} = K_c + C_t$ is the total pressure–flow coefficient. Due to the completely different values of K_q and K_{ce} at $x_v > 0$ and $x_v < 0$, the molecular and hydraulic damping coefficients of the transfer function are also different.

The relevant parameters of the pitch joint are listed in Table 6.

Table 6. The relevant parameters of the pitch joint.

Parameter	Value
A_1	3.1×10^{-3} m^2
A_2	1.5×10^{-3} m^2
n	0.490
K_{q1}	2.333×10^{-3} m$^3/(\text{s} \cdot \text{V})$
K_{q2}	2.917×10^{-5} m$^3/(\text{s} \cdot \text{V})$
K_{ce1}	5.8×10^{-11} m$^3/(\text{s} \cdot \text{Pa})$
K_{ce2}	1.8×10^{-11} m$^3/(\text{s} \cdot \text{Pa})$
β_e	6.9×10^8 N/m^2
V_t	2.5×10^{-3} m^3
m	1150 kg
B_c	5000 N/(m/s)

According to Table 6, the transfer function of a symmetric hydraulic cylinder can be calculated as:

$$Y_{pitch}(s) = \begin{cases} \frac{0.0749}{s\left(\frac{s^2}{76.642^2} + \frac{2\times 0.263s}{76.642} + 1\right)}, x_v > 0 \\ \frac{0.0936}{s\left(\frac{s^2}{76.642^2} + \frac{2\times 0.816s}{76.642} + 1\right)}, x_v < 0 \end{cases} \quad (81)$$

The relevant parameters of the telescopic joint are listed in Table 7.

Table 7. The relevant parameters of the telescopic joint.

Parameter	Value
A_1	2.0×10^{-3} m^2
A_2	9.456×10^{-4} m^2
n	0.482
K_{q1}	1.111×10^{-3} m^3/(s · V)
K_{q2}	1.389×10^{-5} m^3/(s · V)
K_{ce1}	5.8×10^{-11} m^3/(s · Pa)
K_{ce2}	1.8×10^{-11} m^3/(s · Pa)
β_e	6.9×10^{8} N/m^2
V_t	4.5×10^{-3} m^3
m	950 kg
B_c	5000 N/(m/s)

The transfer function of a symmetric hydraulic cylinder can be calculated as:

$$Y_{\text{telescopic}}(s) = \begin{cases} \frac{0.0566}{s\left(\frac{s^2}{39.116^2} + \frac{2\times 0.280s}{39.116} + 1\right)}, x_v > 0 \\ \frac{0.0707}{s\left(\frac{s^2}{39.116^2} + \frac{2\times 0.867s}{39.116} + 1\right)}, x_v < 0 \end{cases} \quad (82)$$

5.3. Model of Servo Valve

Since the natural frequency of the transfer function of the actuator is generally far lower than the frequency of the servo valve, the transfer function of the servo valve is generally regarded as a second-order unit or a first-order unit. It is mainly divided into the following three situations:

- If the response frequency of the servo valve is close to the natural frequency of the hydraulic cylinder, the servo valve can be regarded as a second-order oscillation unit.
- If the response frequency of the servo valve is 3~5 times that of the natural frequency of the hydraulic cylinder, the servo valve can be regarded as an inertial unit.
- If the response frequency of the servo valve is more than 5 times that of the natural frequency of the hydraulic cylinder, the servo valve can be regarded as a proportional unit.

In this study, the transfer function of the servo valve can be defined as:

$$g(s) = \frac{1}{0.035s + 1} \quad (83)$$

Due to the response frequency of the servo valve used in this study being 3–5 times the natural frequency, it is considered as an inertial unit, where 0.035 is the inertia constant of the servo valve.

6. Design of Controller

6.1. Fuzzy PID Controller

The PID controller has a simple structure and strong robustness. However, the parameters of ordinary PID cannot change with the sea state because the sea state changes rapidly. Therefore, fuzzy PID control was adopted here. The three parameters of fuzzy PID are not fixed but are provided in real time according to the external environment to ensure the real-time update of the PID control system. The schematic diagram of fuzzy PID is shown in Figure 13. The accurate input value is fuzzed by the membership function, then the fuzzed results are matched with the established fuzzy rules one-by-one, and finally the matching result is anti-fuzzed to output the accurate value. The output results are transmitted to the PID controller, which controls the controlled object in real time.

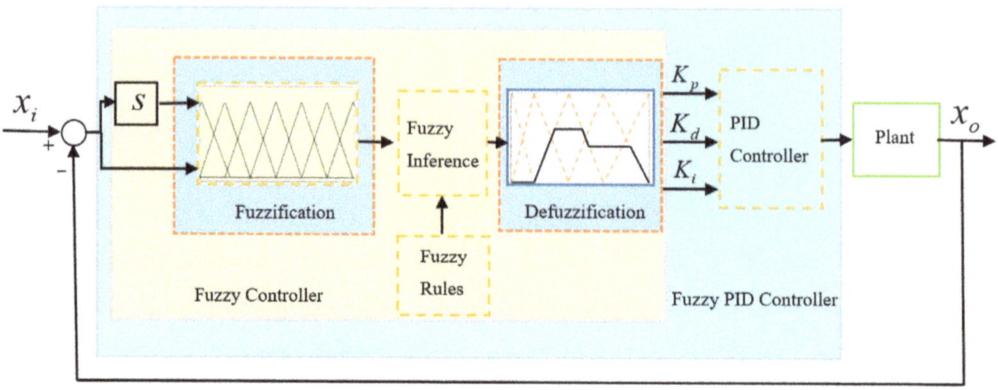

Figure 13. Schematic diagram of fuzzy PID.

In fuzzy PID control, the design of the fuzzy controller plays a vital role, determining whether the PID control is effective. The fuzzy control design of this study was as follows. First, input and output fuzzy subsets were defined, where {NB, NM, NS, ZO, PS, PM, PB} were defined as the input and output fuzzy subsets, representing Negative Big, Negative Medium, Negative Small, Zero, Positive Small, Positive Medium, and Positive Big, respectively.

The error e and the first derivative of the error ec were used as fuzzy input values. The value e range was $(-0.035, 0.035)$, and the value ec range was $(-0.5, 0.5)$. The range of K_p, K_i, and K_d was $(0, 1)$. Then, the appropriate proportion coefficients were selected as the system parameters. Finally, the membership function and fuzzy rules were designed according to the experience of experts.

Moreover, the Mamdani algorithm was used for fuzzy inference, and the centroid algorithm was used for defuzzification. The fuzzy regular surfaces obtained from the above parameters are shown in Figure 14.

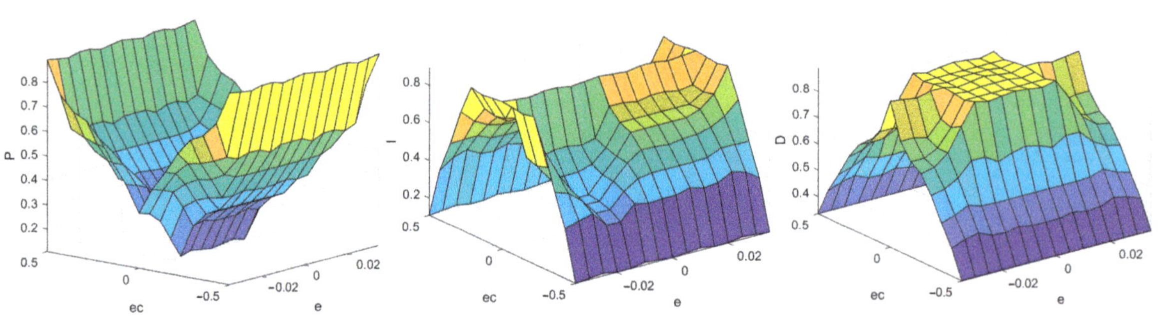

Figure 14. Fuzzy regular surfaces of P, I, and D in fuzzy control. Different colors represent different values, and the larger the value, the warmer the color tone.

From Figure 14, it can be observed that when there is a significant error between the input and output, K_p is increased to achieve rapid error reduction, thereby improving the dynamic performance of the system and reducing K_i and K_d to prevent system instability.

When the error between the input and output is small, K_p should be reduced to prevent excessive overshoot, while K_i and K_d should be increased to improve speed.

6.2. Feedforward Compensation

In order to improve the stability and speed of the system and reduce errors, feedforward compensation was added in three joints. Figure 15 shows the composition structure of the hydraulic control system using the rotary joint as an example. K_f is the transfer function of the angle sensor, $G_p(s)$ is the transfer function of the fuzzy PID controller, K_a is the transfer function of the power amplification unit, $G_v(s)$ is the transfer function of the servo valve, $G_c(s)$ is the transfer function of the hydraulic cylinder, and $G_r(s)$ is the feedforward compensation unit that needs to be designed.

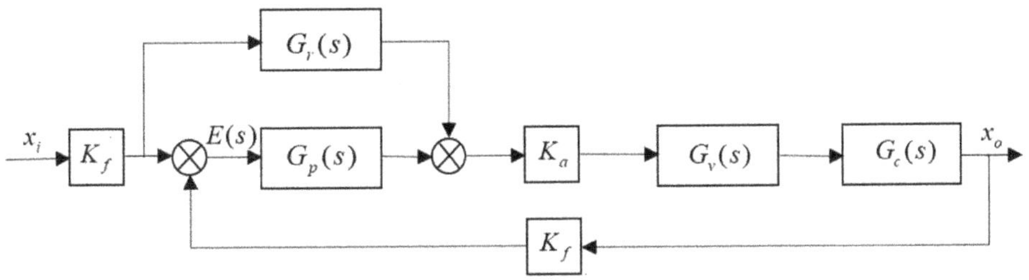

Figure 15. Feedforward compensation of the rotary joint.

The output x_o of the system is obtained from the following equation:

$$x_o = K_a G_v(s) G_c(s) \left(E(s) G_p(s) + K_f G_r(s) x_i \right) \quad (84)$$

where $E(s)$ can be obtained by the following equation:

$$E(s) = K_f(x_i - x_o) \quad (85)$$

It can be inferred that:

$$\frac{x_o}{x_i} = \frac{K_f(G_r(s) + G_p(s))}{\frac{1}{K_a G_v(s) G_c(s)} + K_f G_p(s)} \quad (86)$$

To make x_o equal to x_i, this means:

$$\frac{K_f(G_r(s) + G_p(s))}{\frac{1}{K_a G_v(s) G_c(s)} + K_f G_p(s)} = 1 \quad (87)$$

Then, the following equation can be calculated:

$$G_r(s) = \frac{1}{K_a K_f G_v(s) G_c(s)} \quad (88)$$

However, in reality, implementing Equation (88) is very difficult, especially when it is very complex. Therefore, it is often approximated as a first-order differential unit, which can elevate the control system to a second-order level. Therefore, Equation (88) can be changed to:

$$G_r(s) = \frac{1}{K_a K_f K_v K_c} s \quad (89)$$

where K_v is the molecular coefficient of $G_v(s)$ and $K_c(s)$ is the molecular coefficient of $G_c(s)$.

The compensation unit transfer function of the rotary joint system was obtained according to the above method and the results were compared with the system without feedforward compensation, as shown in Figure 16. The input signal is $x(t) = t$. The error can reach 0.45° without compensation, and after feedforward compensation, the error is less than 0.1°. It is apparent that feedforward compensation accelerates the response speed of the system and greatly reduces system error.

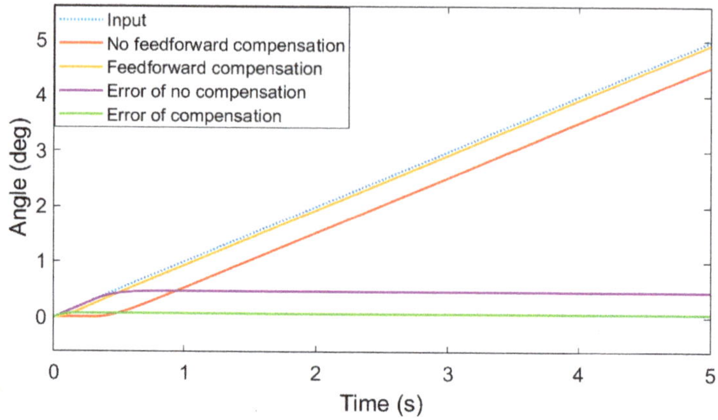

Figure 16. The effect of feedforward compensation under ramp input.

7. Results and Discussion

In this study, the wave period of the wind farm was 6 s, the significant wave height was set as 2.2 m, and the wave direction was 45°. The initial rotary joint angle and initial pitch joint angle were 0, and initial telescopic length was 0.5 m. In this case, the control effects of ordinary PID and fuzzy PID with feedforward compensation were compared.

The errors between the output angle and the expected angle of the ordinary PID and fuzzy PID algorithms are shown in Figure 17. In rotary joint control, the error of PID control was within (−0.66°, 0.65°), while the error for fuzzy PID control can be reduced to (−0.6° 0.44°). In pitch joint control, ordinary PID control displays error within ±6°, while the error for fuzzy PID control can be reduced to (−4°, 5°).

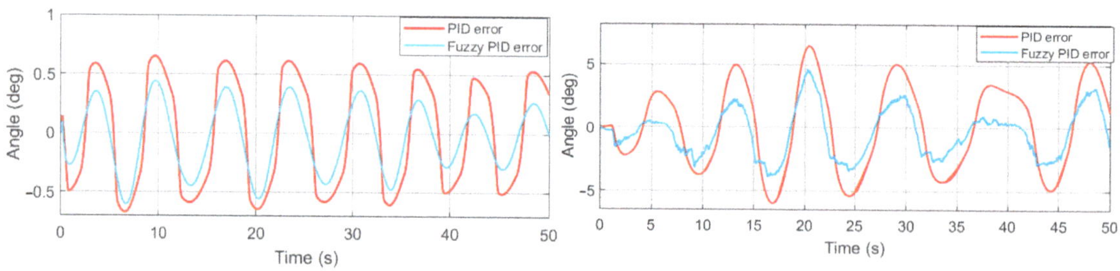

(a) (b)

Figure 17. Angle error comparison: (**a**) angle error of the rotary joint hydraulic cylinder and (**b**) angle error of the pitch joint hydraulic cylinder.

Figure 18 shows a comparison of the position error of the hydraulic cylinder of the ordinary PID and fuzzy PID algorithms. For rotary joint control, the position error of ordinary PID control was controlled within ±7 mm, while fuzzy PID control can reduce the error to (−6.4 mm, 4.8 mm). The error of the fuzzy PID was greater than the error of the PID in a short period of time. For pitch joint control, the error was controlled within

±160 mm by ordinary PID, while for fuzzy PID, the error was reduced to ±80 mm. In the control of the telescopic joint, the error displacement of ordinary PID control was within ±150 mm, while fuzzy PID control can reduce the error to (−100 mm, 80 mm). It is worth noting that Figures 17a and 18a are very similar, because the rotary joint uses a symmetrical hydraulic cylinder as the execution unit, and hydraulic valves, angle sensors, power amplifiers, etc., can be considered as proportional components.

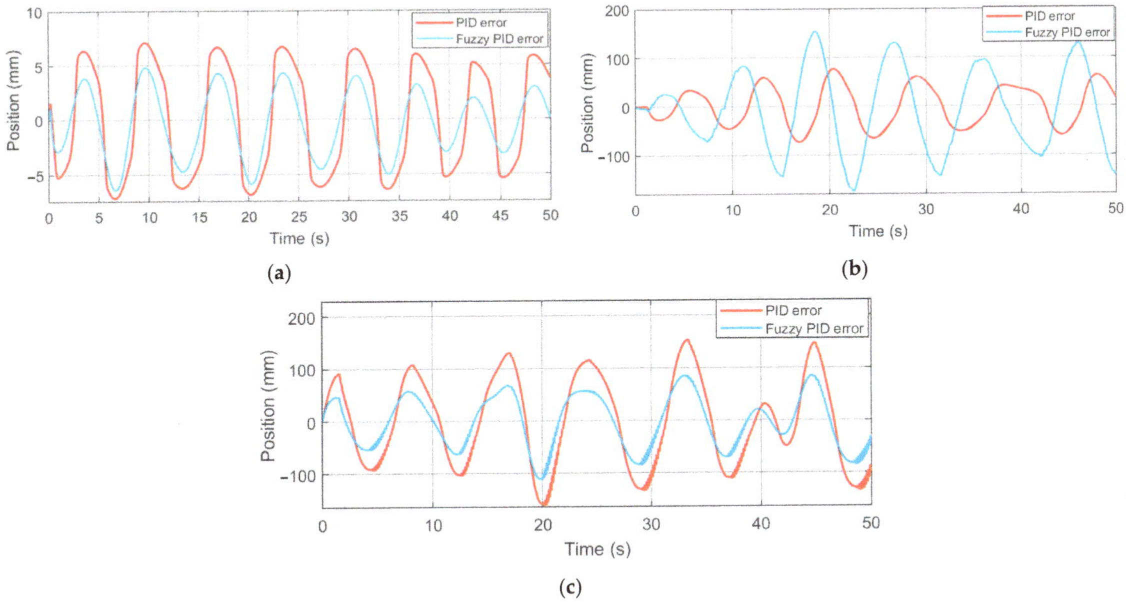

Figure 18. Position error comparison: (**a**) position error of the rotary joint hydraulic cylinder, (**b**) position error of the pitch joint hydraulic cylinder, and (**c**) position error of the telescopic joint hydraulic cylinder.

The position of the TAS end effector at the X-axis, Y-axis, and Z-axis after 50 s of simulation is shown in Figure 19. Before compensation, the maximum position error in the X direction can reach 0.2396 m, the maximum position error in the Y direction can reach 0.2032 m, and the maximum position error in the Z direction can reach 1.2907 m. However, it is obvious that the position error was greatly reduced whether using PID control or fuzzy PID control. For the former, the maximum error of displacement in the X direction was reduced to 0.0526 m, that in the Y direction was reduced to 0.1256 m, and that in the Z direction was reduced to 0.4805 m. For the latter, the maximum error of displacement in the X direction was reduced to 0.0441 m, that in the Y direction was reduced to 0.0967 m, and that in the Z direction was reduced to 0.3492 m.

The roll angle, pitch angle, and heave height were compared before and after compensation here. Since the TAS is not located on the roll and pitch axes, the roll and pitch angles defined here are different from those of the ship. In this study, the roll angle of the TAS, φ_t, is defined as: the angle between the projection of the line connecting the end of the TAS and the center of the B coordinate system in the plane with waves and the z-axis without waves. The pitch angle of the TAS, θ_t, is defined as: the angle between the projection of the line connecting the end of the TAS with and without waves on the y_0Bz_0 plane and the vertical plane, where the end of the TAS is located in the absence of waves. The vertical heave height of the TAS, z_t, is defined as the height difference between the end of the TAS with and without waves. The results calculated based on the above definition are shown in Figure 20.

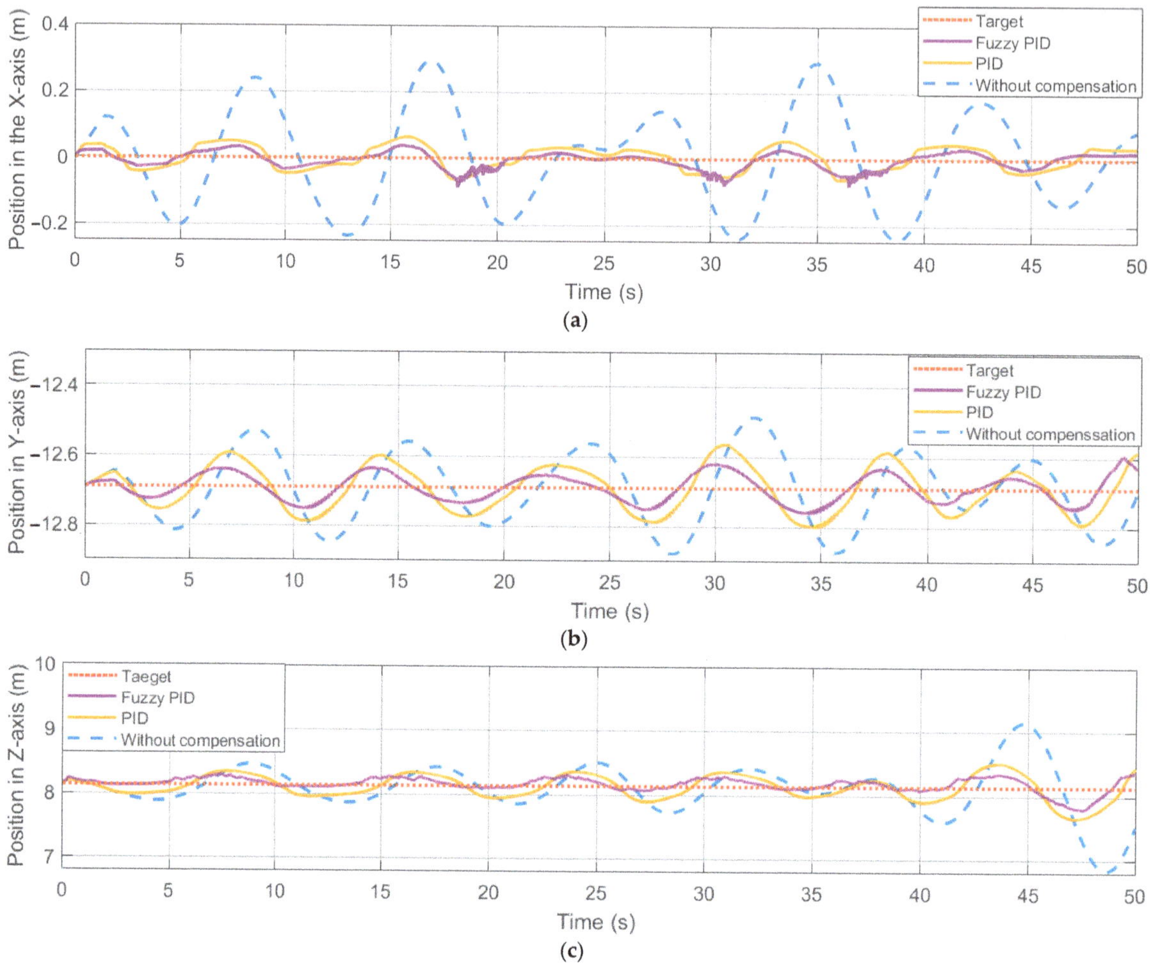

Figure 19. The compensation response at the TAS end effector when the encounter angle is 45 degrees. (**a**) The displacement in the *X* direction, (**b**) displacement in the *Y* direction, and (**c**) displacement in the *Z* direction.

The roll angle of the TAS under the action of waves was between (−2.76°, 4.74°) without compensation, and the difference between the highest and lowest amplitudes was 7.5°. After PID compensation, the roll angle was reduced to ±0.8°, with a reduction of 78.7%. After fuzzy PID compensation, the roll angle was reduced to 0.57°, with a reduction of 84.8%. For the pitch angle, θ_t, the pitch angle of the TAS was between (−0.08°, 0.056°) without compensation, and the difference between the highest and lowest amplitudes was 0.136°. Obviously, the amplitude of the pitch angle is very small compared to the ship's roll angle, due to the different definitions of the pitch angle. After PID compensation, the pitch angle decreased to ±0.022°, with a decrease of 67.6%. After fuzzy PID compensation, the pitch angle decreased to (−0.014°, 0.020°), with a reduction of 75%. For the heave height z_t, the heave height of the TAS was between (−0.665 m, 0.549 m) without compensation and the difference between the highest and lowest amplitudes was 1.214 m. After PID compensation, the heave height was reduced to (−0.297 m, 0.306 m), with a reduction of 50.3%. After fuzzy PID compensation (−0.106 m, 0.214 m), the reduction was 73.6%. Based on the above data comparison, fuzzy PID control can effectively reduce the roll angle, pitch angle, and heave height of the TAS, while maintaining end stability.

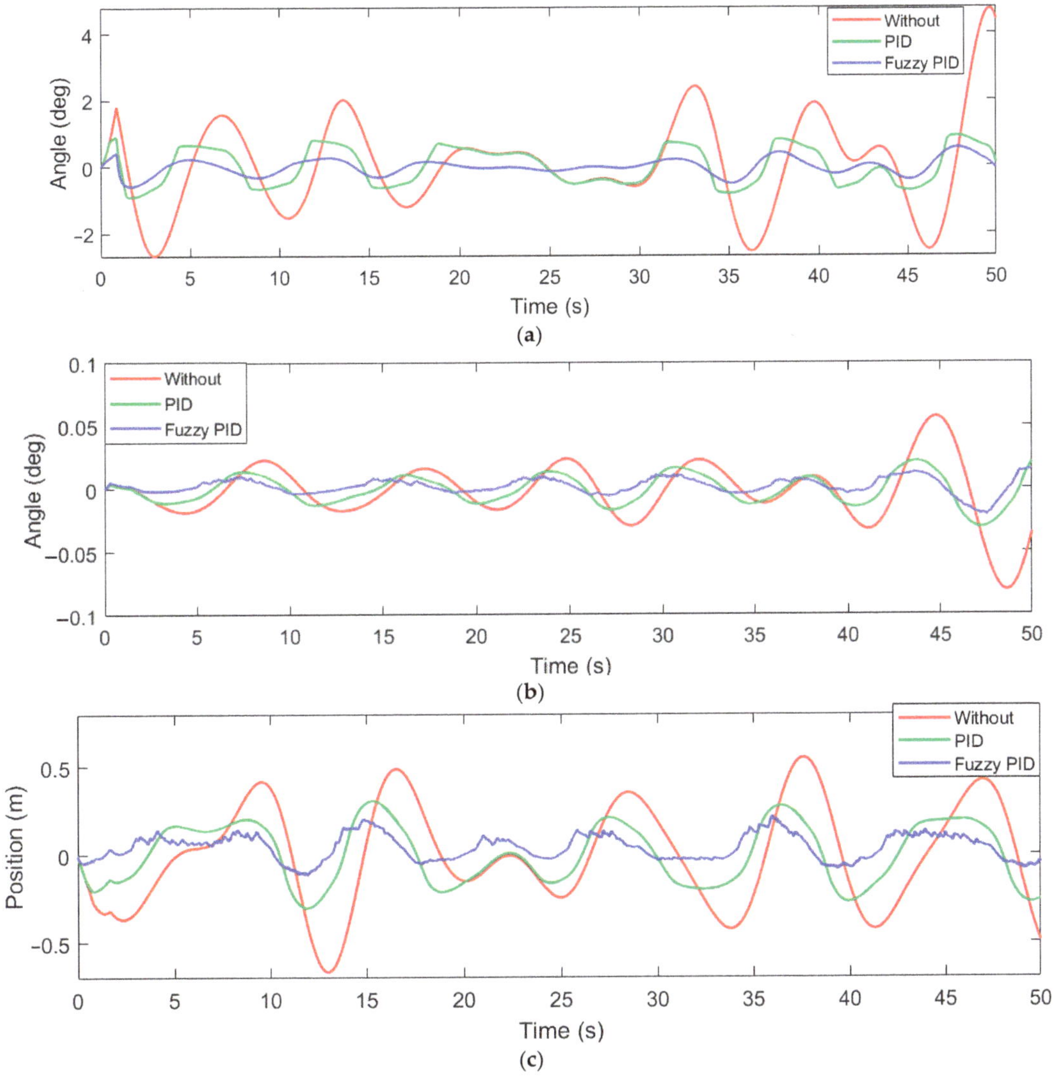

Figure 20. Roll angle, pitch angle, and heave height of the TAS when the encounter angle is 45 degrees. (**a**) Roll angle, (**b**) pitch angle, and (**c**) heave height.

In order to compare the results of the different control methods in more detail, the root mean square error (RMSE) was used to compare the error between the displacement and the ideal output displacement under different conditions to evaluate the performance of the control method. The RMSE formula can be described as:

$$RMSE = \sqrt{\frac{\sum_{i=1}^{n} (d - d_{tar})^2}{n}} \tag{90}$$

where d is displacement with or without compensation, d_{tar} is expected displacement, and n is the number of data points.

The calculation formula for compensation efficiency is as follows:

$$E = \frac{RMSE_{before} - RMSE_{after}}{RMSE_{before}} \quad (91)$$

where $RMSE_{before}$ is the RMSE without compensation, and $RMSE_{after}$ is the RMSE after compensation.

RSME in the X direction, Y direction, and Z direction was calculated without compensation, under PID control and fuzzy PID control. The compensation efficiency in each case was calculated, and the results are shown in Table 8. It is evident that the displacement error was greatly reduced after compensation whether in the X direction, Y direction, or Z direction. Without compensation, the displacement errors in three directions were 0.1039, 0.1226 m, and 0.3316 m, respectively. However, after PID compensation, the displacement errors were reduced to 0.0312 m, 0.0722 m, and 0.1843 m, respectively, and the compensation efficiency was 69.98%, 41.08%, and 44.44%, respectively. After fuzzy PID compensation, the displacement errors were reduced to 0.0180 m, 0.0429 m, and 0.0923 m, respectively. Fuzzy PID compensation can achieve a compensation efficiency of 82.69%, 65.03%, and 72.18% in three degrees of freedom, respectively. The stability of the TAS end effector has been greatly improved. There is no doubt that fuzzy PID control greatly increases the stability of the TAS end effector and has better performance than PID control. To sum up, fuzzy PID can effectively reduce the displacement error of the TAS end effector in the X, Y, and Z directions and enhance the stability of the TAS, and its result is better than that of PID control.

Table 8. RMSE and compensation efficiency in the X direction, Y direction, and Z direction.

Object	X	Y	Z
Without compensation, $RSME_1$ (m)	0.1039	0.1226	0.3316
Compensation with PID, $RSME_2$ (m)	0.0312	0.0722	0.1843
Compensation with fuzzy PID, $RSME_3$ (m)	0.0180	0.0429	0.0923
Compensation efficiency of PID, E_1	69.98%	41.08%	44.44%
Compensation efficiency of fuzzy PID, E_2	82.69%	65.03%	72.18%

In addition, the trajectory of the end effector can be obtained according to the displacement data of the end effector, and its trajectory is shown in Figure 21. The end effector shook violently before compensation, and the motion track seriously deviated from the target position. After PID compensation, the displacement error decreased a lot; that is, the range of the green line in the figure decreased a lot, which indicates that the compensation was effective. Finally, it is worth noting that the range of the track of fuzzy PID control, namely the magenta line, is closer to the target position, which indicates that the effect of fuzzy PID control is better than that of PID control.

When the encounter angle becomes 90, the situation becomes different. The roll, pitch, and heave heights of the TAS at this time are shown in Figure 22. For the roll angle, the size did not increase but decreased instead. The roll angle of the TAS before compensation was (−2.84°, 4.45°), and the difference between the highest and lowest amplitudes was 7.29°, which is 2.8% less than the obtained 7.5° when the encounter angle was 45 degrees. After PID compensation, the amplitude of the roll angle decreased to (−0.923°, 0.892°), with a decrease of 75.1%. After fuzzy PID compensation, the amplitude of the roll angle was reduced to (−0.684°, 0.452°), with a decrease of 84.4%. The decrease in amplitude was not significant compared to when the encounter angle was 45 degrees. For the pitch angle, the amplitude range of the pitch angle before compensation was (−0.075°, 0.071°), and the difference between the highest and lowest amplitudes was 0.136°, which is an 7.4% increase compared to the obtained 0.136° when the encounter angle was 45 degrees. After

PID compensation, the amplitude of the pitch angle decreased to (−0.035°, 0.029°). After fuzzy PID compensation, the amplitude of the pitch angle was reduced to (−0.024°, 0.019°). For the heave height, the range of heave height before compensation was 1.170 m, and the difference between the highest and lowest heave heights was 2.340 m, which is an increase of 89.4% compared to the obtained 1.214 m at an encounter angle of 45 degrees. After PID compensation, the heave height was reduced to (−0.516 m, 0.468 m). After fuzzy PID compensation, the heave height was reduced to (−0.37 m, 0.314 m). Overall, compared to an encounter angle of 45 degrees, the roll angle decreased, while the pitch angle and heave height increased. The reason for this is the same as in Figure 3. The change in the encounter angle to 90 degrees increased the roll angle and heave height of the ship, while the pitch angle decreased. However, in this case, the coordinate direction of the TAS is not consistent with the coordinate direction of the ship, but is 90 degrees different, which increases the ship's roll and causes an increase in the TAS pitch. The decrease in the ship's pitch angle leads to a decrease in the TAS roll angle. The direction of heave is the same, so the heave height increases.

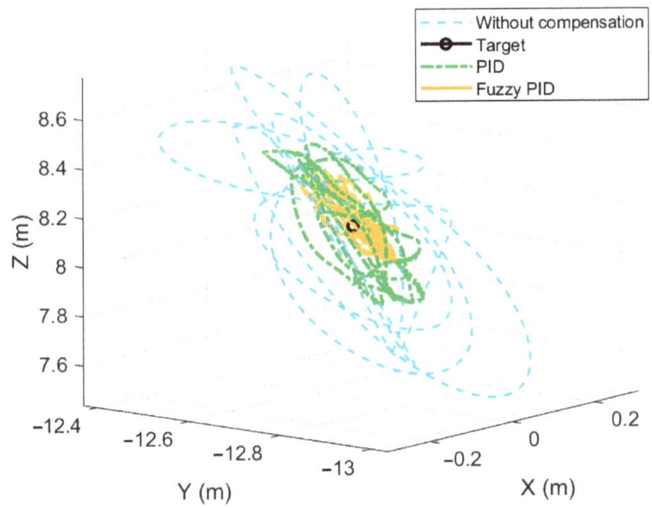

Figure 21. The motion trajectory of the end effector of the TAS.

For the position in the X, Y, and Z directions at an encounter angle of 90 degrees, the results are shown in Figure 23. Before compensation, the maximum position error in the X direction could reach 0.2173 m, the maximum position error in the Y direction could reach 0.2983 m, and the maximum position error in the Z direction could reach 1.4693 m. Obviously, compared to the encounter angle of 45 degrees, the displacement of the TAS on the X-axis was significantly reduced, while the displacement on the Y-axis and Z-axis significantly increased. This is because at the encounter angle of 90 degrees, the increase in ship roll angle increases the displacement of the TAS on the Y-axis and Z-axis, and the decrease in pitch reduces the displacement of the TAS on the X-axis.

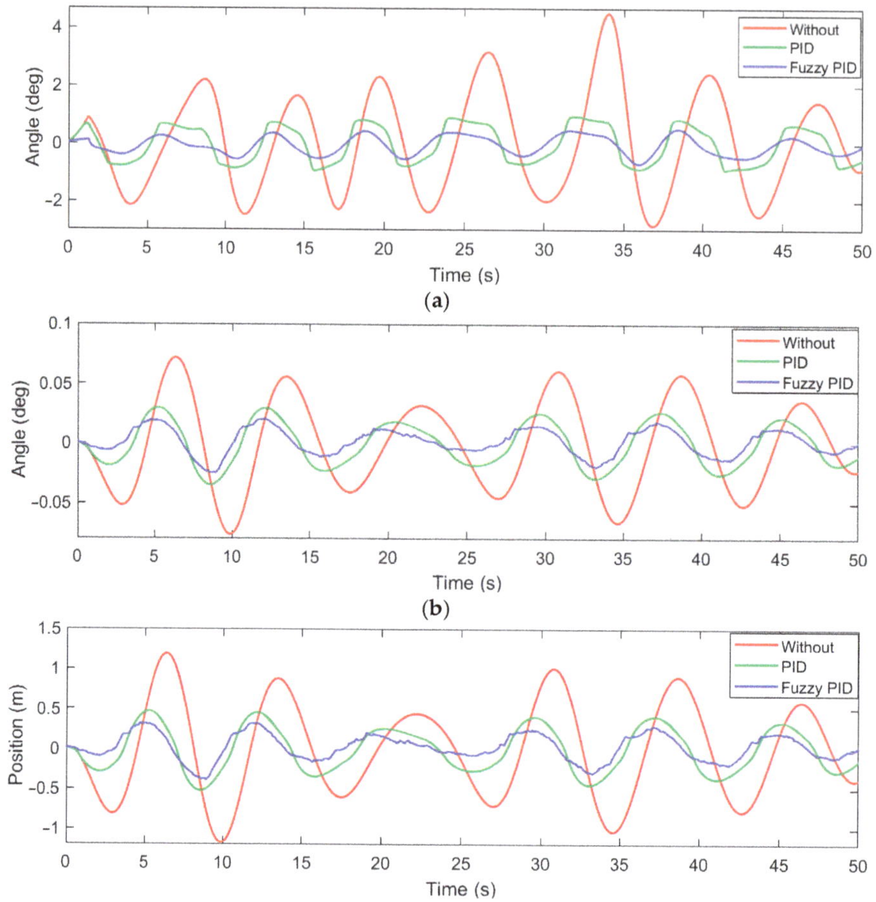

(c)

Figure 22. Roll angle, pitch angle, and heave height of the TAS when the encounter angle is 90 degrees. (**a**) Roll angle, (**b**) pitch angle, and (**c**) heave height.

Moreover, it is obvious that the position error was greatly reduced whether using PID or fuzzy PID control. For the former, the maximum error of displacement in the X direction was reduced to 0.0548 m, that in the Y direction was reduced to 0.1629 m, and that in the Z direction was reduced to 0.5596 m. For the latter, the maximum error of displacement in the X direction was reduced to 0.0447 m, that in the Y direction was reduced to 0.1183 m, and that in the Z direction was reduced to 0.4468 m. The changes in these data are also consistent with the above discussion.

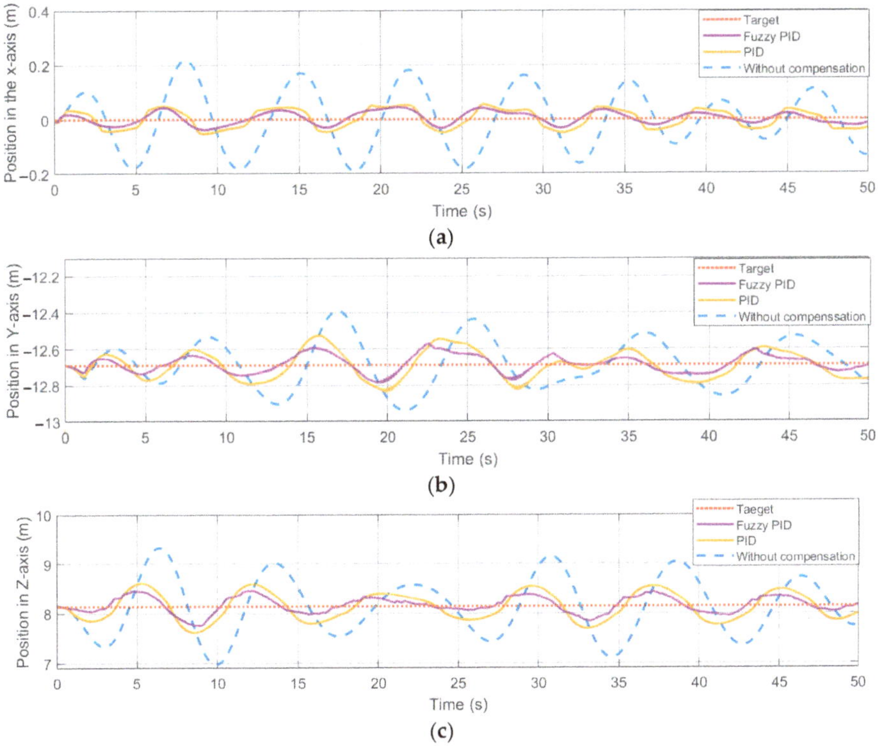

Figure 23. The compensation response at the TAS end effector when the encounter angle is 90 degrees. (**a**) The displacement in the *X* direction, (**b**) displacement in the *Y* direction, and (**c**) displacement in the *Z* direction.

8. Conclusions

A novel turbine access system for the Fujian coastal area was designed to solve the problem of the stability of the end of the TAS under sea conditions with a significant wave height of 2.2 m, including structural design, hydraulic drive design, and control system design. This study conducted the following work:

- A TAS was designed for the sea area near Putian City in Fujian Province, which includes a rotary joint, a pitch joint, and a telescopic joint. The specific compensation range was developed, and the specific length and weight of each joint were designed.
- The D-H parameter table of the TAS was obtained through the D-H parameters method, the kinematic model of the TAS was established, and the reachable motion space of the TAS was displayed; then, the 3-DOFs hydraulic control system model was determined.
- A new stacking compensation method was proposed to compensate for the roll angle, pitch angle, and heave height at the end of the TAS. The effects of ship roll, pitch, and heave on the end of the TAS were analyzed separately and compensation values were calculated. The sum of the compensation values was used as the final compensation value for the TAS.
- A fuzzy PID controller with feedforward compensation was designed and a simulation diagram was built in MATLAB/Simulink to test the feasibility of the design, and then compared with the simulation results of PID. The experimental results showed that the fuzzy PID control with feedforward compensation can achieve maximum compensation efficiencies of 84.8%, 75%, and 73.6% for the roll angle, pitch angle, and heave height of the TAS, respectively, all higher than the 78.7%, 67.6%, and 50.3% achieved under PID control. In addition, the former algorithm can achieve a

maximum compensation efficiency of 82.69% on the *X*-axis, much higher than the 69.98% achieved using PID control.

Moreover, this study achieved good results in roll angle compensation and pitch angle compensation, which can make the error of the TAS end in the *X*-axis and *Y*-axis less than 15 cm. However, the error in the heave direction is still relatively large, ranging from 0.3 m to 0.4 m. Thus, it is necessary to explore new algorithms for heave compensation. In addition, the compensation strategy is subject to physical experiments, and future practical experiments are needed to further verify the effectiveness of the method.

Author Contributions: Conceptualization, J.W., S.Z. and Y.L.; methodology, J.W. and Y.L.; software Y.L., S.Z., Y.S. and Z.W.; validation, Y.L. and J.W.; formal analysis, Y.L. and Y.S.; investigation, Y.L and Z.W.; resources, J.W., S.Z. and J.C.; data curation, Y.L.; writing—original draft preparation, Y.L. writing—review and editing, J.W. and Y.L.; visualization, Y.L. and Y.S.; supervision, J.W. and J.C. project administration, J.W. and Y.L.; funding acquisition, J.W., Y.L., Y.S. and Z.W. All authors have read and agreed to the published version of the manuscript.

Funding: This work was partially supported by the National Natural Science Foundation of China (Grant No. 51875113) and the International Exchange Program of Harbin Engineering University for Innovation-oriented Talents Cultivation.

Institutional Review Board Statement: Not applicable.

Informed Consent Statement: Not applicable.

Data Availability Statement: The data used to support the findings of this study are included within the article and are also available from the corresponding authors upon request.

Conflicts of Interest: The authors declare no conflict of interest.

References

1. Zhang, J.; Wang, H. Development of offshore wind power and foundation technology for offshore wind turbines in China. *Ocean Eng.* **2022**, *266*, 113256. [CrossRef]
2. Faraggiana, E.; Sirigu, M.; Ghigo, A. Conceptual design and optimisation of a novel hybrid device for capturing offshore wind and wave energy. *J. Ocean Eng. Mar. Energy* **2023**, 1–22. [CrossRef]
3. Tang, G.; Lei, J.; Li, F. A modified 6-DOF hybrid serial–parallel platform for ship wave compensation. *Ocean Eng.* **2023**, *280*, 114336. [CrossRef]
4. Wei, Y.; Wang, A.; Han, H. Ocean wave active compensation analysis of inverse kinematics for hybrid boarding system based on fuzzy algorithm. *Ocean Eng.* **2019**, *182*, 577–583. [CrossRef]
5. Cerda Salzmann, D.J. Ampelmann: Development of the Access System for Offshore Wind Turbines. Ph.D. Thesis, Delft University of Technology, Delft, The Netherlands, 2010.
6. Fan, Q.; Wang, X.; Yuan, J. A review of the development of key technologies for offshore wind power in China. *J. Mar. Sci. Eng.* **2022**, *10*, 929. [CrossRef]
7. Hu, B.; Yung, C. Offshore Wind Access Report 2020. Available online: http://resolver.tudelft.nl/uuid:d6cefbf6-2916-42fc-bb3d-b2d713938e09 (accessed on 14 November 2023).
8. Ren, Z.; Verma, A.S.; Li, Y. Offshore wind turbine operations and maintenance: A state-of-the-art review. *Renew. Sustain. Energy Rev.* **2021**, *144*, 110886. [CrossRef]
9. Dai, L.; Stålhane, M.; Utne, I.B. Routing and scheduling of maintenance fleet for offshore wind farms. *Wind. Eng.* **2015**, *39*, 15–30. [CrossRef]
10. Acanfora, M.; Montewka, J.; Hinz, T. On the estimation of the design loads on container stacks due to excessive acceleration in adverse weather conditions. *Mar. Struct.* **2017**, *53*, 105–123. [CrossRef]
11. Southerland, A. Mechanical systems for ocean engineering. *Nav. Eng. J.* **1970**, *82*, 63–74. [CrossRef]
12. Yang, W.; Zhang, Z.; Zhang, A. Research on an active heave compensation system for remotely operated vehicle. In Proceedings of the 2008 International Conference on Intelligent Computation Technology and Automation (ICICTA), Changsha, China, 20–22 October 2008; pp. 407–410.
13. Küchler, S.; Sawodny, O. Nonlinear control of an active heave compensation system with time-delay. In Proceedings of the 2010 IEEE International Conference on Control Applications, Yokohama, Japan, 8–10 September 2010; pp. 1313–1318.
14. Huang, X.; Wang, J.; Xu, Y. Dynamic modeling and analysis of wave compensation system based on floating crane in smart port. In Proceedings of the 2018 International Symposium in Sensing and Instrumentation in IoT Era (ISSI), Shanghai, China, 6–7 September 2018; pp. 1–5.
15. Zhou, M.; Wang, Y.; Wu, H. Control design of the wave compensation system based on the genetic PID algorithm. *Adv. Mater. Sci. Eng.* **2019**, *2019*, 2152914. [CrossRef]

6. Xie, T.; Huang, L.; Guo, Y. Modeling and simulation analysis of active heave compensation control system for electric-driven marine winch under excitation of irregular waves. *Meas. Control* **2022**, *56*, 1004–1015. [CrossRef]
7. Christopher, N. Offshore Access Systems. U.S. Patent 5074382, 24 December 1991.
8. Rong, B.; Rui, X.; Lu, K. Dynamics analysis and wave compensation control design of ship's seaborne supply by discrete time transfer matrix method of multibody system. *Mech. Syst. Signal Process.* **2019**, *128*, 50–68. [CrossRef]
9. Tang, G.; Lu, P.; Hu, X. Control system research in wave compensation based on particle swarm optimization. *Sci. Rep.* **2021**, *11*, 15316. [CrossRef]
0. Cai, Y.; Zheng, S.; Liu, W. Sliding-mode control of ship-mounted Stewart platforms for wave compensation using velocity feedforward. *Ocean Eng.* **2021**, *236*, 109477. [CrossRef]
1. Yin, R.; Xie, W.; Wen, Y. Robust wave compensation controller design for an active hexapod platform with time-varying input delays. *Ocean Eng.* **2023**, *274*, 114084. [CrossRef]
2. Chen, W.; Wang, S.; Li, J. An ADRC-based triple-loop control strategy of ship-mounted Stewart platform for six-DOF wave compensation. *Mech. Mach. Theory* **2023**, *184*, 105289. [CrossRef]
3. Lan, Z.; An, C.; Yi, D. Multivariable fuzzy genetic controller for stabilized platform. In Proceedings of the 2015 34th Chinese Control Conference (CCC), Hangzhou, China, 28–30 July 2015; pp. 783–788.
4. Bai, X.; Turner, J.; Junkins, J. Dynamic analysis and control of a stewart platform using a novel automatic differentiation method. In Proceedings of the AAS Astrodynamics Specialist Conference and Exhibit, Keystone, Colorado, 21–24 August 2006; p. 6286.
5. Liu, J.; Chen, X. Adaptive control based on neural network and beetle antennae search algorithm for an active heave compensation system. *Int. J. Control Autom. Syst.* **2022**, *20*, 515–525. [CrossRef]
6. Chen, B.; Chiang, M. Simulation and experiment of a turbine access system with three-axial active motion compensation. *Ocean Eng.* **2019**, *176*, 8–19. [CrossRef]
7. Huang, Y.; Lin, C.; Chiang, H. Development of an Offshore Turbine Access System with Four-Axial Active Motion Compensation. In Proceedings of the 2022 8th International Conference on Applied System Innovation (ICASI), Nantou, Taiwan, 21–23 April 2022; pp. 127–129.
8. Tang, L.; Chen, Y. Dynamic model of four degree of freedom rope-driven rigid-flexible hybrid wave compensation device. *J. Mech. Sci. Technol.* **2020**, *34*, 1937–1948. [CrossRef]
9. Ampelmann. Available online: http://www.ampelmann.nl/products/e-type/ (accessed on 14 November 2023).
0. Houlder. TAS Steps up to Turbine Access Challenge—Houlder. Available online: https://www.houlderltd.com/tas-turbine-access-system-steps-access-challenge/ (accessed on 14 November 2023).
1. Zhang, B.; Wang, Y. Semi-submersible offshore coupled motion in irregular waves. *J. Marine Sci.* **2020**, *2*, 19–23. [CrossRef]
2. Yang, Z. Statistical properties and spectral analysis of wind waves. *Hydro-Sci. Eng.* **1985**, *3*, 81–98.
3. Guo, J.; Li, G.; Li, B. A ship active vibration isolation system based on a novel 5-DOF parallel mechanism. In Proceedings of the 2014 IEEE International Conference on Information and Automation (ICIA), Hailar, China, 28–30 July 2014; pp. 800–805.
4. Shi, J.; Zheng, J.; Zhang, C. A 39-year high resolution wave hindcast for the Chinese coast: Model validation and wave climate analysis. *Ocean Eng.* **2019**, *183*, 224–235. [CrossRef]
5. Fossen, T.; Perez, T. Marine Systems Simulator (MSS). Available online: https://github.com/cybergalactic/MSS (accessed on 14 November 2023).
6. Li, Y.; Zheng, J. The design of ship formation based on a novel disturbance rejection control. *Int. J. Control Autom.* **2018**, *16*, 1833–1839. [CrossRef]

Journal of
Marine Science and Engineering

MDPI

Article

Design Longitudinal Control System Using Suitable T-Foil Modeling for the Offshore Wind Power Operation and Maintenance Vessel with Severe Sea States

Jia Yuan [1], Zhen Liu [1,*], Hua Geng [1], Songtao Zhang [2,*], Lihua Liang [2] and Peng Zhao [3]

[1] School of Information and Electrical Engineering, Hebei University of Engineering, Handan 056038, China; yuanjia@hebeu.edu.cn (J.Y.); huahua27102710@163.com (H.G.)
[2] College of Intelligent System Science and Engineering, Harbin Engineering University, Harbin 150001, China; liang_hrbjyq@126.com
[3] Department of Transportation and Vehicle Engineering, Tangshan University, Tangshan 063000, China; heuzyyby@163.com
* Correspondence: liuzhen@hebeu.edu.cn (Z.L.); hrbzst@126.com (S.Z.); Tel.: +86-15832040790 (Z.L.); +86-13946032279 (S.Z.)

Abstract: In order to reduce the offshore wind power operation and maintenance vessel motion induced by severe sea states, a suitable stabilizer with the ship based on linear quadratic regulator strategy is proposed in this paper. First of all, the dynamics of the ship motion model are established to study the longitudinal control system. The six degrees of freedom nonlinear motion model and nonlinear coupled longitudinal motion (heave and pitch) model are described in detail in this paper. Secondly, this work presents matching suitability between the T-foil and the operation and maintenance vessel. Therefore, the most suitable installation position and the optimum strut's height of T-foil are determined by meshing the ship hull model, setting the water channel, and a series of corresponding computer fluid dynamic simulation. Following that, the linear quadratic regulator controller is studied with active longitudinal control system based on the suitable T-foil. Furthermore a longitudinal control system is built, including free vessel module and the suitable T-foil stabilizer based proposed controller module. Finally, the simulation results indicate that the designed T-foil and the longitudinal control system are feasible and effective to ensure the heave and pitch motion reduction based on the proposed controller.

Keywords: longitudinal control system; suitable T-foil modeling; heave reduction; pitch reduction; severe sea states

Citation: Yuan, J.; Liu, Z.; Geng, H.; Zhang, S.; Liang, L.; Zhao, P. Design Longitudinal Control System Using Suitable T-Foil Modeling for the Offshore Wind Power Operation and Maintenance Vessel with Severe Sea States. *J. Mar. Sci. Eng.* **2023**, *11*, 2182. https://doi.org/10.3390/jmse11112182

Academic Editor: Kostas Belibassakis

Received: 17 October 2023
Revised: 10 November 2023
Accepted: 14 November 2023
Published: 16 November 2023

1. Introduction

Due to the massive consumption of fossil fuels, global environmental issues cannot be ignored, and seeking the development and utilization of renewable energy has become a research focus. Wind power has become one of the ideal renewable energy options for achieving emission reduction goals. Therefore, in recent years, offshore wind power has played an important role in China's national energy strategy [1,2].

The offshore wind power service business demands its maintenance personnel to perform quickly and effectively, yet the intricate marine environment raises the bar for offshore wind power operation and maintenance (O&M) boats' seakeeping standards. Establishing a suitable hydrofoil stabilizer is a challenging problem due to its characteristics as a catamaran. Therefore, designing the matching control device that effectively reduces the motion of ships due to different wave environments is paramount. During the last decades, the bilge keels, anti-rolling tanks, fin stabilizers, and rudder-roll reductions are widely used in ship stabilizing. The integrated control of rudder and fin stabilizers were used for the reduction of the roll motion based on modern control strategies [3]. Jin et al. investigated how the rudder fin joint control system affected the performance of the ship's sailing by developing a reliable H∞-type state feedback model predictive control [4]. Ren et al. established a ship motion control system using the time-scale decomposition approach, implementing rudder roll reduction, based on a four degrees of freedom (4 DOF) maneuvering motion model (surge, sway, roll, and yaw) [5]. Lee et al. created a ship motion system based on a traditional PID control method, regulating the rudder angle and propeller rotation speed to estimate ship performance accurately and guarantee effective ship operation [6]. Moreover, the new anti-rolling devices have also emerged in recent years. Li et al. carried out numerical models and experiments on the hydrodynamic moments and forces at zero velocity from the roll amplitude, flapping frequency, and aspect ratio perspectives of the flapping fins. The efficiency of flapping fins as actuators at zero velocity was confirmed, and the empirical formulae for estimating the forces and moments were obtained [7]. Song et al. conducted experiments on the proposed zero/low speed fin stabilizer at various speeds and hydrofoil profile parameters to obtain the optimal hydrofoil profile for effectively controlling ship roll [8]. A new anti-rolling equipment called the Magnus stabilizer was obtained based on the Magnus effect by analyzing the aspect ratios and lift/drag hydrodynamic characteristics [9,10].

A large focus has recently been placed on the catamarans. The transverse stability of a catamaran is higher than a monohull ship because the width of the vessel is much wider than that of a mono-hull ship. A catamaran's thin and long demi-hulls can aid in decreasing wave resistance while maintaining fast speeds [11,12]. However, the heave and pitch motion are very obvious because of the above characteristics of the catamarans at high speeds and severe sea states. The heave and pitch motion can threaten the safety of passengers, crew, and cargo. More seriously, the equipment on the ship will appear to have momentary failure. Based on this, it is essential to install control stabilizers on the catamarans [13]. Zong et al. conducted towing tank experiments using three distinct ship models, including a bare hull, a hull with a non-controlled T-foil, and a hull with an actively controlled T-foil. We confirmed that the T-foil is an efficient appendage for improving trimaran seakeeping performance by comparing different testing situations [14]. The work of Zhu's team used the T-foil or trim tab to reduce the motions of fast catamaran and trimaran, and the significant anti-vertical motion results indicated the importance of installing devices [15–17]. Lau et al. restricted the vertical motion of the vessel on a wave piercing catamaran by installing a T-foil and stern tabs, lowering motion sickness incidence (MSI) and enhancing passenger comfort [18]. Based on a hybrid control method, Jiang et al. presented a T-foil that can greatly improve the seakeeping performance of a high-speed trimaran [19].

Although different types of control equipment can be installed on the catamaran to reduce the ship motion, the general conventional control device can not achieve the ideal stabilizing efficiency, or can even have no effect in actual navigation when the ship is in complex and changeable sea conditions. Thus, designing a stabilizer that is suitable for specific ship types is very important and meaningful. According to the different degrees of the ship motion, the stabilizer constrained by the hull is designed to obtain the best matching parameters, in order to achieve the best stabilization effect. The case studies in [20] elaborated on the control of longitudinal motion of high-speed mono-hull ships, with a focus on analyzing the matching between T-foil and the hull, as well as the stabilization effect on the hull. Finally, an optimized model of T-foil was obtained. Liang et al. proposed a suitable T-foil by analyzing the optimum installation location and the strut's height. Results show that the fitted T-foil is at an advantage over the heave and pitch reduction, and the seakeeping of the wave-piercing catamaran is improved effectively [21].

However, a fixed or suitable stabilizer can suppress ship motion within a certain range, and these control devices cannot provide timely feedback in complex marine environments. The ship ride control system (RCS) has been offered as a solution to the fixed hydrofoil constraint, and several solutions have been investigated by researchers. Davis et al. investigated an 86 m high-speed catamaran with active T-foils and stern flaps. They compared movements estimated through a rapid, fixed-frame, time-based domain strip hypothesis that took the T-foil and two flaps into account. The acceleration levels at various locations were computed [22]. A data-based linear quadratic Gaussian (LQG) controller for reference tracking was established using the Markov parameters, and a detailed study was conducted on tensegrity morphing airfoils [23]. The active T-foil and two flaps were used for catamaran- or trimaran-based Kalman filter methods. The feasibility and effectiveness of the devices installed on multi-hull ship were proved by the results from their simulations [24,25]. Liang et al. demonstrated a nonlinear feedback system for controlling the fin stabilizer. The controller and feedback error decoration were used to create a type of nonlinear feedback control. The results showed that the performance of controller was not affected and the energy was saved in rough sea [26]. The reference [27] investigated a robust fuzzy controller design problem for discreditably and perturbed nonlinear ship fin stabilizing system. To demonstrate the feasibility and usefulness of the suggested robust fuzzy control approach, a simulation for managing a discrete-time nonlinear ship fin stabilizing system was presented. Zhang et al. suggested a model predictive control strategy for anti-pitching the multi-hull without violating the control input limitations, and simulations and tests demonstrated the control efficacy [28]. Cao et al. suggested a succinct and robust controller for the particular vessels that has a strong pitching reduction impact, anti-interference ability, and energy-saving effect. The controller's efficacy was validated by theoretical analysis and simulated trials under varied scenarios [29].

According to a study of previous studies, most controllers are based on discrete devices that are not confined by the hull. The actual situation is that when the control device is installed on the hull, its lift characteristics will be affected by the hull. Our proposition takes into consideration both the installation position and the strut's height of T-foil [30] Therefore, the longitudinal control system for suitable T-foil models constrained by the hull is proposed in this paper. The controller's efficiency is validated by theoretical research and simulated studies in extreme sea situations. The main contributions of this study are as follows:

- The two degrees of freedom (2 DOF) motion (heave and pitch) model discussed in this paper contains coupled motion model, which is more general and more complex than the systems with one degree of freedom motion. The existence of coupled longitudinal motion makes the design process of ride control system more complicated;
- The most suitable installation position and the optimum strut's height of T-foil are determined by meshing the ship hull model, setting the water channel, and a series of corresponding computer fluid dynamic (CFD) simulations at different high sea state numbers (SSN) and speeds;
- The attitude control system for ship multiple degrees of freedom (M DOF)-coupled motion is established, which increases the complexity of system modeling and analysis due to the transformation of the system model from a bare hull model to a model with control devices. The proposed control strategy in this paper effectively suppresses the ship's heave and pitch motion.

The remainder of the study is organized as follows. In Section 2, the mathematical models of six degrees of freedom (6 DOF) for ships and 2 DOF motion for the offshore wind power O&M vessel are described, in which disturbing and control forces and moments are elaborated. Section 3 details the matchability simulation of T-foil for the O&M vessel using the CFD software. Section 4 presents the linear quadratic regulator (LQR) strategy for the offshore wind power O&M vessel with the T-foil stabilizer and the longitudinal control system is built. Moreover, the simulation results demonstrate the efficacy of the suggested controller. In Section 5, some results in various sea conditions and speeds are provided to demonstrate the practicality of the suggested control strategy. The concluding remarks are provided in Section 6.

2. Dynamics of Ship Motion

2.1. Six Degrees of Freedom Model

Based on Newton's laws, the studies of marine vehicles' dynamics are divided into two parts: kinematics, which only treats geometric aspects of motion, and kinetics, which are also divided into two parts: rigid-body dynamics, which consist of vehicle inertia matrix, and mechanics, which are the analyses of the forces and moments causing motion [31].

Figure 1 shows a diagram of a ship coordinate system based on the notations of above picture. The following vectors can describe the general motion of a ship in 6 DOF:

$$\begin{aligned} \eta &= [\eta_1^T, \eta_2^T]^T, \eta_1 = [x, y, z]^T, \eta_2 = [\phi, \theta, \psi]^T \\ \nu &= [\nu_1^T, \nu_2^T]^T, \nu_1 = [u, v, w]^T, \nu_2 = [p, q, r]^T \\ \tau &= [\tau_1^T, \tau_2^T]^T, \tau_1 = [X, Y, Z]^T, \tau_2 = [K, M, N]^T \end{aligned} \quad (1)$$

where a ship's motion descriptions are surge, sway, heave, roll, pitch, and yaw, which is a ship's 6 DOF motion. $\eta_1 = [x, y, z]^T$ denotes the displacement of a ship in the x, y, and z directions, respectively, $\eta_2 = [\phi, \theta, \psi]^T$ denotes the rotation angle around the x, y, and z axes, respectively, $\nu_1 = [u, v, w]^T$ denotes the translational velocity in the x, y, and z directions, respectively, $\nu_2 = [p, q, r]^T$ denotes the rotational velocity around the x, y, and z axes, respectively, and $\tau_1 = [X, Y, Z]^T$ and $\tau_2 = [K, M, N]^T$ denote the external force and moment [32].

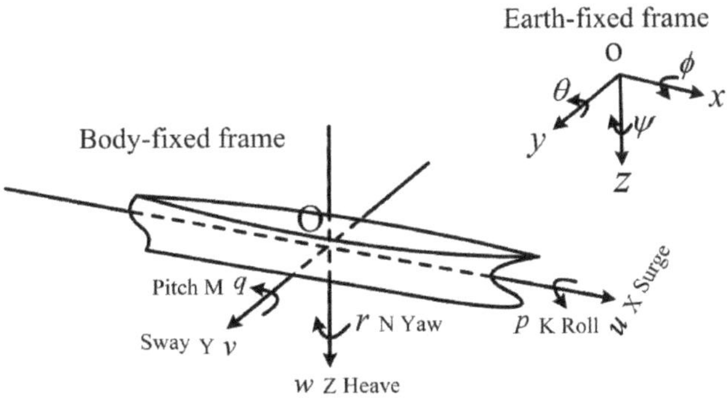

Figure 1. Body-fixed and earth-fixed frame of a ship.

By using the basic principles of dynamics, the general model structure [31] can be conveniently expressed as:

$$M\dot{\nu} + D(\nu)\nu + g(\eta) = \tau_{waves} + \tau_{control} \quad (2)$$

where M represents the state of inertia matrix and $D(\nu)$ represents the fluid damping matrix, which includes potential damping, friction between layers, wave slide damping, and vortex-related damping, $g(\eta)$ is the gravitational force and moment vector, τ_{waves} is the wave interference force and moment vector, and $\tau_{control}$ is the force and moment vector of control device. When highlighted, $\dot{\eta} = J(\eta)\nu$, $J(\eta)$ is the coordinate transform that connects the translational and rotational velocities of body-fixed and earth-fixed frames, $J_1(\eta_2)$ represents a transformation matrix connected to the Euler angle, and $J_2(\eta_2)$ represents a transformation matrix connected to the angular velocity transformation. The expressions of $J(\eta)$, $J_1(\eta_2)$, and $J_2(\eta_2)$ are as follows:

$$J(\eta) = \begin{bmatrix} J_1(\eta_2) & 0_{3\times3} \\ 0_{3\times3} & J_2(\eta_2) \end{bmatrix},$$

$$J_1(\eta_2) = \begin{bmatrix} \cos\theta\cos\psi & \sin\phi\sin\theta\cos\psi - \cos\phi\sin\psi & \cos\phi\cos\psi\sin\theta + \sin\phi\sin\psi \\ \cos\theta\sin\psi & \sin\theta\sin\theta\sin\psi + \cos\theta\cos\psi & \cos\phi\sin\theta\sin\psi - \sin\phi\cos\psi \\ -\sin\theta & \sin\phi\cos\theta & \cos\phi\cos\theta \end{bmatrix},$$

$$J_2(\eta_2) = \begin{bmatrix} 1 & \sin\phi\tan\theta & \cos\phi\tan\theta \\ 0 & \cos\phi & -\sin\phi \\ 0 & \frac{\sin\phi}{\cos\theta} & \frac{\cos\phi}{\cos\theta} \end{bmatrix}.$$

2.2. Coupled Equations of Heave and Pitch Motion

Due to the complex and changeable wave disturbances, it is very difficult to solve the hydrodynamic coefficients of the 6 DOF model. In this paper, according to the characteristics of the O&M vessel, motion in surge, sway, roll, and yaw can often be neglected in comparison with the vertical motion, when the ship is traveling at great speeds in head seas [33]. According to the 6 DOF motion model (2), when the origin of the body-fixed frame is located at the center of gravity (CG), the axis of the body-fixed frame is selected on the ship's inertia spindle, and the smaller hydrodynamic parameters and higher order elements are neglected. The nonlinear coupled equations in 2 DOF based on T-foil reduction model are described as follows [34,35]:

$$\begin{aligned} &(m - Z_{\dot{w}})\dot{w} - Z_w w - Z_{w|w|}w|w| - Z_{\dot{q}}\dot{q} - Z_q q \\ &- Z_{q|q|}q|q| + mg\cos\phi\cos\theta = Z_{waves} + Z_{T\text{-}foil} \\ &(I_{yy} - M_{\dot{q}})\dot{q} - M_q q - M_{q|q|}q|q| - M_{\dot{w}}\dot{w} - M_w w \\ &- M_{w|w|}w|w| - \rho g\nabla\overline{GM_L} = M_{waves} + M_{T\text{-}foil} \end{aligned} \quad (3)$$

where Z and M denote forces in the Z direction and moments with relation to the Y axis respectively, m is the ship mass, I_{yy} is the inertia about the Y axis, g is the acceleration of gravity, ρ is the water density, ∇ is the displaced volume of the ship, $\overline{GM_L}$ is the longitudinal metacentric height, and $Z_{\dot{w}}, Z_w, \ldots,$ and $M_{w|w|}$ are the coefficients of the ship's fluid dynamics and are defined as follows: $Z_{\dot{w}} = \frac{dZ}{d\dot{w}}$, $Z_w = \frac{dZ}{dw}, \ldots,$ and $M_{w|w|} = \frac{dM}{dw|w|}$ where the terms associated with movements or angles are the hydrodynamic restorative coefficients, the terms associated with speeds are the dynamical damping coefficients and the ones associated with accelerated velocities are the fluid inertia coefficients. A ship's hydrodynamic characteristics are sophisticated and change depending on sea state, speed and so on. As a result, they may be acquired using the formula estimation technique the computer simulation approach, and the towing tank test methodology. Z_{waves} and M_{waves} are the disturbing forces and moments, respectively, caused by wave disturbances and $Z_{T\text{-}foil}$ and $M_{T\text{-}foil}$ are the control forces and moment,m respectively, provided by the anti-heaving and pitching stabilizer T-foil.

2.3. Wave-Induced Forces and Moments

Waves are incredibly complicated in actuality, and they are the product of irregular random waves. Under the premise of linear theory, complex sea states formed by irregular waves are supplied by adding a huge number of essentially autonomous routine contributions with stages of randomness [36].

Power spectral density (PSD), often known as the wave spectrum, is widely used to define sea conditions. The following expression is used in this research to model fully formed oceans with limitless depth, no swell, and unlimited fetch [37].

$$S(\omega) = \frac{8.1 \times 10^{-3} g^2}{\omega^5} \exp\left(\frac{-3.11}{\omega^4 h_{1/3}^2}\right) \quad (4)$$

where g is still the gravitational acceleration, ω is the wave frequency, and $h_{1/3}$ is the significant wave height, and it is defined as the average of one-third of the greatest wave amplitude measurements.

The constant wave amplitude of each regular wave component ζ_i and the resultant wave $\zeta(t)$ are calculated by:

$$\zeta_i = \sqrt{2S(\omega_i)\Delta\omega} \tag{5}$$

and

$$\zeta(t) = \sum_{i=1}^{n} \zeta_i \cos(\omega_i t + \varphi_i) \tag{6}$$

where the subscript i denotes the ith regular wave, $\Delta\omega$ denotes the wave frequency spacing, φ denotes the random phase angle, and it is chosen as a random variable with an average distribution on the interval $[0, 2\pi]$.

The disruptive forces Z_{waves} and moments M_{waves} in this study are formed using 91 regular parts of waves. They are explained as follows:

$$\begin{aligned} Z_{waves}(t) &= \sum_{i=1}^{91} Z_i \cos(\omega_i t + \varphi_i) \\ M_{waves}(t) &= \sum_{i=1}^{91} M_i \cos(\omega_i t + \varphi_i) \end{aligned} \tag{7}$$

where Z_i and M_i are the heave forces and pitch moments produced by the ith wavelet ζ_i disturbance to the hull, respectively.

2.4. T-Foil Lift

An anti-heaving and pitching stabilizer, T-foil, is installed in the bow and it is used to control the longitudinal motion of the O&M vessel. The T-foil is a typical lift hydrofoil. One of the most important characteristics of lift hydrofoil is the pressure difference between upper and lower surfaces of hydrofoil. This difference is produced by changing the attack angle of the hydrofoil. The force analyses are shown in Figure 2.

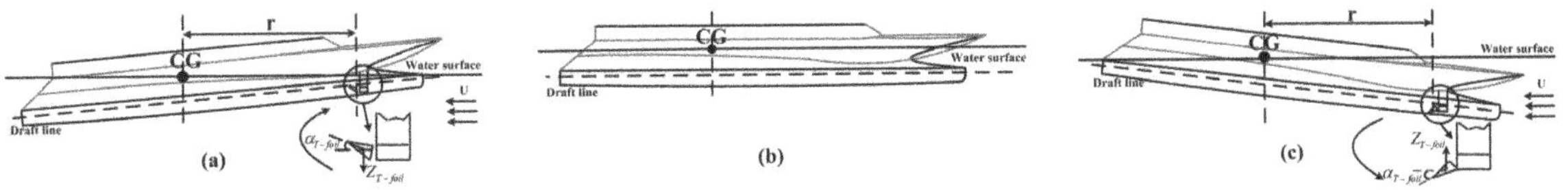

Figure 2. The force analyses of T-foil. (**a**) Life bow state. (**b**) Balance state. (**c**) Submerged bow state.

According to Section 2.2, when the O&M vessel sails at high speeds in head seas, heave and pitch motion are obviously violent. As can be seen from Figure 2, when the ship is in the lift bow state (see Figure 2a), the T-foil located in the bow should provide a downward force to restore the balance of the ship (see Figure 2b). At this point, the attack angle of T-foil needs to be swung from bottom to top to provide the downward force. On the contrary, when the ship is in the submerged bow state (see Figure 2c), the directions of the T-foil's lift and flap swinging are all in opposition to the above conclusion.

Following that, the control forces $Z_{T\text{-}foil}$ and moments $M_{T\text{-}foil}$ can be expressed as:

$$\begin{aligned} Z_{T\text{-}foil} &= \frac{1}{2}\rho A U^2 C_L(\alpha_{T\text{-}foil}) \\ M_{T\text{-}foil} &= Z_{T\text{-}foil} \cdot r \end{aligned} \tag{8}$$

where ρ is the liquid density, A is the T-foil's area, U is the ship's speed, r is the distance between the pressure center of the T-foil and the CG of the ship, $\alpha_{T\text{-}foil}$ is the attack angle of T-foil's flap, and it can rotate in $-15°$–$15°$ with respect to the horizontal, and $C_L(\alpha_{T\text{-}foil})$ is the lift coefficient and it can be obtained by CFD simulation software when the T-foil is in different attack angles.

However, there is a mutual constraint between the T-foil and the hull, so the location and the strut's height of T-foil will be considered as the factors affecting the lift coefficient. Therefore, it will analyze the matchability of T-foil for the O&M vessel from the optimization design of T-foil's location and its strut's height in the next section.

3. Matchability Analysis of T-Foil for the O&M Vessel

3.1. CFD Simulation of the O&M Vessel

The offshore wind power O&M vessel is more conducive to navigate at high speeds and reduce the rolling motion, as the ship has two thin and long demi-hulls and a good length and width ratio, and its bow is sharpened (see Figure 3). Based on the three main principles of similarity (geometric similarity, kinematic similarity, and dynamic similarity) [38,39], the original computer simulation hull model will be reduced at a certain scale. This is all for the simulation time-saving and simulation efficiency-improving. The parameters of the original vessel model and its reduced model are listed in Table 1.

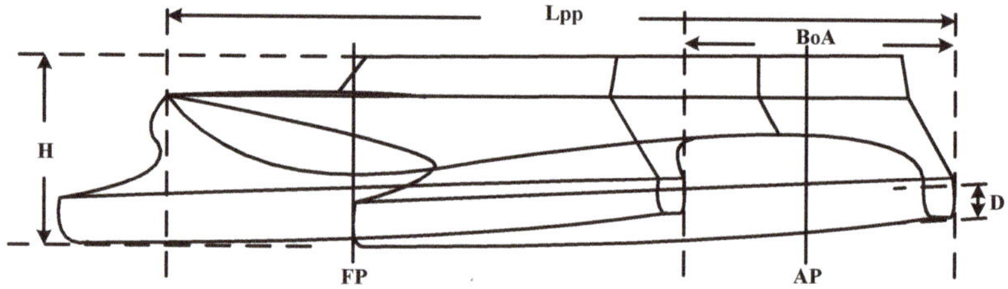

Figure 3. A diagram of the O&M vessel.

Table 1. Parameters of the O&M vessel.

Parameters	Sign	Original Model	Reduced Model	Unit
Length	L_{pp}	90.33	10.04	m
Beam overall	BoA	25.96	2.88	m
Height	H	13.16	1.46	m
Designed draft	D	2.60	0.29	m
Displacement	∇	734.54	1.01	ton

It is necessary to build a precise ship mesh model for the ship simulation. Figure 4 shows the grids of the reduced the O&M vessel model. There are more than one hundred thousand grid units, including triangular and quadrilateral mesh shapes. The deformation degree of the grid model is a crucial parameter in the simulation model. If it is reduced although the calculation accuracy is improved, it is a great test of computer performance. If it is increased, it saves time and reduces accuracy. Therefore, it needs to be repeatedly verified through trial and error methods to ultimately determine the grid deformation degree of the ship as 0.05.

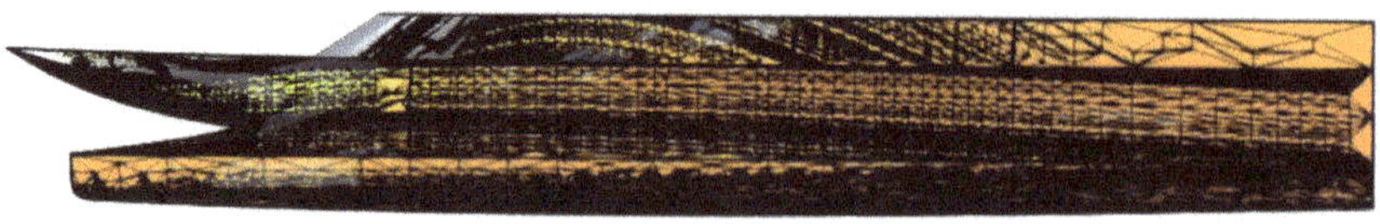

Figure 4. Mesh model of the O&M vessel.

The mesh model needs to be put in a fluid domain to simulate. Figure 5 shows the parameter settings for the entire fluid domain. In order to determine the accuracy of the simulation data, the dimension design principle of the fluid domain is as follows:

- The length of a fluid domain is about eight times the simulation model's length;
- The height is about eleven times the model's height;

- The width is about three times the model's width.

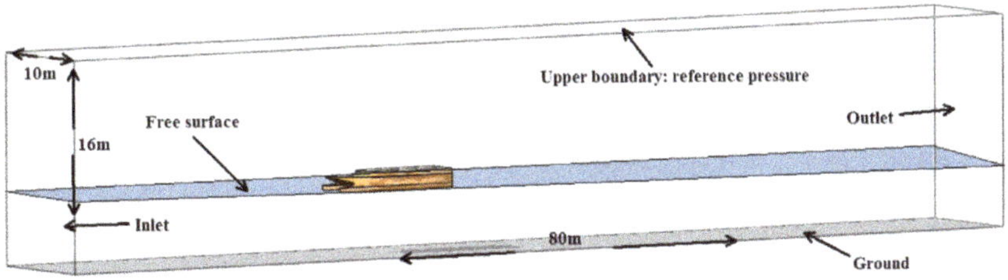

Figure 5. Parameter settings of fluid domain.

Therefore, the dimensions of the reduced O&M vessel model's fluid domain are 80 m × 16 m × 10 m. As seen in Figure 5, the flow velocity is set at the inlet surface (the negative direction of *X*), and the positive direction of *X* is the outlet surface, the upper boundary (the positive direction of *Y*) is reference pressure, the wall condition is used for the ground surface (the negative direction of *Y*), and the final two sidewalls (the positive and negative direction of *Z*) apply the periodic boundary condition.

When the wave heights increase with the increase of sea states, as shown in Figure 6, the wave crest positions are consistent because of the same speeds. Relatively, the wave heights are consistent when the wave crest positions move faster with the increase of the ship speeds, as shown in Figure 7.

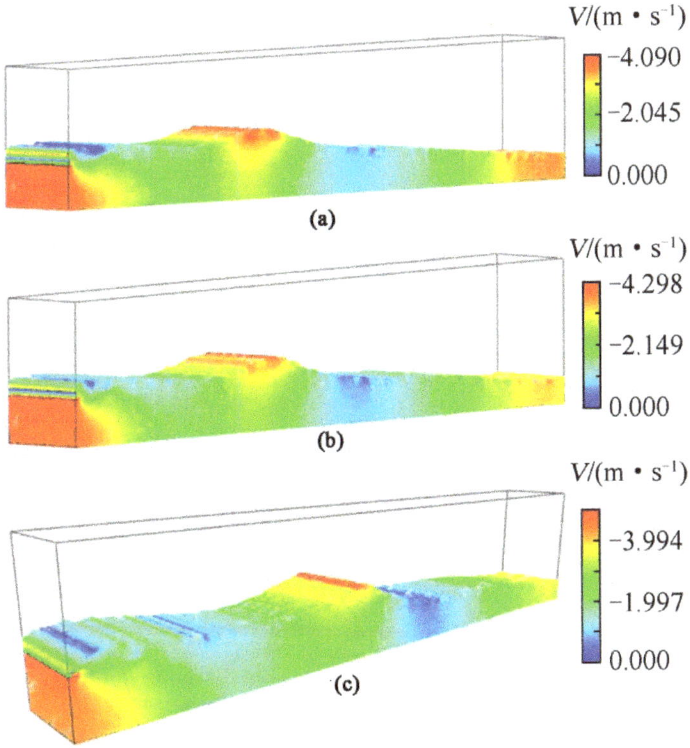

Figure 6. Wave contour of free surface for different sea states (t = 4.6 s). (**a**) SSN4. (**b**) SSN5. (**c**) SSN6.

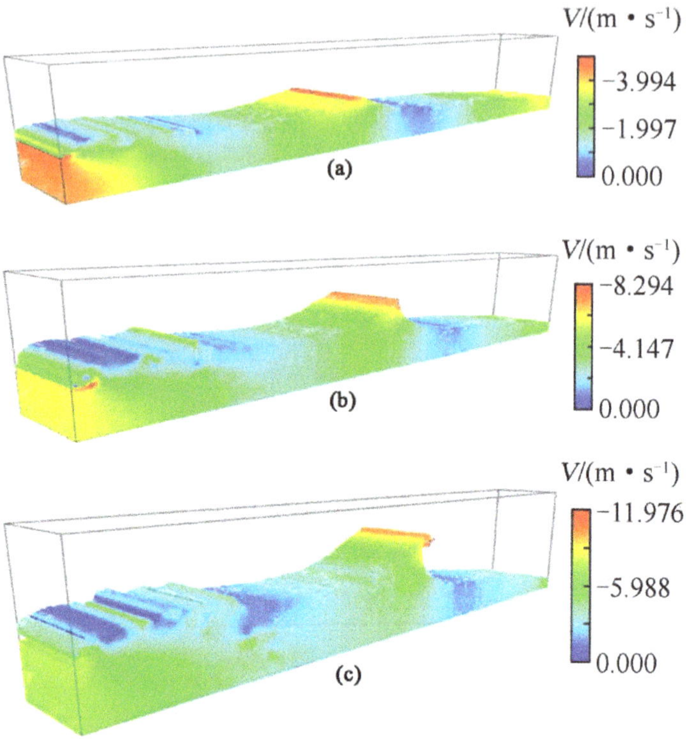

Figure 7. Wave contour of free surface for different speeds (t = 4.6 s). (**a**) 20 kts. (**b**) 30 kts. (**c**) 40 kts.

3.2. Optimization Analysis of T-Foil's Location and Strut's Height

According to Section 2.4, when the suitable T-foil for the offshore wind power O&M vessel from the optimization design of T-foil's location and its strut's height is found the optimal lift coefficient of T-foil will be obtained. Based on CFD simulation, the T-foil's location can be preliminarily judged according to the vessel's attitudes and its wetted area in the following figures [21].

Figures 8 and 9 show the longitudinal attitudes of the O&M vessel in one motion period at different sea conditions and speeds. It is evident that with the rise of the sea state there is a small upward trend for the ship's heave and pitch motion. However, with the increase of the speed, the heave and pitch motion of the ship fluctuate greatly. Therefore the change of the speed has a greater influence on the ship motion. Furthermore, with the increase of the sea state and speed, the vertical motion becomes more and more obvious.

Illumined by [40], the simulation results of different positions of a ship hull show that the foil positioned at the bow is by far the most effective in reducing the longitudinal motion of the vessel. According to Figures 10 and 11, the lifting bow covers 10–44% of the ship length in different sea states and speeds. Therefore, the preliminary judgment of the T-foil's location is obtained and the T-foil should be installed in the distances of 5 m–35 m from the forward perpendicular (FP) of the ship, which is consistent with the above-mentioned literature. Moreover, the final installation position of the T-foil depends on the lift coefficient simulation at a certain location. Figure 12 shows the different locations of T-foil and they are simulated by CFD software at intervals of 5 m (see Figure 13).

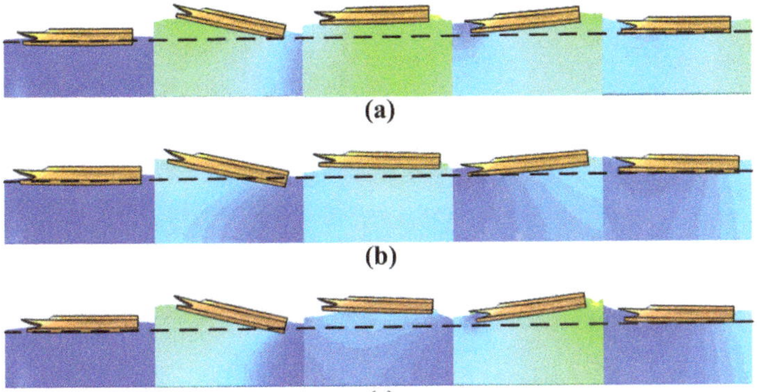
(a)
(b)
(c)

Figure 8. The O&M vessel motion at different sea states. (**a**) SSN4. (**b**) SSN5. (**c**) SSN6.

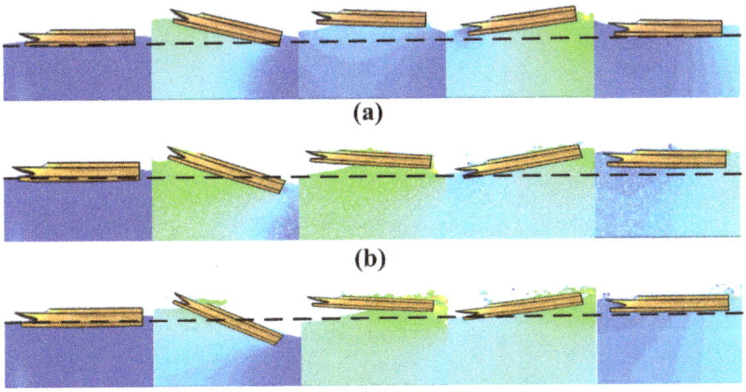
(a)
(b)
(c)

Figure 9. The O&M vessel motion at different speeds. (**a**) 20 kts. (**b**) 30 kts. (**c**) 40 kts.

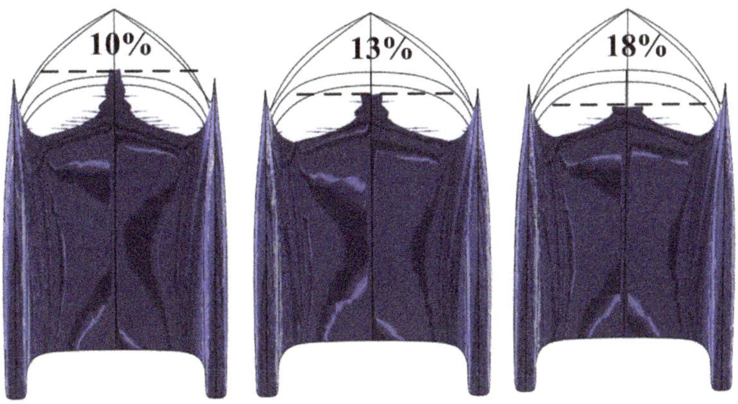

(a) (b) (c)

Figure 10. The wetted area of the O&M vessel for different sea states (t = 4.6 s). (**a**) SSN4. (**b**) SSN5. (**c**) SSN6.

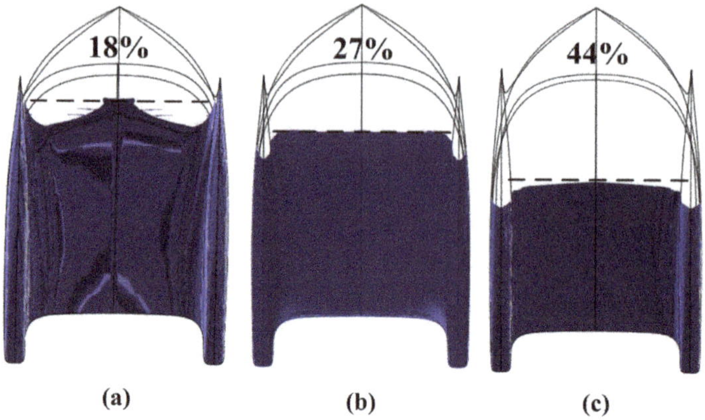

Figure 11. The wetted area of the O&M vessel for different speeds (t = 4.6 s). (**a**) 20 kts. (**b**) 30 kts. (**c**) 40 kts.

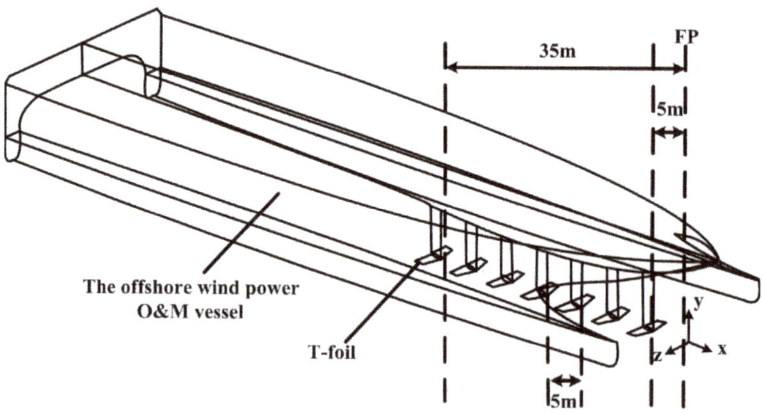

Figure 12. Installation schematic of T-foil.

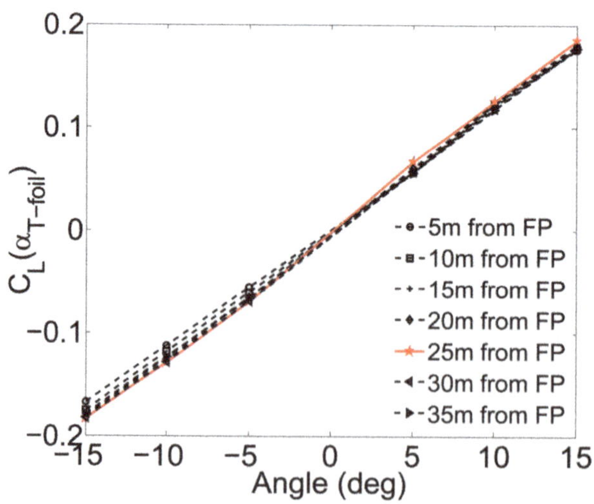

Figure 13. Lift coefficients of T-foil in different installation positions.

Here, the selection of fluid viscosity models needs to be based on the different flow rates in actual situations, which can be divided into laminar flow states with low flow rates and turbulent flow states with high speeds. However, in actual water conditions, the most common fluid flow is often in an unstable and disordered chaotic state with random changes in flow characteristics, namely the turbulent state. The motion equation of turbulence is:

$$\rho\left(\frac{\partial \bar{u}_i}{\partial t}+\bar{u}_j\frac{\partial \bar{u}_i}{\partial j}\right)=\rho m_i-\frac{\partial \bar{p}_i}{\partial i}+\frac{\partial}{\partial j}\left(\mu\frac{\partial \bar{u}_i}{\partial j}-\rho\overline{u'_i u'_j}\right)\quad (i,j=x,y,z) \tag{9}$$

where $\bar{u}_i$ and $\bar{u}_j$ are the time-averaged instantaneous velocity of turbulence on the x , y, and z axes, $\bar{u}'_i$ and $\bar{u}'_j$ are the mean of the fluctuating velocity of turbulence, $\bar{p}_i$ is the time average of instantaneous pressure, m_i is the mass component on the x, y, and z axes, and $\mu = \rho \cdot \nu$, ν is the coefficient of fluid viscosity term.

Figure 13 shows the lift coefficient curves when the different T-foil's attack angles are fixed. It can be seen that when the distance between T-foil and FP is 25 m, the lift coefficient is obviously optimal, and when the distance is beyond this position, the lift coefficient begins to decrease. Therefore, 25 m is the best position to install T-foil for the O&M Vessel.

Depending on the installation of the T-foil's at 25 m, its strut heights can only be made the optional range from 2.55 m–5.15 m. From Figure 14, based on practicability and safety considerations, the biggest strut's height is 5.15 m, and meanwhile the T-foil's main foil is at the same level as the bottom of the demi-hull of the O&M Vessel. The smallest strut's height is 2.55 m, and meanwhile the T-foil's main foil is at the same level as the draft line. Figure 15 shows the lift coefficients of different strut's heights at intervals of 0.65 m. It can be seen that 3.85 m is the optimal option because the lift coefficient of this height is best.

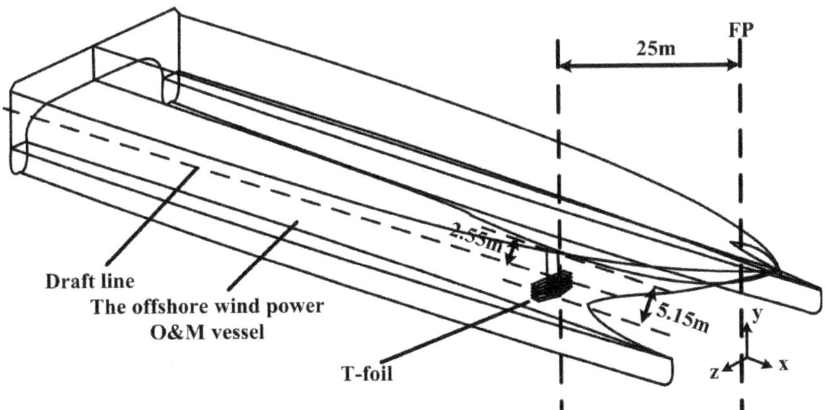

Figure 14. Selections of T-foil's strut height.

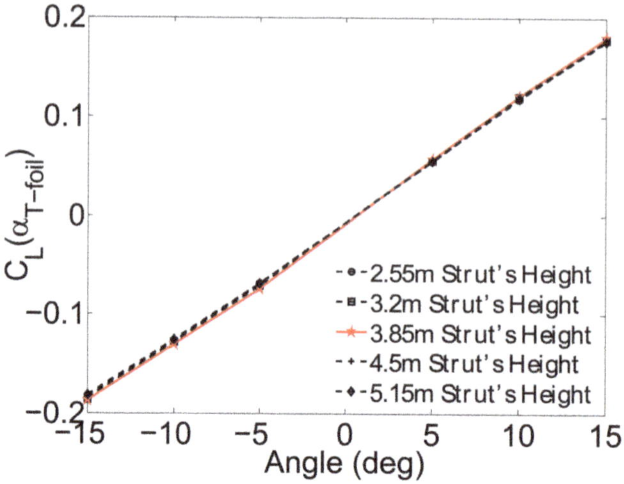

Figure 15. Lift coefficients of T-foil in different strut's heights.

3.3. Characteristics of T-Foil

According to the results of the above optimization analysis, the suitable T-foil model for this O&M vessel is shown in Figure 16 and the parameter details are listed in Table 2 [30].

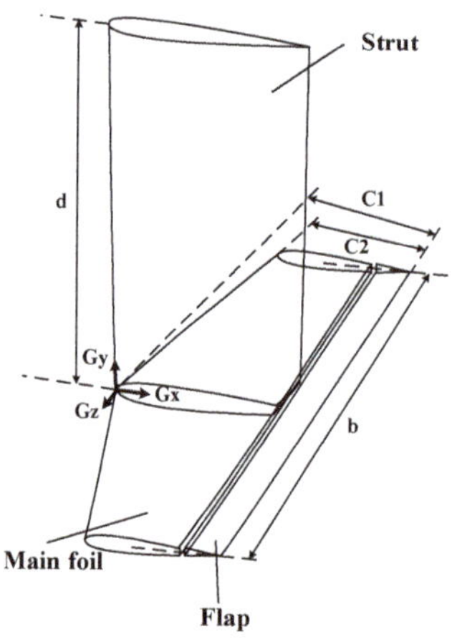

Figure 16. T-foil model.

Table 2. Parameters of T-foil.

Parameters	Sign	Value	Unit
From FP of the O&M vessel	-	25	m
Strut's height	d	3.85	m
Wingspan of main foil	b	4.80	m
Chord length of root	C1	1.50	m
Chord length of tip	C2	1.00	m
Aspect ratio	$\frac{b}{c}$	3.84	-
Main foil type	-	NACA0012	-
Strut type	-	NACA16021	-
Rotation angle range	$\alpha_{T\text{-}foil}$	−15–15	deg

Figure 17 shows the optimal lift coefficient of the suitable T-foil for the ship. The fitting equation is described as follows:

$$C_L(\alpha_{T\text{-}foil}) = p_1\alpha_{T\text{-}foil} + p_2 \tag{10}$$

where $p_1 = 0.01239$, $p_2 = -0.0055$.

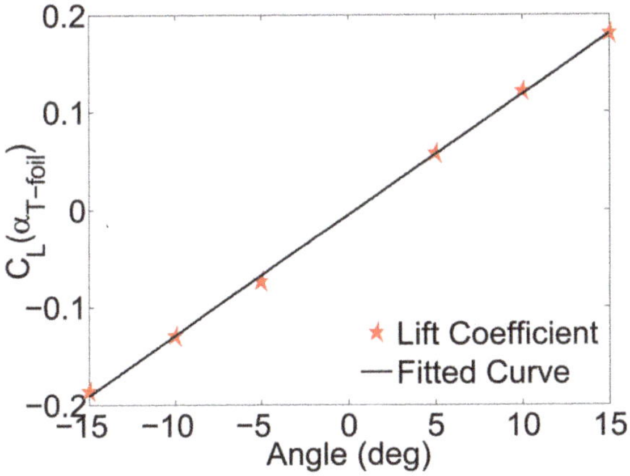

Figure 17. Lift coefficient fitting curve.

4. LQR Controller Design for the O&M Vessel with T-Foil

4.1. Linear Quadratic Regulator Model for the O&M Vessel with T-Foil

Linear quadratic regulator (LQR) [41,42] is a potent technical mean which serves complex systems such as multiple input and multiple output (MIMO) system. The longitudinal control system for the O&M vessel is a MIMO system. In the design of LQR, the optimal control law u^* can be obtained, and then, the performance index J is optimized. The performance index J is defined as follows:

$$J = \frac{1}{2}\int_0^{\infty}(x^TQx + u^TRu)dt \tag{11}$$

where $Q \geq 0$ and $R > 0$ are quadratic symmetrical matrices. They represent respective weights on state vector x and control (input) vector u, which satisfy the linearized anti-heaving and pitching system of this study.

$$\begin{aligned} \dot{x} &= Ax + B_2u + B_1d \\ y &= Cx \end{aligned} \tag{12}$$

where $x = [w, q, z, \theta]^T$ is state variable, $y = x$ is output variable, d is wave disturbance signal, A is the system matrix, B_1 is the wave disturbance matrix, B_2 is the control matrix, and $C = I_{4\times4}$ is the output matrix. The expressions of A, B_1, B_2 and the variables within each matrix are as follows:

$$A = \begin{bmatrix} -A_a^{-1}B_b & -A_a^{-1}C_c \\ I_{2\times2} & 0_{2\times2} \end{bmatrix}, B_1 = \begin{bmatrix} A_a^{-1} \\ 0_{2\times2} \end{bmatrix}\begin{bmatrix} Z_d \\ M_d \end{bmatrix}, B_2 = \begin{bmatrix} A_a^{-1} \\ 0_{2\times2} \end{bmatrix}\begin{bmatrix} Z_u \\ M_u \end{bmatrix},$$
$$[A_a] = \begin{bmatrix} m - Z_{\dot{w}} & Z_{\dot{q}} \\ M_{\dot{w}} & I_{yy} - M_{\dot{q}} \end{bmatrix}, [B_b] = \begin{bmatrix} Z_w & Z_q \\ M_w & M_q \end{bmatrix}, [C_c] = \begin{bmatrix} Z_z & Z_\theta \\ M_z & M_\theta \end{bmatrix}.$$

Determine the control law u^* that minimizes the performance index J under the following LQR design steps.

4.2. Solve the Matrix Algebraic Riccati Equation (ARE)

The Hamilton–Jacobi–Bellman theory is developed in this study to solve the LQR optimal control issue because the performance indicator J is convex.

Introducing the Hamilton function:

$$H(x,\lambda,u) = \frac{1}{2}(x^T Q x + u^T R u) + \lambda^T (Ax + B_2 u) \tag{13}$$

with the necessary canonical equation and extreme condition:

$$\dot{x} = \frac{\partial H}{\partial \lambda} = Ax + B_2 u \tag{14}$$

$$\dot{\lambda} = -\frac{\partial H}{\partial x} = -Qx - A^T \lambda \tag{15}$$

$$\frac{\partial H}{\partial u} = Ru + B_2^T \lambda = 0 \tag{16}$$

From Equation (16), the optimal control u^* is obtained:

$$u^* = -R^{-1} B_2^T \lambda \tag{17}$$

Assumption 1. *A linear relationship between λ and x can be obtained from Equations (14) and (15) that can be written as:*

$$\lambda = Px \tag{18}$$

And differentiating between both sides of Equation (18) with respect to the time, we then obtain:

$$\dot{\lambda} = \dot{P}x + P\dot{x} = (\dot{P} + PA - PB_2 R^{-1} B_2^T P)x \tag{19}$$

Thus, the preceding Equation (19) becomes as follows by simultaneous Equations (15) and (18).

$$\dot{\lambda} = (\dot{P} + PA - PB_2 R^{-1} B_2^T P)x = -Qx - A^T P x \tag{20}$$

Remark 1. *P tends to a constant matrix in the infinite time, meanwhile, first derivative of P tends to zero that is $\dot{P} \to 0$.*

Then, according to Equation (20), for any x, a sufficient condition for optimal control is that the matrix P must be satisfied:

$$PA + A^T P - PB_2 R^{-1} B_2^T P + Q = 0 \tag{21}$$

where P is the solution of Riccati Equation (21) and it is the symmetric positive definite matrix

4.3. Obtain the Optimal Control

Combining Equations (17) and (18), the optimal control law u^* is obtained:

$$u^* = -R^{-1} B_2^T P x^* \tag{22}$$

where x^* is the optimal state.

Remark 2. *Letting $R^{-1} B_2^T P = K$, and finally the optimal control law u^* can be rewritten as:*

$$u^* = -Kx^* \tag{23}$$

and then, it is found that the optimal control can be determined once P has been solved.

4.4. Obtain the Minimum Value of Performance Index J

From Equation (24), the minimum value of performance index J can be obtained:

$$J^* = \frac{1}{2}x^{*T}Px^* \tag{24}$$

When the O&M vessel is sailing at 20 kts and SSN6 in head seas, the system matrix A, the wave disturbance matrix B_1, and the control matrix B_2 are as follows:

$$A = \begin{bmatrix} -1.0673 & -13.8683 & -14.7318 & 8.4357 \\ 0.0322 & -0.8313 & 0.2239 & -15.3339 \\ 1 & 0 & 0 & 0 \\ 0 & 1 & 0 & 0 \end{bmatrix},$$

$$B_1 = \begin{bmatrix} 0.0086 & -2.8 \times 10^{-5} \\ -1.5 \times 10^{-4} & 2.1 \times 10^{-5} \\ 0 & 0 \\ 0 & 0 \end{bmatrix}, B_2 = \begin{bmatrix} 24.8660 \\ 1.6297 \\ 0 \\ 0 \end{bmatrix}.$$

According to the above steps, $Q = diag(1, 1, 10, 100)$ and $R = 10$ are chosen as the weighting coefficients to trade off the reduction of heave and pitch motion. Therefore, the solution of Riccati equation P and the gain K are as follows:

$$P = \begin{bmatrix} 22.6724 & 1.4870 & 0.4301 & 0.0257 \\ 1.4870 & 0.2513 & 0.0382 & 0.0024 \\ 0.4301 & 0.0382 & 0.8184 & 0.0634 \\ 0.0257 & 0.0024 & 0.0634 & 0.0145 \end{bmatrix}, K = \begin{bmatrix} 0.2742 & 0.7323 & 0.5560 & 0.6765 \end{bmatrix}.$$

Proof of stability analysis. The LQR controller's stability analysis is as follows [10]. □

Select the following Lyapunov function:

$$V(x) = x^T Px \tag{25}$$

where for the $\forall x \neq 0$, there is a $V(x) > 0$, and for the $\forall \|x\| \to \infty$, there is a $V(x) \to \infty$. P remains a symmetric positive definite solution of Equation (21). The first-order time derivative of $V(x)$ is:

$$\dot{V}(x) = \dot{x}^T Px + x^T P\dot{x} \tag{26}$$

Substituting Equations (12), (21) and (23) into Equation (26) yields:

$$\begin{aligned} \dot{V}(x) &= [(A - B_2R^{-1}B_2^TP)x]^T Px + x^T P(A - B_2R^{-1}B_2^TP)x \\ &= x^T(A^T - P^TB_2(R^{-1})^TB_2^T)Px + x^TPAx - x^TPB_2R^{-1}B_2^TPx \\ &= x^TA^TPx - x^TP^TB_2(R^{-1})^TB_2^TPx + x^TPAx - x^TPB_2R^{-1}B_2^TPx \\ &= x^T(A^TP - P^TB_2(R^{-1})^TB_2^TP + PA - PB_2R^{-1}B_2^TP)x \\ &= x^T(-Q - P^TB_2(R^{-1})^TB_2^TP)x \\ &= -x^T(Q + PB_2R^{-1}B_2^TP)x < 0 \end{aligned} \tag{27}$$

The suggested controller stabilizes the system described in Equation (12) using the Lyapunov stability theorem [43].

5. Results and Discussion

5.1. Simulation Results Using LQR Method

Figure 18 shows the anti-heaving and pitching motion system of the O&M vessel. There are two main sections—free O&M vessel block and T-foil active ride control block in the overall block diagram. The section labeled "O&M vessel without T-foil" in dashed frame is a free vessel model, namely that only waves induce the vessel's heave and pitch motion. Another labeled "Longitudinal control system of the O&M vessel with T-foil" is a model of the stabilizer which contains the free vessel model and the longitudinal control system composed of stabilizer and proposed controller. The input signal u^* can be obtained by the aforementioned controller design as the input attack angle of T-foil, and then control forces and moments of T-foil can help to decrease heave and pitch motion.

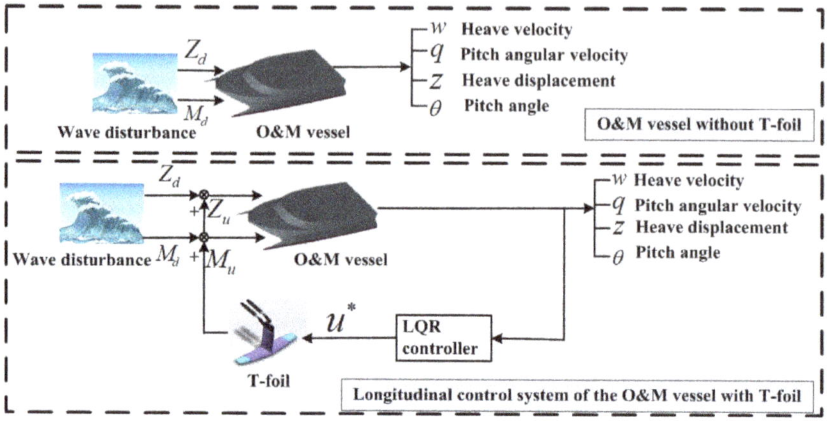

Figure 18. Longitudinal control system of the O&M vessel with T-foil.

The responses of heave and pitch motion are obtained owing to the above building of ride control platform. Taking into account the damage caused by the vessel's motion to passengers and freight [13], the pitch reduction is preferred in this paper. Figures 19 and 20 show that the T-foil active ride control is feasible and effective at high sailing speed and high sea state for the O&M Vessel. Moreover, as can be shown, pitch reduction outperforms heave. The anti-heaving forces and pitch moments have a better compensation for wave-induced disturbing forces and moments.

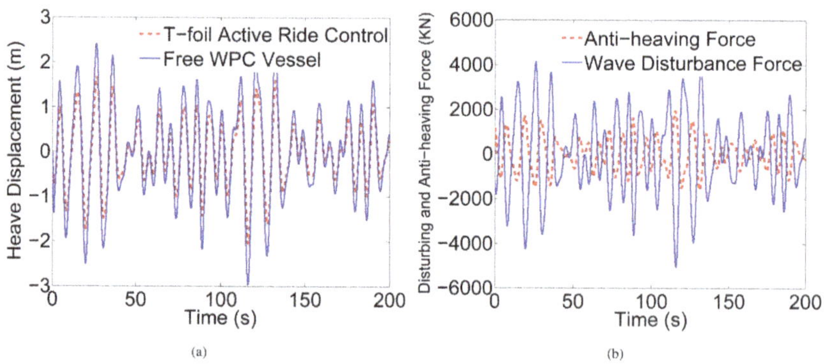

Figure 19. Simulation results of heave at 20 kts and SSN6 in head seas. (**a**) Heave reduction. (**b**) Heave disturbing forces and T-foil control forces.

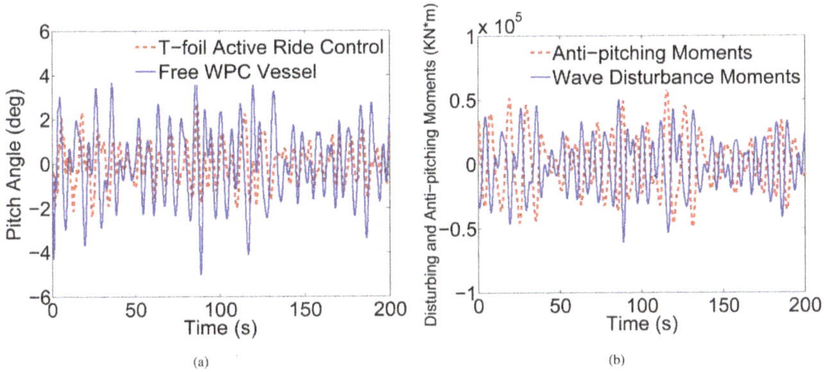

Figure 20. Simulation results of pitch at 20 kts and SSN6 in head seas. (**a**) Pitch reduction. (**b**) Pitch disturbing moments and T-foil control moments.

5.2. T-Foil Stabilization under Multiple Operating Conditions

The heave and pitch motion responses at different sea states and speeds are shown in Figures 21 and 22. It can be seen that with the increase of sea state, the amplitude of heave and pitch motion increases obviously. However, the increase of the speed does not have much effect on heave and pitch motion. Therefore, we can come to the conclusion that the change of sea state has a greater impact on ship motion compared with the change of speed.

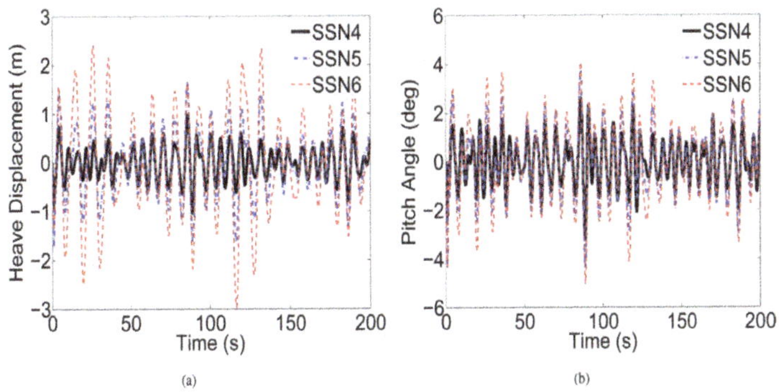

Figure 21. Seakeeping performance of the O&M vessel at different sea states. (**a**) Heave. (**b**) Pitch.

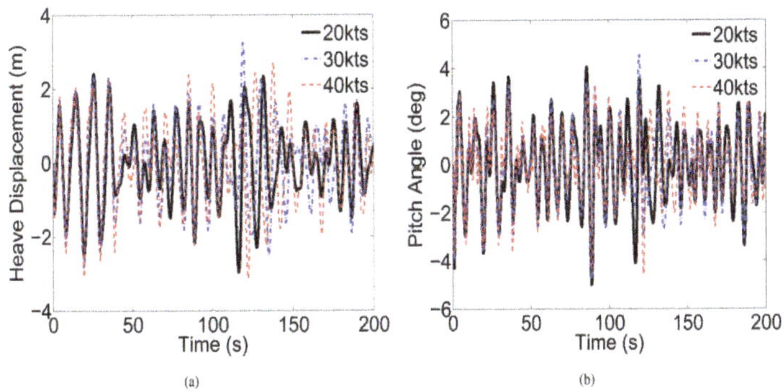

Figure 22. Seakeeping performance of the O&M vessel at different speeds. (**a**) Heave. (**b**) Pitch.

To investigate the effectiveness of longitudinal control system with T-foil for the O&M vessel, the following evaluation formula was selected to calculate the anti-heaving and pitching effect.

$$E_{PE} = \frac{HPB - HPA}{HPB} \times 100\% \quad (28)$$

where E_{PE} is anti-heaving and pitching effect, HPA and HPB represent the root mean square (RMS) of the heave displacement and pitch angle with and without the T-foil active longitudinal control system, respectively [44].

This paper pays attention to the situations of high sea states and speeds for the offshore wind power O&M vessel. The simulation results of the anti-heaving and pitching effect at the considered same speed 20 kts and different high sea states (SSN4, SSN5, and SSN6), and the same sea state SSN6 and different high sailing speeds (20 kts, 30 kts, and 40 kts) with encounter angle 180° are depicted in Figure 23. As is shown in Figure 23, the anti-heaving effect of T-foil active longitudinal control system is above 20% and most of them are above 30% at different sea states and speeds. This result satisfies the reduction index of the O&M vessel. However, the anti-heaving effect at 40 kts and SSN6 just reaches 20.18% because of the worst sea environment in this paper. This further illustrates that it is difficult to control the attitude of ship at high sea states and speeds. In addition, the anti-pitching effect is above 40% at different sea states and speeds. There is 10% points higher than that of heave, which confirms the description of Figures 19 and 20. It can be concluded that the T-foil active longitudinal control system with the proposed controller is promising and effective for the heave and pitch reduction of the O&M vessel.

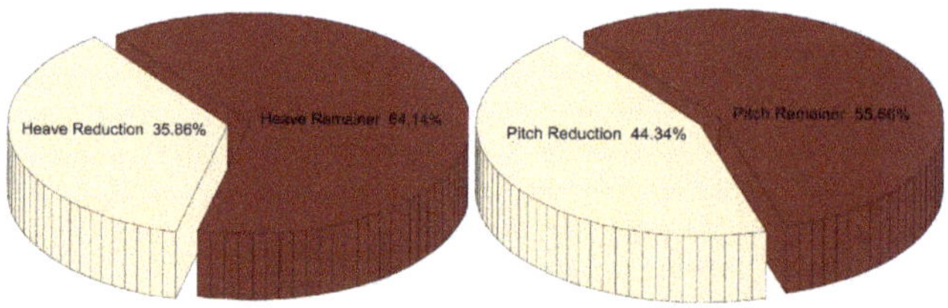

(a)

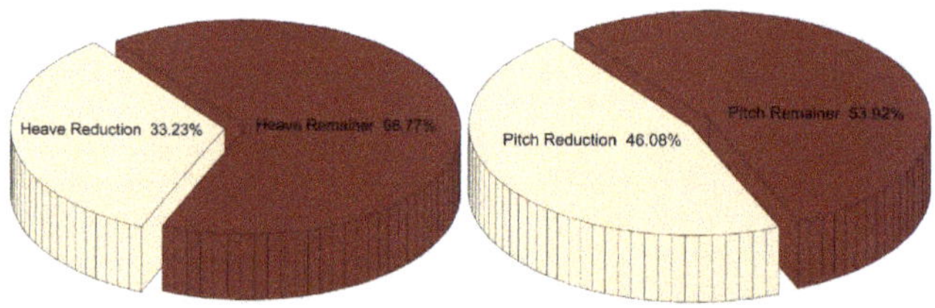

(b)

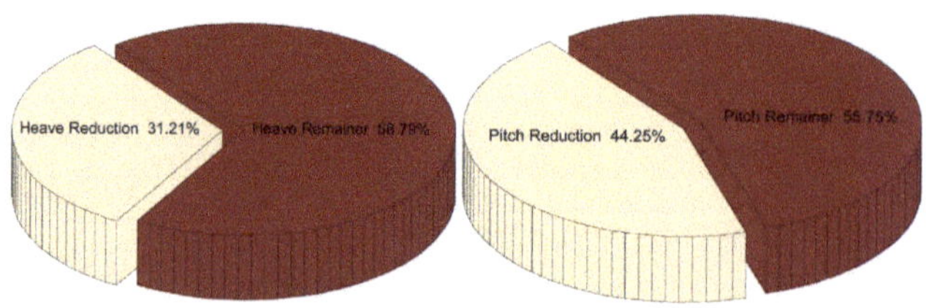

(c)

Figure 23. *Cont.*

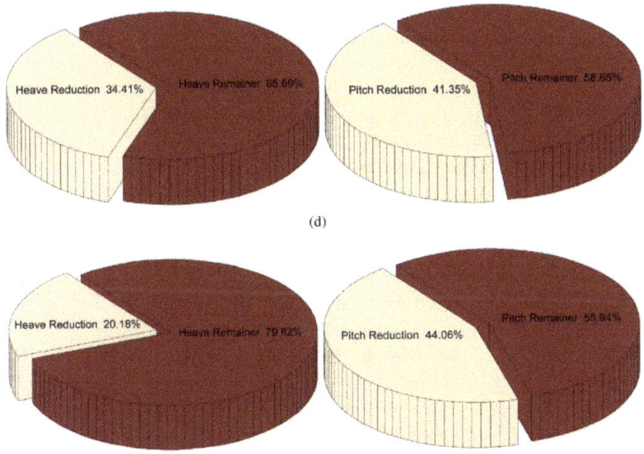

(d)

(e)

Figure 23. Heave and pitch reductions for the O&M vessel in different situations. (**a**) Reduction for SSN4, 20 kts. (**b**) Reduction for SSN5, 20 kts. (**c**) Reduction for SSN6, 20 kts. (**d**) Reduction for SSN6, 30 kts. (**e**) Reduction for SSN6, 40 kts.

6. Conclusions

A suitable stabilizer is the core of the ride control system for a ship. In this paper, a reduced scale ship simulation model was used to find the range of the installation of T-foil preliminarily. Following that, through the CFD simulation of different positions, the optimum T-foil's installation was determined. Next, the optimal strut's height of T-foil can be obtained by CFD software in different height situations after the determination of the position. Therefore, the most suitable T-foil was obtained for a certain catamaran. In addition, this paper introduces a 2 DOF motion model with detailed disturbing and control forces and moments. Based on the ship motion model, a longitudinal control system with LQR controller was established, including free vessel block and the suitable T-foil active ride control block. In conclusion, the simulation results of heave and pitch reduction indicate that the longitudinal control system-based suitable T-foil stabilizer and stated controller is both practicable and efficient to ensure the O&M vessel motion reduction.

Future work will be focused on extension of the present models (e.g., 4 DOF model and 6 DOF model) and will add a motion sickness incidence model to measure passenger comfort in the final evaluation method. The implementation of real ship model experiments can further improve the motion model and controller design methods.

Author Contributions: Conceptualization, J.Y. and L.L.; methodology, J.Y. and S.Z.; software, J.Y. and P.Z.; validation, Z.L.; formal analysis, Z.L.; investigation, J.Y. and Z.L.; resources, J.Y.; data curation, H.G., S.Z. and P.Z.; writing—original draft preparation, J.Y.; writing—review and editing, Z.L.; visualization, J.Y.; supervision, Z.L., H.G. and L.L.; project administration, J.Y.; funding acquisition, J.Y., Z.L. and H.G. All authors have read and agreed to the published version of the manuscript.

Funding: This research was funded by Science and Technology Project of Hebei Education Department grant number QN2021034, the Natural Science Foundation of Hebei Province of China grant number F2019402419 and the Innovation Fund Project of Hebei University of Engineering grant number SJ230100695, SJ220100317.

Institutional Review Board Statement: Not applicable.

Informed Consent Statement: Not applicable.

Data Availability Statement: The authors confirm that the data supporting the findings of this study are available within the article.

Conflicts of Interest: The authors declare no conflict of interest.

Abbreviations

The following abbreviations are used in this manuscript:

O&M	Operation and maintenance
4 DOF	Four degrees of freedom
MSI	Motion sickness incidence
RCS	Ride control system
LQG	Linear quadratic Gaussian
2 DOF	Two degrees of freedom
CFD	Computer fluid dynamic
SSN	Sea state numbers
M DOF	Multiple degrees of freedom
6 DOF	Six degrees of freedom
LQR	Linear quadratic regulator
CG	Center of gravity
PSD	Power spectral density
PM	Pierson-Moskowits
FP	Forward perpendicular
MIMO	Multiple input and multiple output
ARE	Algebraic riccati equation
RMS	Root mean square

References

1. Zhang, S.; Zhao, P.; Li, C.; Song, Z.; Liang, L. Study on the Accessibility Impact of Anti-Rolling Tank on the Offshore Wind O&M Gangway. *J. Mar. Sci. Eng.* **2023**, *11*, 848.
2. Fan, Q.; Wang, X.; Yuan, J.; Liu, X.; Hu, H.; Lin, P. A Review of the Development of Key Technologies for Offshore Wind Power in China. *J. Mar. Sci. Eng.* **2022**, *10*, 929. [CrossRef]
3. Liang, L.; Cheng, Q.; Li, J.; Le, Z.; Cai, P.; Jiang, Y. Design of the roll and heel reduction controller on ship's turning motion. *Ocean Eng.* **2023**, *284*, 115093. [CrossRef]
4. Jin, Z.; Liu, S.; Jin, L.; Chen, W.; Yang, W. Model based robust predictive control of ship roll/yaw motions with input constraints. *Appl. Sci.* **2020**, *10*, 3377. [CrossRef]
5. Ren, R.; Zou, Z.; Wang, J. Time-scale decomposition techniques used in the ship path-following problem with rudder roll stabilization control. *J. Mar. Sci. Eng.* **2021**, *9*, 1024. [CrossRef]
6. Lee, S.; Hwang, S.; Kim, H.; Hyun, Y.; Lee, S.; Paik, K. A numerical study on the hydrodynamic performance of a tanker in bow sea conditions depending on restraint conditions. *J. Mar. Sci. Eng.* **2023**, *11*, 1726. [CrossRef]
7. Li, M.; Shan, L.; Xu, G.; Xu, W.; Wang, J. Experimental and numerical study of the hydrodynamics of a flapping fin at zero speed. *Ocean Eng.* **2023**, *288*, 115945. [CrossRef]
8. Song, J.; Zhao, P.; Liang, L.; Ji, M. Force modeling of zero/low-velocity fin stabilizer and hydrofoil profile optimization. *Ocean Eng.* **2020**, *213*, 107635. [CrossRef]
9. Deniz, O. Performance of a Magnus effect-based cylindrical roll stabilizer on a full-scale Motor-yacht. *Ocean Eng.* **2020**, *218*, 108247.
10. Liang, L.; Jiang, Y.; Zhang, Q.; Le, Z. Aspect ratio effects on hydrodynamic characteristics of Magnus stabilizers. *Ocean Eng.* **2020**, *216*, 107699. [CrossRef]
11. Liang, L.; Yuan, J.; Zhang, S.; Zhao, P. Design a software real-time operation platform for wave piercing catamarans motion control using linear quadratic regulator based genetic algorithm. *PLoS ONE* **2018**, *13*, e0196107. [CrossRef] [PubMed]
12. Li, J.; Li, Z.; Wu, Y.; Xiong, X.; Li, Z.; Xiong, W. Research on high-speed catamaran motion reduction with semi-active control of flexible pontoon. *J. Mar. Sci. Eng.* **2023**, *11*, 1747. [CrossRef]
13. Esteban, S.; Giron-Sierra, J.; Andres-Toro, D.; Cruz, J.; Riola, J. Fast ships models for seakeeping improvement studies using flaps and T-Foil. *Math. Comput. Model.* **2005**, *41*, 1–24. [CrossRef]
14. Zong, Z.; Sun, Y.; Jiang, Y. Experimental study of controlled T-foil for vertical acceleration reduction of a trimaran. *J. Mar. Sci. Technol.* **2019**, *24*, 553–564. [CrossRef]
15. Zhu, Q.; Ma, Y. A design of T-foil and trim tab for fast catamaran based on NSGA-II. *J. Hydrodyn.* **2020**, *32*, 161–174. [CrossRef]
16. Zhu, Q.; Ma, Y. Design of H∞ anti-vertical controller and optimal allocation rule for catamaran T-foil and trim tab. *J. Ocean Eng. Mar. Energy* **2019**, *5*, 205–216. [CrossRef]
17. Ma, Y.; Zhu, Q. Fast trimaran anti-longitudinal motion control system based on active disturbance rejection control with controller tuning. *J. Mar. Sci. Technol.* **2022**, *27*, 1045–1064. [CrossRef]
18. Lau, C.; Ali-Lavroff, J.; Holloway, D.; Shabani, B.A.; Mehr, J.; Thomas, J. Influence of an active T-foil on motions and passenger comfort of a large high-speed wave-piercing catamaran based on sea trials. *J. Mar. Sci. Technol.* **2022**, *27*, 856–872. [CrossRef]
19. Jiang, Y.; Bai, J.; Liu, S.; Zong, Z.; Li, P. Experimental investigation of T-foil hybrid control strategy for ship motion reduction in head seas. *Ocean Eng.* **2022**, *243*, 110251. [CrossRef]

20. Esteban, S. Modelado y Control del Movimiento Longitudinal de un Ferry de Alta Velocidad. Ph.D. Thesis, Universidad Complutense de Madrid, Madrid, Spain, 2002.
21. Liang, L.; Yuan, J.; Zhang, S.; Liu, Y. Simulation study on wave piercing catamarans based on T-foil optimization model. *J. Huazhong Univ. Sci. Tech.* **2018**, *46*, 7–12.
22. Davis, M.; Watson, N.; Holloway, D. Wave response of an 86m high speed catamaran with active T-foils and stern tabs. *Int. J. Marit. Eng.* **2003**, *145*, 15–34.
23. Shen, Y.; Chen, M.; Skelton, R. Markov data-based reference tracking control to tensegrity morphing airfoils. *Eng. Struct.* **2023**, *291*, 116430. [CrossRef]
24. Chota, T.;Takanori, H.; Youhei, T. Numerical analysis of wake wash reduction for catamaran with hydrofoils. *Appl. Ocean Res.* **2023**, *135*, 103556.
25. Liu, Z.; Zheng, L.; Li, G.; Yuan, S.; Yang, S. An experimental study of the vertical stabilization control of a trimaran using an actively controlled T-foil and flap. *Ocean Eng.* **2021**, *219*, 108224. [CrossRef]
26. Liang, C.; Zhang, X. Concise and economical control implemented on ship fin stabilizer system based on nonlinear feedback algorithm. *J. Mar. Sci. Technol.* **2021**, *26*, 88–96. [CrossRef]
27. Chang, W.; Chang, C.; Lin, Y.; Du, J. Discrete-time robust fuzzy control synthesis for discretized and perturbed ship fin stabilizing systems subject to variance and pole location constraints. *J. Mar. Sci. Technol.* **2021**, *26*, 201–215. [CrossRef]
28. Zhang, J.; Liu, Z.; Sun, T.; Bu, M. Explicit stochastic model predictive control for anti-pitching a high-speed multihull. *Appl. Ocean Res.* **2022**, *119*, 102917. [CrossRef]
29. Cao, T.; Zhang, X. Nonlinear Decoration Control Based on Perturbation of Ship Longitudinal Motion Model. *Appl. Ocean Res.* **2023**, *130*, 103412. [CrossRef]
30. Sicard, B. Non-Magnetic Pitch and Heave Stabilizing T-Foil. Master's Thesis, Royal Institute of Technology University, Stockholm, Sweden, 2002.
31. Fossen, T. *Guidance and Control of Ocean Vehicles*; John Wiley and Sons: Hoboken, NJ, USA, 1994.
32. Chin, C.; Lin, W.; Lin, J. Experimental validation of open-frame ROV model for virtual reality simulation and control. *J. Mar. Sci. Technol.* **2018**, *23*, 267–287. [CrossRef]
33. Prestero, T. *Verification of a Six-Degree of Freedom Simulation Model for the REMUS Autonomous Underwater Vehicle*; MIT Press: Cambridge, MA, USA, 2001.
34. Tiwari, B.; Sharma, R. Design and analysis of a variable buoyancy system for efficient hovering control of underwater vehicles with state feedback controller. *J. Mar. Sci. Eng.* **2020**, *8*, 263. [CrossRef]
35. Dai, Y.; Liu, L.; Feng, S. On the identification of coupled pitch and heave motions using opposition-based particle swarm optimization. *Math. Probl. Eng.* **2014**, *2014*, 784049. [CrossRef]
36. Perez, T.; Blanke, M. *Simulation of Ship Motion in Seaway*; Technical Report; The University of Newcastle: Callaghan, Australia, 2005.
37. Price, W.; Bishop, R. *Probabilistic Theory of Ship Dynamics*; Chapman and Hall: London, UK, 1974.
38. Wang, Z.; Yuan, T.; Kong, X.; Wu, W. A universal similarity method and design procedure for buckling assessment of stiffened plates under compression load on real ships. *Thin-Walled Struct.* **2022**, *181*, 110025. [CrossRef]
39. Khomyakov, A.; Elyukhina, I. Complete dynamic similarity for sea trials and towing tank experiments by means of polymer drag reduction. *Ocean Eng.* **2019**, *178*, 31–37. [CrossRef]
40. Sclavounos, P.; Borgen, H. Seakeeping analysis of a high speed monohull with a motion control bow hydrofoil. *J. Ship Res.* **2004**, *48*, 77–117. [CrossRef]
41. Lv, Y.; Li, H. Strong fixed-time dynamic inverse adaptive LQR integrated control strategy for dynamic positioning of ship. *Ocean Eng.* **2023**, *288*, 115969. [CrossRef]
42. Moradi, M.; Bayat, F.; Charmi, M. A salient object detection framework using linear quadratic regulator controller. *J. Vis. Commun. Image Represent.* **2021**, *79*, 103259. [CrossRef]
43. Merabet, A. Adaptive sliding mode speed control for wind energy experimental system. *Energies* **2018**, *11*, 2238. [CrossRef]
44. Yuan, J. Research on Ride Control System for Wave Piercing Catamarans Based T-Foil and Stern Flaps. Ph.D. Thesis, Harbin Engineering University, Harbin, China, 2019.

MDPI
St. Alban-Anlage 66
4052 Basel
Switzerland
www.mdpi.com

Journal of Marine Science and Engineering Editorial Office
E-mail: jmse@mdpi.com
www.mdpi.com/journal/jmse

www.ingramcontent.com/pod-product-compliance
Lightning Source LLC
Chambersburg PA
CBHW040823250926
51850CB00007BA/1245
9783725811410